# Environmental Division 2014

AA003404

Core Programming Area at the 2014 AIChE Spring Meeting and 10th Global Congress on Process Safety

## New Orleans, Louisiana, USA
## 30 March – 3 April 2014

ISBN: 978-1-63439-071-2

**Printed from e-media with permission by:**

Curran Associates, Inc.
57 Morehouse Lane
Red Hook, NY 12571

proceedings.com

**Some format issues inherent in the e-media version may also appear in this print version.**



Printed by Curran Associates, Inc. (2014)

For permission requests, please contact AIChE
at the address below.

AIChE
3 Park Avenue
New York, NY 10016-5991

Phone: (203) 702-7660
Fax: (203) 775-5177

www.aiche.org

**Additional copies of this publication are available from:**

Curran Associates, Inc.
57 Morehouse Lane
Red Hook, NY 12571 USA
Phone: 845-758-0400
Fax: 845-758-2634
Email: curran@proceedings.com
Web: www.proceedings.com

# TABLE OF CONTENTS

# 354012 Possible vs. Practical: Engineers Must Lead the Development of Practical Technologies

**Abstract:**

Chemical Engineering is need today more than ever. There is a strong societal desire for sources of energy and feedstocks that can sustainably support a high standard of living for all. Thermodynamic, material and economic challenges stand in the way. Most of the general public does not have the basic understanding of the energy flows, chemical transformations, and the scale of the systems that support our current way of life. These concepts are fundamental to chemical engineering. Chemical engineers should be on the vanguard of separating hype from truly practical options yet occationally we let society down. This talk will present several examples where the fundamentals of chemical engineering can be used to access the practicality of proposed solutions to energy and water issues, as well as explain why the venture capital model struggles in the materials market.

# 352696 Hydrocarbons from Biogas Via Biogas Reforming, Syngas Cleaning and Catalytic Conversion

**Abstract:**

This research work focused on taking advantage of biogas as a cleaner, sustainable and renewable feedstock to produce liquid hydrocarbons. In this approach, biogas is first reformed to high quality synthesis gas ($H_2$ + CO) followed by Fischer-Tropsch synthesis (FTS). The Sustainable Energy Research Center (SERC) at Mississippi State University has developed and tested a continuous process over a pilot plant system to demonstrate the technology of biomass to liquid (BTL) fuels. The process includes biomass gasification, syngas conditioning and cleaning, and FTS steps. Currently, the typical gaseous product components from the gasifier are 20% $H_2$, 19% CO, 12% $CO_2$, and 47% $N_2$ if using wood chips as the feedstock. The $N_2$ and $CO_2$ contents are too high, the total syngas ($H_2$ + CO) content is only about 40%, and $H_2$/CO ratio of the bio-syngas is less than 1. The biomass-derived product gas is not ideal for the hydrocarbon synthesis. Exceptional product gas is produced from biogas (Table 1), it contains high concentrated syngas (CO+ $H_2$ > 90%), low impurities and $CO_2$, as well as $N_2$,$H_2$/CO ratio is between 1 and 2 which makes it perfect production of liquid hydrocarbon by FTS. The liquid products generated by FTS can be readily upgraded to transportation fuels since they are primarily composed of hydrocarbons initially.

Table 1 Product gas composition for simultaneous dry and partial oxidation (using air) of biogas.

| Temperature | $H_2$ (%) | CO (%) | $N_2$ (%) | $CH_4$ (%) | $CO_2$ (%) | $H_2$/CO |
|---|---|---|---|---|---|---|

| (°C) | | | | | | |
|---|---|---|---|---|---|---|
| **750** | 48.96 | 40.82 | 6.33 | 1.75 | 2.01 | 1.2 |
| **775** | 49.45 | 41.22 | 6.27 | 1.3 | 1.74 | 1.2 |
| **800** | 50.06 | 42.2 | 6.05 | 0.96 | 1.12 | 1.19 |
| **825** | 50.57 | 42.7 | 6.03 | 0.57 | 0.78 | 1.18 |
| **850** | 50.75 | 43.11 | 5.93 | 0.4 | 0.25 | 1.18 |

## 351056 Research for an Energy Future

**Abstract:**

Global demand for electrical energy is increasing even as environmental concerns are changing the ways in which we produce it. Southern Company is a leading US producer of electricity and is dedicated to developing the latest technologies of pollution control, energy production, and energy efficiency, and bringing them to economic feasibility. This talk will highlight efforts in all these areas, including Southern Company's major research projects, research centers, and commercial operations. The talk will primarily focus on Southern Company's roles in developing pollution control technology, including acid gas scrubbers, selective catalytic reduction of NOx, and mercury control (including the Mercury Research Center and the High-Sulfur Pilot Baghouse Research Center). $CO_2$ capture will be discussed including a 25-MW demonstration plant and ongoing work at the National Carbon Capture Center, located at Alabama Power's Gaston Plant. Changing energy production methods including commercial operations of coal gasification, wind, and solar will also be discussed, as well as the new Solar Power Research Center. Efforts to develop technologies which conserve water in power production will be highlighted by a discussion of the Water Research Center located at Georgia Power's Bowen Plant. Finally, ongoing research to improve energy efficiency for our customers will be highlighted, as well as electric transportation research which would have major environmental benefits.

# Sustainability and Environmental Impacts of Biofuel

*Shijie Liu*

***Department of Paper and Bioprocess Engineering***
***SUNY College of Environmental Science and Forestry***

1

Imagine a renewable world

2

# Outline

- **The Hype of Biomass**
- **What is Sustainability?**
- **Biomass Crop**
- **Ethanol: Good or Bad**
- **Ethanol: a Platform chemical**
- **Ethanol: Biomass conversion efficiency**
- **Conclusion**

3

# The Hype in Biomass

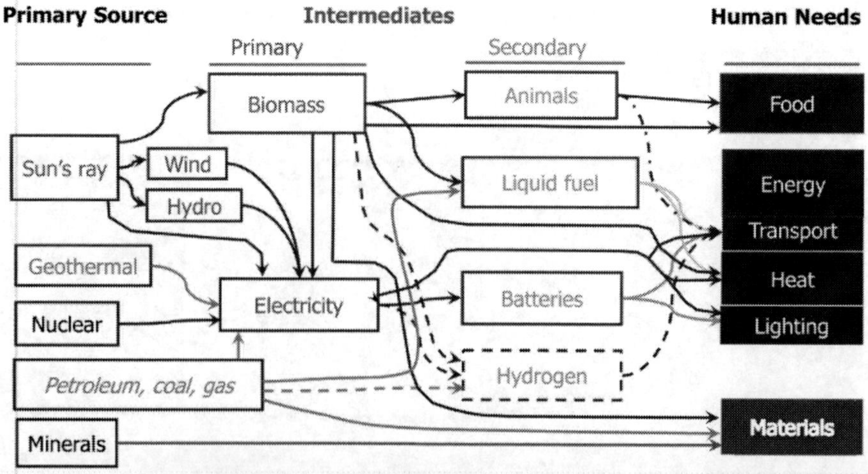


4

## Where do we get our chemicals today?

- Petroleum, coal and natural gas
  - Dominant source of energy for human use
  - Materials use - plastics
- Biomass
  - Food
  - Heating use on the steep declining slope
  - Materials – dwellings, furniture
- Solar energy
  - Limited use in heating and lighting

5

## What is Renewability?

- Merriam-Webster:
  - 1. Renewable – capable of being renewed
  - 2. Renewable – capable of being replaced by natural ecological cycles or sound management practices
- Point on a complete cycle

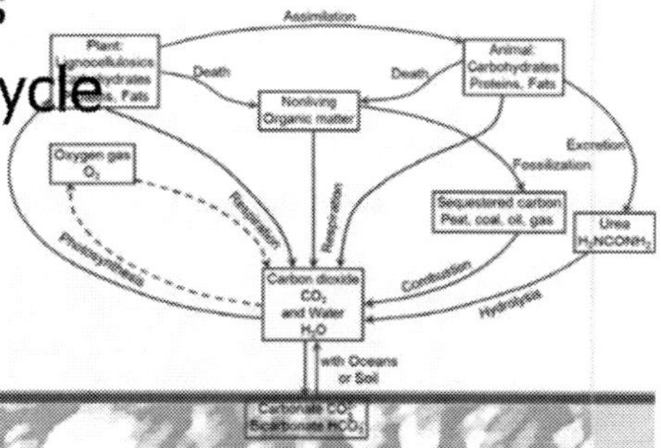

## What is Sustainability?

- Merriam-Webster:
  - 1. Sustainable – capable of being sustained
  - 2. Sustainable – of, relating to, or being a method of harvesting or using a resource so that the resource is not depleted or permanently damaged.
- Non-trivial Steady State

(S. Liu, Bioprocess Engineering: Kinetics, Biosystems, Sustainability, and Reactor Design, Elsevier, 2012)

## Biomass

- Agricultural crop
  - Cereal grain; Tubers
  - Straws
  - Grass
- Forest biomass
  - Trees
- Long rotation crops
  - Fruits, nuts
  - "Short-rotation" woody biomass

8

# Is Forest Biomass Renewable?

- Manomet Center for Conservation Sciences report "Biomass Sustainability and Carbon Policy Study" also know as the Massachusetts Biomass Report (June 2010)
  - Woody biomass utilization sending more carbon into the atmosphere than fossil
  - Equal setting of forest biomass use with fossil use will occur after 32 years IF ...
  - Forest biomass carbon neutrality will not occur after 90 years

9

# Massachusetts' report

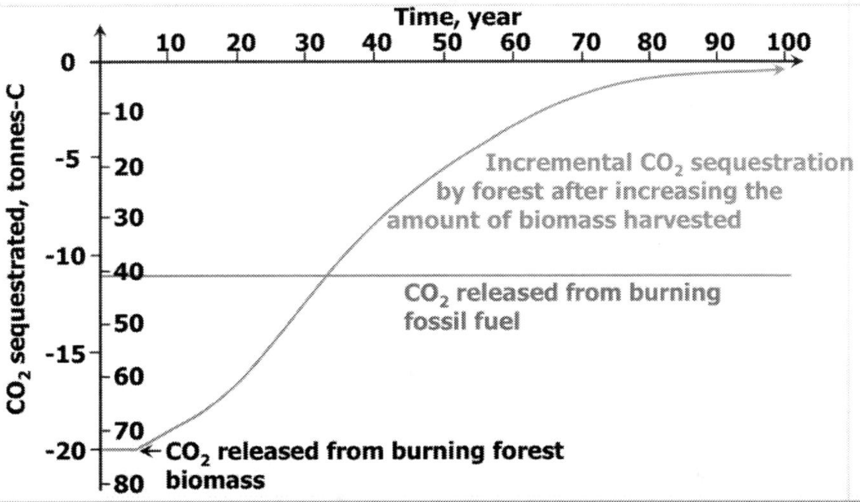


10

# Biomass: Good or Bad?

- Fossil source
  - Energy dense: low $CO_2$ emission
- Biomass, $CO_2$ is $CO_2$
  - Less energy content (already oxygenated) - bulky
  - Not accounting: Forest disturbance – only half storage grows back when converting an untouched forest to a managed forest and equal amount underground in jeopardy
  - More $CO_2$ in atmosphere at sustainability
  - Hydrocarbon emission, e.g. isosprene

11

# Sustainability – Biomass Production

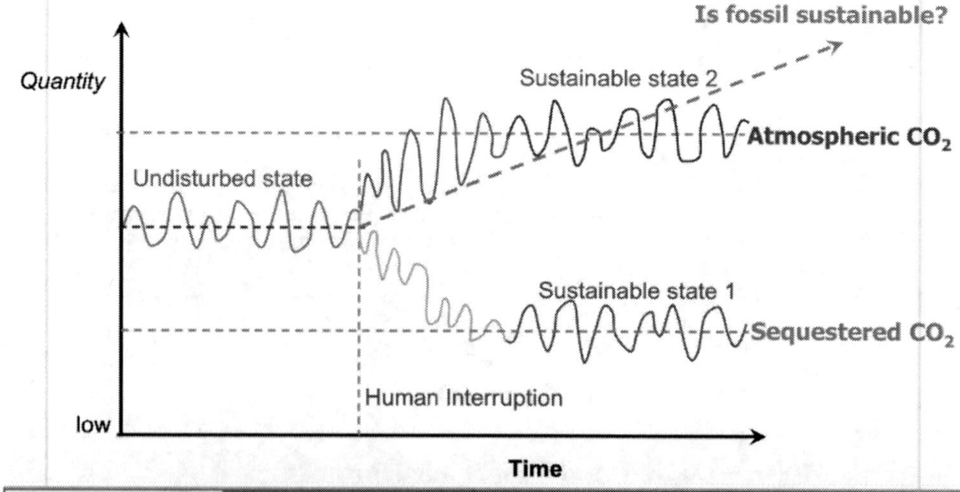

# What is the Magic Biomass (Crop)?

**Table 1. Recycle times and productivities of chemical/energy feedstock (Liu, 2010).**

| Feed stock | Recycle time | Standing Biomass tons/ha | Biomass Production tons/(ha·year) |
|---|---|---|---|
| Algae | 1 month | 0.9 | 11.25 |
| Agricultural crops | 3 month - 1 year | 4.5 | 2.93 |
| Temperate grasses | 1 year | 7.2 | 2.70 |
| Savanna | 1 year | 18.0 | 4.05 |
| Shrubs | 1 – 5 years | 27.0 | 3.15 |
| Tropical forest | 5 – 25 years | 202.5 | 9.90 |
| Trop. season forest | 5 – 25 years | 157.5 | 7.20 |
| Boreal forest | 25 – 80 years | 90.0 | 3.60 |
| Temperate deciduous | 10 – 50 years | 135.0 | 5.40 |
| Temperate evergreen | 10 – 80 years | 157.5 | 5.85 |
| *Oil, gas and coal* | *280 million years* | $(38.4\times10^{21}$ J) | (0) |

# Standing Biomass ($\sim X_\infty$)

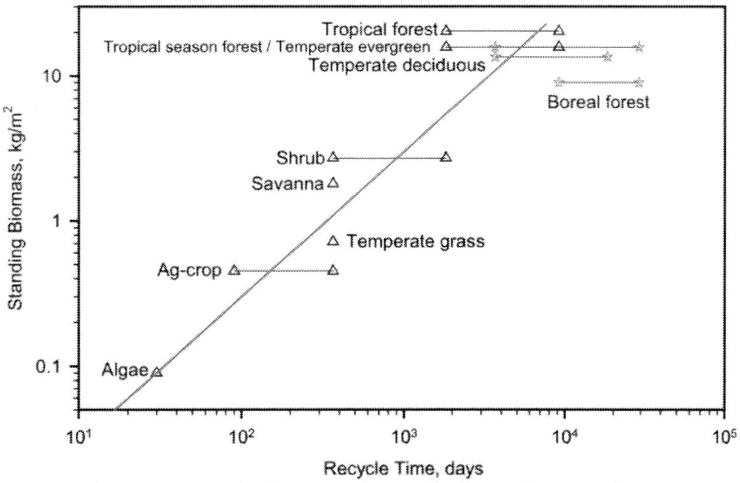


14

# Biomass crop selection

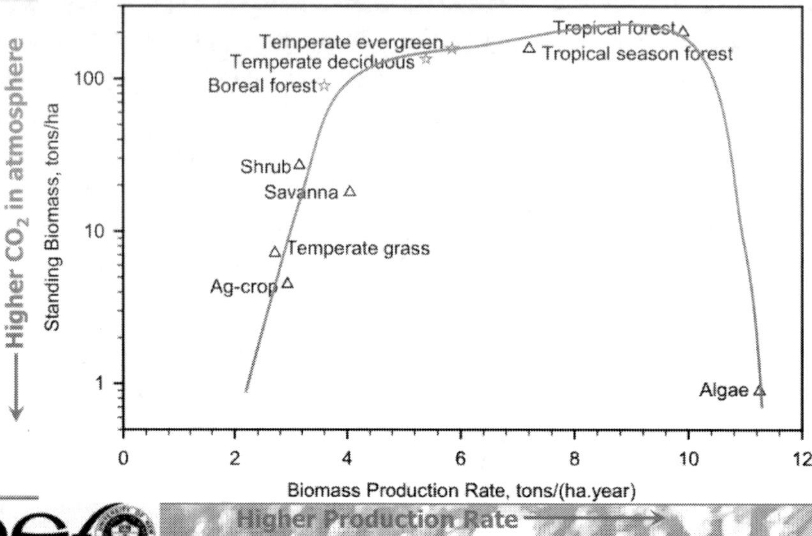


# Biomass Crop Selection

- Productivity: rate of supply
- Standing biomass: $CO_2$ sequestration
- Water, Soil condition, Terrain
- Neighboring community
- Food needs, why "corn" is a better biomass
- Production cost
- There is no magic crop

# Integrated Biorefinery:

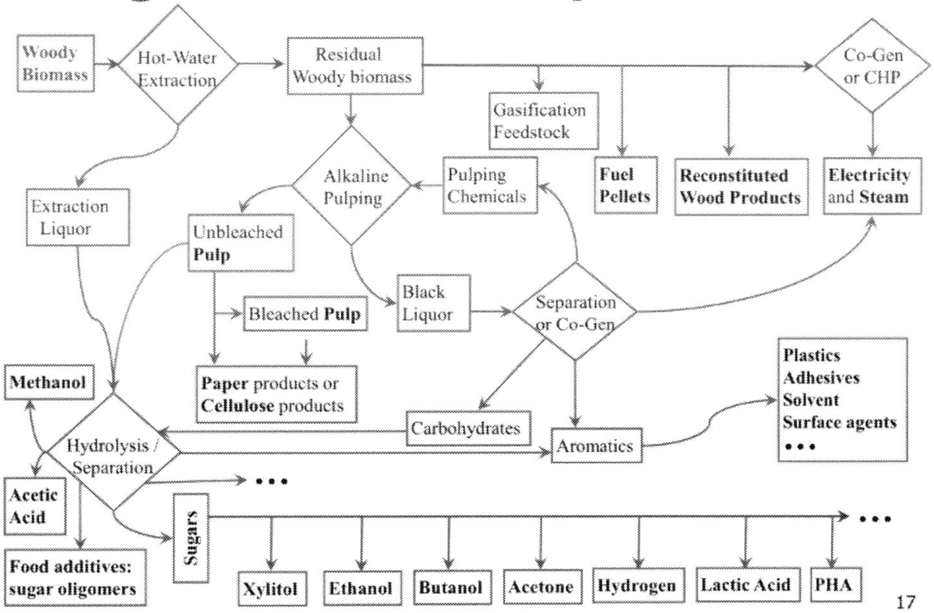


17

## Ethanol: Good or Bad

- Gasoline additive: E10, going for E15
  - Commercial
  - Infrastructure established, and production facility design mature for new ones
  - Blending at distribution centers
  - Transportation
- Cereal grain use
- Physical and chemical properties

# Ethanol: Platform chemical

- Jet fuel
- Bioplastics or bioploymers
- Hydrocarbons

$$CH_3CH_2OH \longrightarrow CH_2=CH_2 + H_2O$$
$$2CH_3CH_2OH \longrightarrow CH_2=CH\text{-}CH=CH_2 + H_2O + 2H_2$$

Wang, Y. and Liu, S. J. Bioproc. Eng. Biorefinery, 1(1), 33-43.

# Ethanol: Biomass conversion

- Theoretical yield: 51% on sugars
  - Anaerobic fermentation, can tolerate $O_2$
  - Fermentation to high concentrations
- As compared with
- Butanol – low yield, low concentration
- PHA – low yield, high recovery cost

# Fermentation: Wood extract Hydrolysate

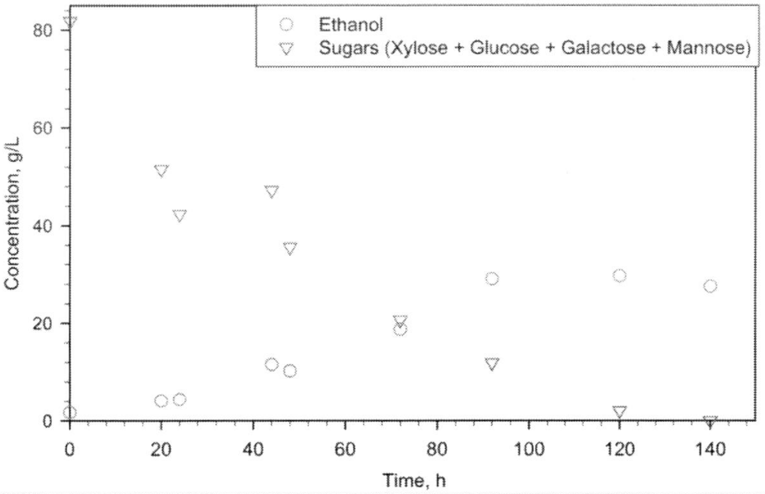


# Butanol fermentation

**Glucose in feed: 45 g/L**

| | | |
|---|---|---|
| Acetone | 5.2 | g/L |
| Ethanol | 1.8 | g/L |
| Butanol | 9.8 | g/L |
| Total | 16.8 | g/L |

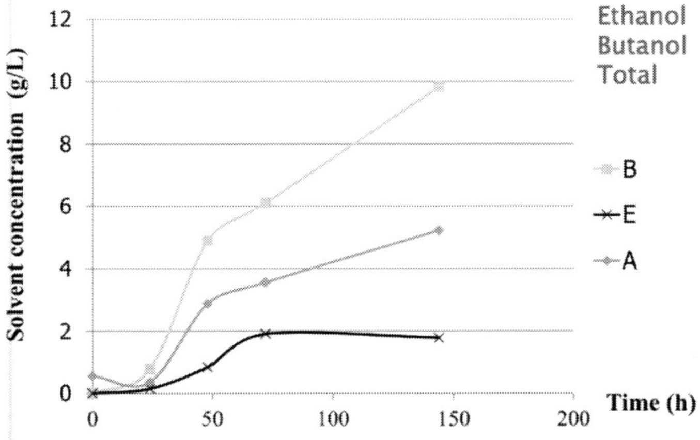

# Butanol fermentation

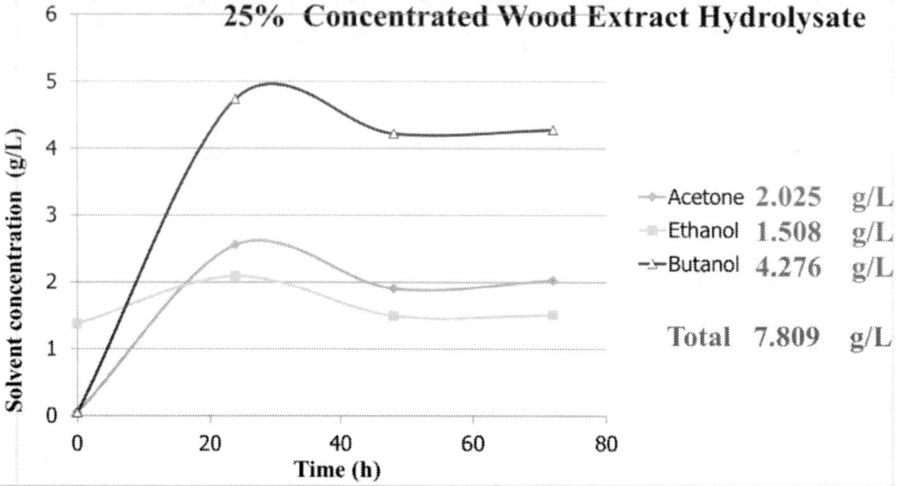


# Conclusions

- Future depends on biomass, despite a hard choice to make: $CO_2$ is $CO_2$
- Biomass is both renewable and sustainable
- Ethanol is a good choice for continuing developments in renewable chemical and energy sectors
- Ethanol is a platform chemical, from which products know today from petroleum can be produced

24

Thanks!

# 352830 Air Permitting of a Major Capital Project at a Chemical Plant in Louisiana -

**Abstract:**

*ABSTRACT for 2014 AIChE Spring Meeting – New Orleans iE Sprin Modn*

***Authors:***

*Steve David, Senior Manager of Air Strategy, Mosaic Corporation*

*Peter H. Anderson, Principal, Geosyntec Consultants*

***Air Permitting of a Major Capital Project at a Chemical Plant in Louisiana -*** As part of preparing a prevention of significant deterioration (PSD) and Part 70 (Title V) permit application for a proposed $1 billion expansion at a chemical plant a number of technical analyses were completed to demonstrate that the Project would meet federal and state air pollution control requirements. Included as part of this effort, was implementing a comprehensive permitting strategy to evaluate control technology options to minimize regulatory impacts and executing a plan to reduce critical non-technical risks to support an expedited permitting process. This paper will provide insights on how to permit a complex expansion of a chemical plant and cover the challenging topics of determining best available control technology (BACT) for greenhouse gases (GHG) and $CO_2$ emissions and carrying-out air dispersion modeling to demonstrate compliance with national ambient air quality standards (NAAQS) and state toxic air pollutant ambient air standards.

# Theoretical Screening of Mixed Solid Sorbent for Applications to $CO_2$ Capture Technology

**Yuhua Duan**[1]
National Energy Technology Laboratory, United States Department of Energy, Pittsburgh, Pennsylvania 15236, USA

## Abstract

Since current technologies for capturing $CO_2$ to fight global climate change are still too energy intensive, there is a critical need for development of new materials that can capture $CO_2$ reversibly with acceptable energy costs. Accordingly, solid sorbents have been proposed to be used for $CO_2$ capture applications through a reversible chemical transformation. By combining thermodynamic database mining with first principles density functional theory and phonon lattice dynamics calculations, a theoretical screening methodology to identify the most promising $CO_2$ sorbent candidates from the vast array of possible solid materials has been proposed and validated. The calculated thermodynamic properties of different classes of solid materials versus temperature and pressure changes were further used to evaluate the equilibrium properties for the $CO_2$ adsorption/desorption cycles. According to the requirements imposed by the pre- and post- combustion technologies and based on our calculated thermodynamic properties for the $CO_2$ capture reactions by the solids of interest, we were able to screen only those solid materials for which lower capture energy costs are expected at the desired pressure and temperature conditions. Only those selected $CO_2$ sorbent candidates were further considered for experimental validations. The *ab initio* thermodynamic technique has the advantage of identifying thermodynamic properties of $CO_2$ capture reactions without any experimental input beyond crystallographic structural information of the solid phases involved. Such methodology not only can be used to search for good candidates from existing database of solid materials, but also can provide some guidelines for synthesis new materials. In this presentation, we apply our screening methodology to mixing solid systems to adjust the turnover temperature to help on developing $CO_2$ capture Technologies.

## I. Introduction

Carbon dioxide is one of the major combustion products which once released into the air can contribute to global climate change.[1-4] One approach to solve such environmental problems is to capture and sequester the $CO_2$. Current technologies for capturing $CO_2$ including solvent-based (amines) and CaO-based materials are still too energy intensive. Hence, there is critical need for development of new materials that can capture and release $CO_2$ reversibly with acceptable energy costs. In particular, solid oxide sorbent materials have been proposed for capturing $CO_2$ through a reversible chemical transformation leading primarily to formation of carbonate products. Solid sorbents containing alkali and alkaline earth metals have been reported in several previous studies

[1] Tel. 412-386-5771, Fax 412-386-5920, E-mail: yuhua.duan@netl.doe.gov

to be promising candidates for $CO_2$ sorbent applications due to their high $CO_2$ absorption capacity at moderate working temperatures.[5-8]

During past few years we developed a theoretical methodology to identify promising solid sorbent candidates for $CO_2$ capture by combining thermodynamic database searching with *ab initio* thermodynamics obtained based on first-principles density functional theory (DFT) and lattice phonon dynamics.[5-7, 9-13] As shown in Fig.1, the primary outcome of our screening scheme is a list of promising $CO_2$ sorbents with optimal energy usage.

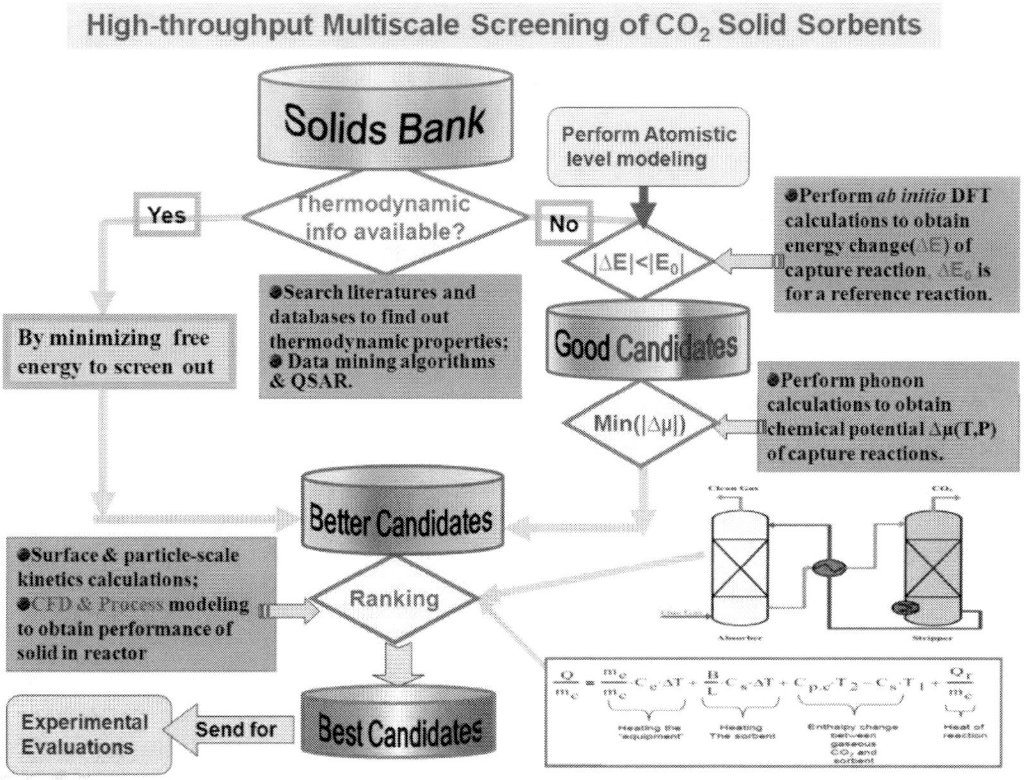


**Fig. 1.** Schematic of our screening methodology.

At a given $CO_2$ pressure, the turnover temperature ($T_t$) of an individual solid capture $CO_2$ reaction is fixed. Such $T_t$ may be outside the operating temperature range ($\Delta T_o$) for a particularly capture technology. In order to adjust $T_t$ to fit the practical $\Delta T_o$, in this study, we demonstrate that by mixing different types of solids it's possible to shift $T_t$ to the range of practical operating conditions.

## II. Calculation Methods for Mixed Solid Sorbents

The complete description of the computational methodology together with relevant applications can be found in our previous publications.[5-7, 9-16] The $CO_2$ capture reactions of solids can be expressed generically in the form (for convenient description, we normalized the reaction to 1 mole of $CO_2$)

$$\sum_{Ri} n_{Ri}\text{Solid}\_R_i + CO_2 \leftrightarrow \sum_{Pj} n_{Pj}\text{Solid}\_P_j \qquad \text{(a)}$$

where $n_{Ri}$, $n_{Pj}$ are the numbers of moles of reactants ($R_i$) and products ($P_j$) involved in the capture reactions. We treat the gas phase $CO_2$ as an ideal gas. By assuming that the difference between the chemical potentials ($\Delta\mu^o$) of the solid phases of reactants ($R_i$) and products ($P_j$) can be approximated by the difference in their total energies ($\Delta E^{DFT}$), obtained directly from DFT calculations, and the vibrational free energy of the phonons and by ignoring the PV contribution terms for solids, the variation of the chemical potential ($\Delta\mu$) for reaction (a) with temperature and pressure can be written as[5-7, 9-16]

$$\Delta\mu(T,P) = \Delta\mu^0(T) - RT\ \ln\frac{P_{CO_2}}{P_0} \qquad (1)$$

where

$$\Delta\mu^0(T) \approx \Delta E^{DFT} + \Delta E_{ZP} + \Delta F^{PH}(T) - G^0_{CO_2}(T) \qquad (2)$$

Here, $\Delta E^{DFT}$ is the DFT energy difference between the reactants and products of the reaction (a), $\Delta E_{ZP}$ is the zero point energy difference between the reactants and products and can be obtained directly from phonon calculations. $\Delta F^{PH}$ is the phonon free energy change excluding zero-point energy (which is already counted into the $\Delta E_{ZP}$ term) between the solids of products and reactants. $P_{CO_2}$ is the partial pressure of $CO_2$ in the gas phase and $P_0$ is the standard state reference pressure taken to be 1 bar. The heat of reaction ($\Delta H^{cal}(T)$) can be evaluated through the following equation

$$\Delta H^{cal}(T) = \Delta\mu^0(T) + T[\Delta S_{PH}(T) - S_{CO_2}(T)] \qquad (3)$$

where $\Delta S_{PH}(T)$ is the difference of entropies between product solids and reactant solids. The free energy of $CO_2$ ($G^0_{CO_2}$) can be obtained from standard statistical mechanics,[6, 7, 13] and its entropy ($S_{CO_2}$) can be found in the empirical thermodynamic databases.[17]

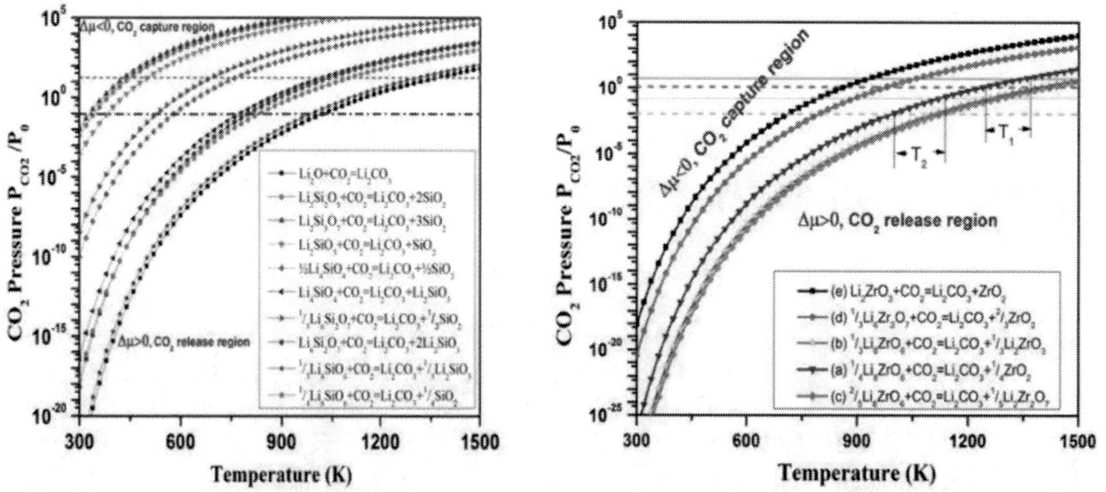


(a) $Li_2O/SiO_2$ (b) $Li_2O/ZrO_2$

**Fig. 2.** Contour plots of the calculated chemical potential ($\Delta\mu$) versus temperature and the $CO_2$ pressure (*P* plotted in logarithmic scale) for the $CO_2$ capture reactions. Only $\Delta\mu$=0 curve is shown explicitly. For each reaction, above its $\Delta\mu$=0 curve, their $\Delta\mu$<0, which means the sorbents absorb $CO_2$ and the reaction goes forward, whereas below the $\Delta\mu$=0 curve, their $\Delta\mu$>0, which indicates $CO_2$ start to be released and reaction goes backward with regeneration of the sorbents.

# III. Results and Discussions

When we mix two solids *A* and *B* to form a new sorbent *C*, the turnover temperature of the newly resulted system ($T_C$) is located between those of *A* and *B* ($T_A$, $T_B$). Here it was assumed that *A* is a strong $CO_2$ sorbent while *B* is a weak $CO_2$ sorbent and $T_A>T_B$. Also, we assumed that the desired operating temperature $T_O$ is between $T_A$ and $T_B$ ($T_A>T_O>T_B$). Now, depending on the properties of *A* and *B*, we have typically three scenarios to synthesize the mixing sorbent *C*:

**3.1. $T_A>>T_B$ and the *A* component is the key part to capture $CO_2$.**

An example of this case is represented by $Li_2O$. This is a very strong $CO_2$ sorbent which forms $Li_2CO_3$. However, its regeneration from $Li_2CO_3$ only can occur at very high temperature ($T_A$). In order to move its $T_A$ to lower temperatures, one can mix some weak $CO_2$ sorbents (such as $SiO_2$, $ZrO_2$). Fig. 2 shows the relationship of chemical potential, $P_{CO2}$, and T of the $CO_2$ capture reactions by the mixed $Li_2O/SiO_2$ and $Li_2O/ZrO_2$ solids with different mixing ratios. Fig. 3 shows the turnover temperature and the $CO_2$ capture capacity of $Li_2O/SiO_2$ mixture versus the ratio of $Li_2O/SiO_2$.[5-7, 9, 15, 18-20]

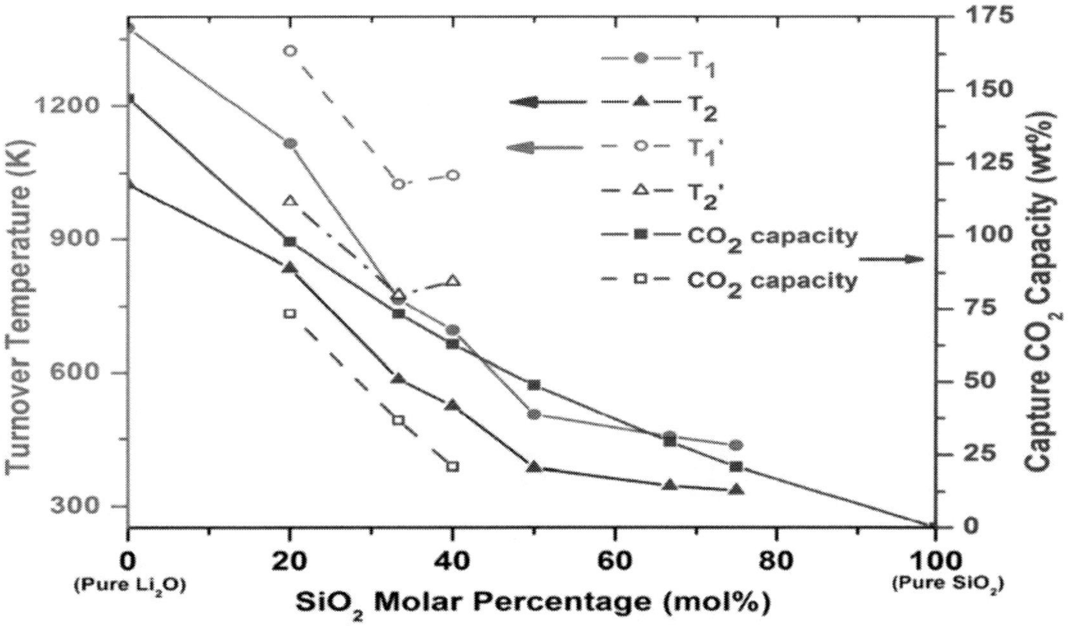


**Fig. 3** The dependence of the turnover temperatures defined in the text and of $CO_2$ capture capacity on molar percentage of $SiO_2$ in the silicates for which calculations are reported here. $T_1$ and $T_1$' are the turnover temperatures under pre-combustion conditions with $CO_2$ partial pressure at 20 bars, while $T_2$ and $T_2$' are the turnover temperatures under post-combustion conditions with $CO_2$ partial pressure at 0.1 bar. The solid lines indicate to convert lithium silicates to $SiO_2$ and $Li_2CO_3$ ($T_1$, $T_2$). For those $Li_2O$-rich lithium silicates ($Li_8SiO_6$, $Li_4SiO_4$, $Li_6Si_2O_7$) capturing $CO_2$, the data shown in dash lines indicate convert them to $Li_2SiO_3$ and $Li_2CO_3$ ($T_1$', $T_2$'). The corresponding $CO_2$ capture capacities are plotted with open blue squares.

From Figs. 2 and 3, one can see that after mixing $Li_2O$ and $SiO_2$ (or $ZrO_2$) with different $Li_2O/SiO_2$ (or $Li_2O/ZrO_2$) ratios, the $T_C$ of the newly formed *C* compound (silicate or zirconate) is lower than $T_A$ of pure $Li_2O$ and could be close to the $\Delta T_o$ range to fit the practical needs.

### 3.2 $T_A >> T_B$ and B component is the key part to capture $CO_2$

In this case, since $T_B$ is lower than $T_O$, by mixing *A* into *B* will increase the turnover temperature $T_C$ of the *C* solid to values closer to $T_o$. For example, pure MgO has a very high theoretical $CO_2$ capture capacity. However, its turnover temperature (250 °C) is lower than the required temperature range of 300-470 °C used in warm gas clean up technology and its practical $CO_2$ capacity is very low, and therefore, it cannot be used directly as a $CO_2$ sorbent in this technology.[21-23] As shown in Fig. 4, by mixing alkali metal oxides $M_2O$ (M=Na, K, Cs, Ca) or carbonates ($M_2CO_3$) into MgO, the corresponding newly formed mixing systems have higher turnover temperatures making them useful as $CO_2$ sorbents through the reaction $MgO + CO_2 + M_2CO_3 = M_2Mg(CO_3)_2$.[21, 23]

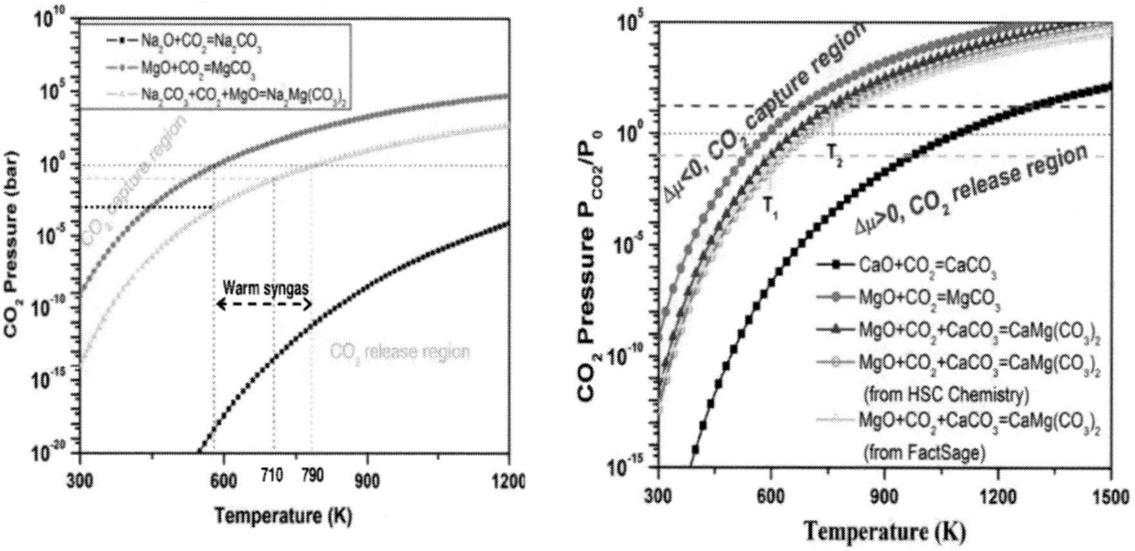


(a) $MgO/Na_2CO_3$ ($Na_2O$) (b) $MgO/CaCO_3$ ($CaO$)

**Fig. 4.** Plots of the calculated chemical potentials versus $CO_2$ pressures and temperatures for the $CO_2$ capture reaction by MgO to form double salts.

### 3.3 $T_A$ and $T_B$ are close to each other and both A and B are active for $CO_2$ capture

In this case, both *A* and *B* components are active to capture $CO_2$, and the $CO_2$ capacity of the mixture is the summation of those of *A* and *B*. As we know another potential advantage of mixing solids is to increase the surface area of the solids in order to have faster reaction rates. Such a mixing scenario doesn't show too much advantage in shifting the capture temperature, but may enhance the kinetics of the capture process and eventually make the mixtures more efficient. Although there is no such report in literature, we think such an attempt is worthwhile and are working on several doped systems. Figure 5 shows a case of Na-/K- doped $Li_2ZrO_3$ capture $CO_2$ through the following reaction (b):[24]

$$LiMZrO_3 + CO_2 \leftrightarrow \frac{1}{2}Li_2CO_3 + \frac{1}{2}M_2CO_3 + ZrO_2 \qquad (M = Na, K) \qquad \text{(b)}$$

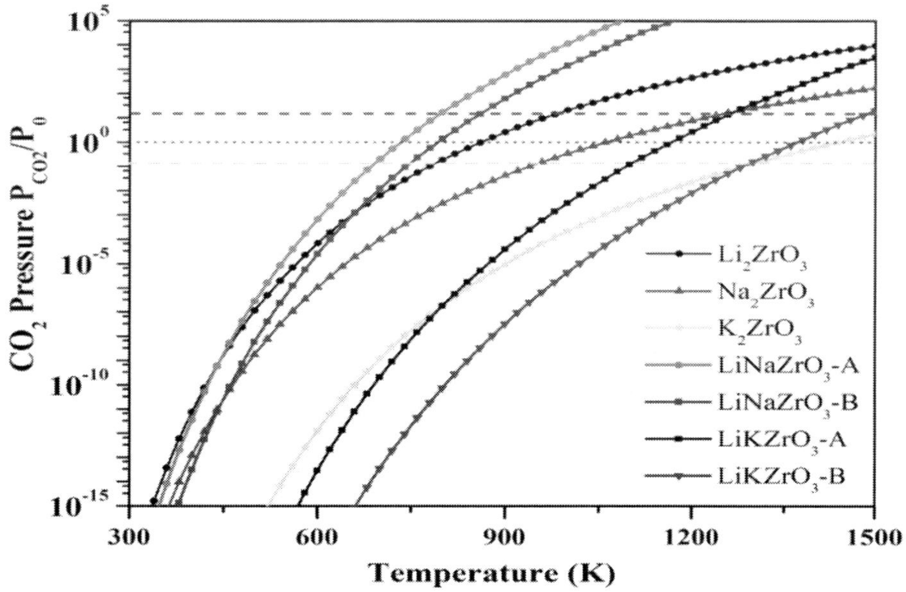


**Fig. 5**. The contour plotting of calculated chemical potentials versus $CO_2$ pressures and temperatures of the $CO_2$ capture reactions by $LiMZrO_3$ and $M_2ZrO_3$. Y-axis plotted in logarithm scale. Only the $\Delta\mu=0$ curve is shown explicitly. For each reaction, above its $\Delta\mu=0$ curve, their $\Delta\mu<0$, which means the solids absorb $CO_2$ and the reaction goes forward, whereas below the $\Delta\mu=0$ curve, their $\Delta\mu>0$, which means the $CO_2$ start to release and the reaction goes backward to regenerate the sorbents.

All of the reactions are thermodynamically favorable for $CO_2$ sorption over a quite wide range of temperatures and reduced $CO_2$ pressures. But a well preforming $CO_2$ solid sorbent should not only easily absorb $CO_2$ in the first half cycle but also easily release $CO_2$ from the product ($M_2CO_3$ and $ZrO_2$ for example) in the second half cycle. Table 1 shows the calculated thermodynamical properties and turnover temperatures of $LiMZrO_3$ capture $CO_2$ reaction (b).

From Fig.5 and Table 1, one can see that they are not good sorbents for capturing $CO_2$ for pre-combustion technology. However, they could be used for high-temperature post-combustion $CO_2$ capture with $T_2$=685K, 745 K for $LiNaZrO_3$ (phases _A and _B) and $T_2$=1095K, 1285 K for $LiKZrO_3$ (phases _A and _B) respectively. When compared to $Li_2ZrO_3$, the K substituted system $LiKZrO_3$ does not decrease the turnover temperature of the $CO_2$ capture reaction. In other words, although the $LiKZrO_3$ has a stronger driving force to capture $CO_2$, with improved $CO_2$ capture capability when compared to $Li_2ZrO_3$, its regeneration only can happen at higher temperature. Therefore, $LiKZrO_3$ is not a good $CO_2$ sorbent, since it requires more energy to regenerate. In contrast, the Na-substituted sorbent system $LiNaZrO_3$ does have a more compatible turn over temperature, making it a possible good candidate as a $CO_2$ sorbent. Such conclusions are in good agreement with experimental findings.[25-27]

In summary, from the calculated thermodynamic properties of $LiMZrO_3$ reacting with $CO_2$, we found that the $CO_2$ sorbent performance of $LiNaZrO_3$ is better than that of $Li_2ZrO_3$, but the performance of $LiKZrO_3$ is much worse since its regeneration temperature is higher. Therefore, the Na doped $Li_2ZrO_3$, $LiNaZrO_3$, is a better $CO_2$ sorbent for post-combustion capture technology. Our calculated thermodynamic results showed that the K doped $Li_2ZrO_3$, $LiKZrO_3$, didn't gain any improvements over $Li_2ZrO_3$ on its overall $CO_2$ capture performance.

**Table 1**. The weight percentage of $CO_2$ capture, the calculated energy change $\Delta E^{DFT}$, the zero-point energy changes $\Delta E^{ZP}$ and the thermodynamic properties ($\Delta H$, $\Delta G$) of the $CO_2$ capture reactions. (unit: kJ/mol). The turnover temperatures ($T_1$ and $T_2$) of the reactions of $CO_2$ capture by solids under the conditions of pre-combustion ($P_{CO2}$=20 bar) and post-combustion ($P_{CO2}$=0.1 bar) are also listed.

| reaction | absorbing $CO_2$ Wt% | $\Delta E^{DFT}$ | $\Delta E^{ZP}$ | $\Delta H$ (T=300K) | $\Delta G$ (T=300K) | Turnover T (K) | |
|---|---|---|---|---|---|---|---|
| | | | | | | $T_1$ | $T_2$ |
| $Li_2ZrO_3+CO_2 \leftrightarrow Li_2CO_3+ZrO_2$ [a] | 28.75 | -146.648 | 11.311 | -158.562 | -103.845 | 1000 | 780 |
| $LiNaZrO_3$_A$+CO_2 \leftrightarrow ½(Li_2CO_3+Na_2CO_3)+ZrO_2$ | 26.01 | -152.936 | 7.069 | -176.666 | -110.892 | 805 | 685 |
| $LiNaZrO_3$_B$+CO_2 \leftrightarrow ½(Li_2CO_3+Na_2CO_3)+ZrO_2$ | 26.01 | -167.872 | 6.934 | -191.526 | -126.477 | 865 | 745 |
| $LiKZrO_3$_A$+CO_2 \leftrightarrow ½(Li_2CO_3+K_2CO_3)+ZrO_2$ | 23.75 | -264.115 | 6.006 | -287.513 | -225.611 | 1275 | 1095 |
| $LiKZrO_3$_B$+CO_2 \leftrightarrow ½(Li_2CO_3+K_2CO_3)+ZrO_2$ | 23.75 | -311.604 | 7.080 | -332.612 | -272.410 | hT[b] | 1285 |
| $Na_2ZrO_3+CO_2 \leftrightarrow Na_2CO_3+ZrO_2$ [a] | 23.76 | -140.862 | 2.236 | -158.327 | -114.121 | 1275 | 925 |
| $K_2ZrO_3+CO_2 \leftrightarrow K_2CO_3+ZrO_2$ [a] | 20.24 | -223.158 | 5.813 | -238.490 | -187.884 | hT[b] | 1285 |

[a] from Refs.[15, 19, 20].
[b] hT means the temperature is higher than our temperature range (1500K)

## IV. Conclusions

The obtained results showed that by changing the mixing ratio of solid *A* and solid *B* to form mixed solid *C* it's possible to shift the turnover $T_t$ of the newly formed solid *C* to fit the practical $CO_2$ capture technologies. When mixing $SiO_2$ or $ZrO_2$ into the strong $Li_2O$ sorbent, one can obtain a series of lithium silicates (or zirconates) with $T_t$ lower than that of pure $Li_2O$. By mixing oxides ($Na_2O$, $K_2O$, CaO) or their corresponding carbonates into MgO, the obtained mixtures exhibit different thermodynamic behaviors and their $T_t$ are higher than that of pure MgO. Such results can be used to provide insights for designing new $CO_2$ sorbents. Therefore, although one single material taken in isolation might not be an optimal $CO_2$ sorbent to fit the particular needs to operate at specific temperature and pressure conditions, by mixing or doping two or more materials to form a new material, our results showed that it is possible to synthesize new $CO_2$ sorbent formulations which can fit the industrial needs. Our results also show that computational modeling can play a decisive role for identifying materials with optimal performance.

## Acknowledgement

The author thanks Drs. D. C. Sorescu, D. Luebke, H. W. Pennline, B. Y. Li, H. Pfeiffer, J. K. Johnson, B. Zhang, K. Zhang, D. King, K. Parlinski for their kind help and collaborations.

## References


1 B. Y. Li, Y. Duan, D. Luebke, and B. Morreale, Applied Energy **102**, 1439 (2013).

2 D. Aaron and C. Tsouris, Separation Science and Technology **40**, 321 (2005).

3 M. R. Allen, D. J. Frame, C. Huntingford, C. D. Jones, J. A. Lowe, M. Meinshausen, and N. Meinshausen, Nature **458**, 1163 (2009).

4 R. S. Haszeldine, Science **325**, 1647 (2009).

5 Y. Duan, D. Luebke, and H. W. Pennline, International Journal of Clean Coal and Energy **1**, 1 (2012).

6 Y. Duan and D. C. Sorescu, Phys Rev B **79**, 014301 (2009).

7 Y. Duan and D. C. Sorescu, J Chem Phys **133**, 074508 (2010).

8 Y. Duan, in *Proceedings of 7th~12th Ann. Conf. on Carbon Capture, Sequestration & Utilization*, Pittsburgh, 2008-2013).

9 Y. Duan and K. Parlinski, Phys Rev B **84**, 104113 (2011).

10 Y. Duan, B. Zhang, D. C. Sorescu, and J. K. Johnson, J. Solid State Chem. **184**, 304 (2011).

11 Y. Duan, D. R. Luebke, H. W. Pennline, B. Y. Li, M. J. Janik, and J. W. Halley, Journal of Physical Chemistry C **116**, 14461 (2012).

12 Y. Duan, B. Zhang, D. C. Sorescu, J. K. Johnson, E. H. Majzoub, and D. R. Luebke, Journal of Physics-Condensed Matter **24**, 325501 (2012).

13 B. Zhang, Y. Duan, and J. K. Johnson, J Chem Phys **136**, 064516 (2012).

14 Y. Duan, Phys. Rev. B **77**, 045332 (2008).

15 Y. Duan, J Renew Sustain Ener **3**, 013102 (2011).

16 Y. Duan, J Renew Sustain Ener **4**, 013109 (2012).

17 M. W. J. Chase, J. Phys. Chem. Ref. Data, **Monograph 9**, 1 (1998).

18 Y. Duan, Physical Chemistry Chemical Physics **15**, 9752 (2013).

19 Y. Duan, H. Pfeiffer, B. Y. Li, I. C. Romero-Ibarra, D. C. Sorescu, D. R. Luebke, and J. W. Halley, Physical Chemistry Chemical Physics **15**, 13538 (2013).

20 Y. Duan, J Renew Sustain Ener **4**, 013109 (2012).

21 K. L. Zhang, X. H. S. Li, Y. Duan, D. L. King, P. Singh, and L. Y. Li, International Journal of Greenhouse Gas Control **12**, 351 (2013).

22 K. L. Zhang, X. H. S. Li, W. Z. Li, A. Rohatgi, Y. Duan, P. Singh, L. L. Y., and D. L. King, Advanced Materials Interfaces (2014) accepted.

23 Y. Duan, K. L. Zhang, X. H. S. Li, D. L. King, B. Y. Li, L. F. Zhao, and Y. H. Xiao, Aerosol and Air Quality Research, DOI: 10.4209/aaqr.2013.05.0178 (2014).

24 Y. Duan, ScienceJet, submitted (2014).

25 M. Y. Veliz-Enriquez, G. Gonzalez, and H. Pfeiffer, J Solid State Chem **180**, 2485 (2007).

26 H. Pfeiffer, E. Lima, and P. Bosch, Chemistry of Materials **18**, 2642 (2006).

27 H. Pfeiffer, C. Vazquez, V. H. Lara, and P. Bosch, Chemistry of Materials **19**, 922 (2007).

# 351628 Utilization of Carbonic Anhydrase-Displaying Escherichia Coli in the Foam Bioreactor to Capture and Sequester Carbon Dioxide

**Abstract:**

In recent years, global warming as the result of greenhouse gas emissions, especially that of $CO_2$, has become of great concern. Many technologies have been proposed and employed to capture post-combustion $CO_2$. Conventional methods such as absorption and adsorption have the problems of regeneration of spent solvents and media requiring additional energy input and expensive operating costs. For these reasons, the carbonic anhydrase enzyme (CA), which quickly catalyzes the conversion of $CO_2$ into bicarbonate ion and a proton, has been studied for use in carbon capture and sequestration (CCS) technologies. This process would solve the limitations of current carbon capture technologies because it readily converts $CO_2$ into environmentally friendly products at mild conditions. Furthermore, upon $CO_2$ removal, the aqueous bicarbonate can be precipitated and separated as calcium carbonate (bio-mineral) as a value-added product. However, current CA-driven carbon capture remains expensive: the purified enzyme itself costs in the range of thousands of dollars per gram; the enzyme works best in slightly alkaline conditions, so some buffering capacity is necessary to combat the protons it generates; and CA must be replaced over time as it gradually loses its activity. The major goals of this project are to employ genetically modified *E. coli* to continuously produce and display CA on their cell membranes and to use these bacteria in the Foam Bioreactor to capture and sequester $CO_2$. The bacteria incubated in auto inducible media (AIM) have shown successful growth and enzyme expression as well as high activity and foaming capacity in the Foam Bioreactor. Because of the huge gas-to-liquid surface area in the Foam Bioreactor, the $CO_2$ in gas phase was quickly transferred to many CA-displaying bacteria, allowing the CA to convert $CO_2$ to bicarbonate ions and protons. Under *ex situ* pH control and an inlet stream of 0.5-2% $CO_2$ at 1 L/min, the increased cell concentrations enhanced the removal efficiency and elimination capacity of $CO_2$. A decrease in gas flow rate from 1.0 to 0.4 L/min at an inlet stream of 0.5-2% $CO_2$ also enhanced the removal efficiency of $CO_2$. Future work will include the effect of tertiary amine solutions (i.e., methyl diethanolamine) to enhance mass transfer and overall removal efficiency of $CO_2$, the *in situ* pH control in the foam for optimal CA activity inside the reactor, and the mathematical modeling of CA-driven carbon capture in the Foam Bioreactor.

# 352818 Purification of Waste Carbon Dioxide from Ethylene Glycol Plants and its' Commerical Application

**Abstract:**

Reduction of waste and convert them to the valuable product is the key challenge in today's competitive petrochemical business. Large amount of the carbon dioxide is treated as waste and vented to atmosphere from ethylene glycol plant. Carbon dioxide (CO2) is one of the byproducts produced during the manufacturing process of ethylene glycol. Instead, it may be used as one of the raw material in urea & methanol process, soda ash, baking powder etc. Moreover, it can be liquefy up to food grade or can be utilized for dry Ice, cold drink, beverage preparation, even as preservative etc. These studies will emphasis on opportunity to reduce carbon dioxide emission to atmosphere and utilization in ecofriendly manner by conversion into competitive petrochemical products. This paper will also describes technology in detail, how to recover & purified the waste carbon di oxide to meet the specification of commercial & food grade as a raw material. The purification of waste CO2 will provide two fold benefit. First, there will be reduction in the large CO2 venting & subsequent reduction of pollution & global warming. Secondly, availability of raw materials from waste stream, which will reducing the production cost.

# 352865 Fuel Based Approach for CO2 Emission Factor Estimation

**Abstract:**

Global warming is an issue facing the world nowadays and $CO_2$ emissions are contributing directly to this issue. There are many ways to quantify and predict $CO_2$ emissions from different sources. Combustion process is considered to be the major source of $CO_2$ emissions. There are many published $CO_2$ emission factors in order to find the released emissions. In this study, $CO_2$ emission factors from fuel specifications for oil and petrochemical industry is used and compared with the published data. It was found that $CO_2$ emissions vary with quality of fuel and operating condition. For example, the average $CO_2$ emission factor for gas turbine was calculated as 114 lb/MMbtu. This emission factor has a good match with published emission factor of EPA and CAPP data.

# 350799 Adsorption of CO2 and H2S from Natural Gas by Activated Fly Ash

**Abstract:**

Waste oil fly ash (OFA), a by- product of fossil fuel combustion from power generating plants, was successfully convert to activated carbon using both chemical and physical treatment. A combination of three acids (Sulfuric, nitric and phosphoric acids) was mixed at different volume ratios and concentrations with the ash, followed by physical treatment at 990 oC while $CO_2$ is introduced at flow rate of 1 L/min. The produced AC was characterized according to its surface area and morphology. BET surface area has increased from less than 1 $m^2/g$ for untreated ash to 375 $m^2/g$ after both chemical and physical activation. Then it is utilized in the removal of $CO_2$ and $H_2S$ from natural gas stream. The maximum adsorption capacity of the treated ash was found to be 240 mg/g and 175 mg/g, respectively. The adsorption rate increased with the increase in flow rate and concentration. The results indicated the potential of using waste OFA in acidic gases removal from downstream.

# 351274 Energy Systems Model: Development of a Discrete-Event Simulation Tool to Facilitate Operating Decisions

**Abstract:**

The Dow Energy business is responsible for managing and providing power, steam and other utilities to some of Dow's largest sites located around the world. At certain sites, 100% of the steam requirements are generated on-site, and power is additionally supplied to the external grid and any disruption to the steam and energy supply would severely impact operations. An energy plant typically contains gas and steam turbines, boilers, synchronous condensers, and transformers. Each of these assets has the potential to experience unplanned shutdowns, in addition to scheduled maintenance shutdowns. It is important to ensure that steam and power requirements are still met, despite the uncertainties associated with asset shutdowns. Interrupted steam and power supply can cause short-term problems in meeting customer commitments and damage customer confidence in Dow's supply chain .This drives the need for a model that effectively captures energy system operations so that data based decisions on effective and robust operating strategies and capital investments can be made. We present an approach that would enable evaluation of the reliability of steam and power supply under various conditions (e.g. the removal/addition of new assets) through computationally inexpensive simulations.

Our model development includes (1) extracting asset reliability information from asset specific historical data and used in the model to accurately depict unplanned shutdowns; and (2) understanding interactions between the various energy assets in order to accurately represent the energy plant from a steam and power perspective. The model is based on discrete event simulations (DES) and has been developed in ExtendSim®. Our model output focuses on statistics that would be pertinent to operating decisions and includes: equipment reliability, steam and power supply reliability. We additionally focus on developing a tool that is user-friendly to promote its use, longevity, and ability to easily leverage to other sites in the energy business. We then demonstrate the use of the tool on a hypothetical case study that aims to drive key business decisions.

In summary, the modeling approach we present involves the development of a discrete event simulation-based tool that can be used to facilitate operating decisions by quantitatively characterizing a site's exposure to the risk of not obtaining the required steam and/or power supply. The tool would be used to drive key proactive energy decisions that include process development and enhancement related to steam and power assets.

# Oxygenated Compounds as Reducing Agent for $NO_x$ Removal by Selective Non-Catalytic Reduction (SNCR) Method

AIChE 2014 Spring Conference

Maoqi (Mark) Feng, Ph.D.

**Southwest Research Institute®**

March 31, 2014

# Agenda

- Introduction
- Background
- Objectives
- Technical Approach
- Test Results
- Summary and Conclusions

# Introduction

- $NO_x$ are of regulated emissions from stationary sources
  - Clean Air Act (CAA), Section 111(b) requires to use "*best demonstrated technology* (BDT)"
  - California requires up to **75%** reduction in $NO_x$ emissions
  - Texas Natural Resources Conservation Commission (TNRCC) adopted new rules for Houston-Galveston Area (HGA) - **90%** reduction for major industrial sources

- Examples of Stationary Sources
  - Coal-fired power plants
  - Waste incinerators
  - Process heaters

3

# Technical Background (1 of 4)

- Commonly used $NO_x$ reduction technologies
  - Flue Gas Recirculation (FGR)
    - ✓Similar to EGR in automotive $NO_x$ control
    - ✓FGR $\Rightarrow O_2 \Downarrow \Rightarrow$ Combustion Temperature $\Downarrow \Rightarrow NO_x \Downarrow$
  - Selective Catalytic Reduction (S**C**R)
    - ✓$4NO + 4NH_3 + O_2 \Rightarrow 4N_2 + 6H_2O$ (Major Reaction)
    - ✓$NO + NO_2 + 2NH_3 \Rightarrow 2N_2 + 3H_2O$
  - Selective Noncatalytic Reduction (S**NC**R)
    - ✓$4NO + 4NH_3 + O_2 \Rightarrow 4N_2 + 6H_2O$

4

# Technical Background (2 of 4)

➢ Comparison of selected technologies

In-combustion

| Methods | Need a Catalyst? | Temperature, °C | NOx Efficiency | Relative Cost |
|---|---|---|---|---|
| FGR | No | N/A | 10~25% | 0.85 |
| SCR | Yes | 300~450 | 70~90% | 3.5 |
| SNCR | No | 800~1050 | 25~50% | 1.0 |

Post-combustion

5

# Technical Background (3 of 4)
## – SNCR Overview

➢ **Key Terminologies**
- $NH_3$-to-$NO_x$ Ratio (ANR): Typically 2
- $NH_3$ slip

$$4NO + 4NH_3 + O_2 \Rightarrow 4N_2 + 6H_2O$$

➢ **Operating Conditions**
- Temperature (850~1100°C)
- Residence time

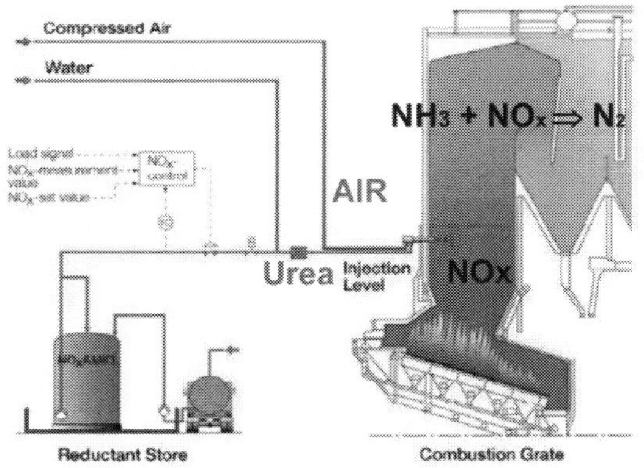

6

## Technical Background (4 of 4)

**– Remaining Issues with SNCR**

- Relatively low $NO_x$ conversion efficiency
  - Flow distribution
    - ✓Reaction efficiency
    - ✓$NH_3$ oxidation
  - Limited residence time
    - ✓Narrow temperature window
    - ✓Typically less than 1.0 second

7

## Objectives

- Improve the current SNCR technology
  - $NO_x$ efficiency ↑ 10-15%
  - Reducing Agent to $NO_x$ ratio = 2 - 4

- Use oxygenated compounds as reducing agent to replace urea (or $NH_3$)

8

## Reaction Conditions

- Simulated flue gas and SNCR conditions

| Parameter | Concentration |
|---|---|
| $NO_x$ | 200, 400, 600 ppm |
| $O_2$ | 10% |
| $H_2O$ | 6% |
| $CO_2$ | 8% |
| $N_2$ | Balance |
| Reducing Agent | $RA/NO_x = 3$ |
| *($SO_2$)* | *0, 200, 500 ppm* |
| *(HCs)* | *0, 200, 400, 600 ppm* |
| Space Velocity: | Residence time 1 s |
| Temperature: | 800, 900, 1000°C |

9

## Reaction Setup

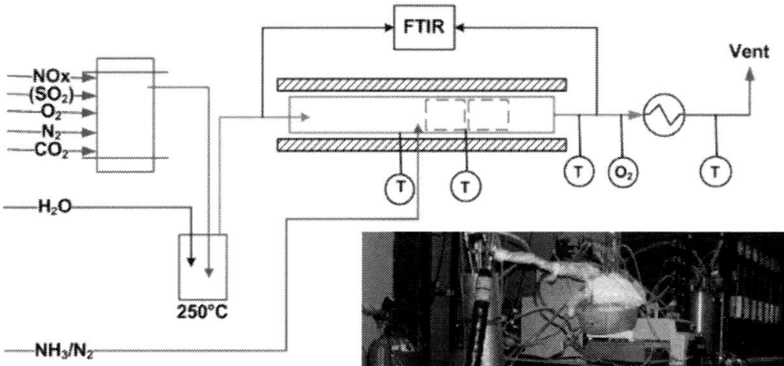

10

## Advantages of Structured Packing

- A solution to combine two functions into one "structure"
  - Improved flow distribution ⇒
    - ✓$NO_x$ efficiency ⇑
    - ✓Reducing agent oxidation ⇓
  - Increased surface area ⇒
    - ✓$NO_x$ efficiency ⇑

11

## Structure Packing

- Three Types of "Structures"
  - Ceramic packing
  - Non-metalic structures
  - Clay cakes

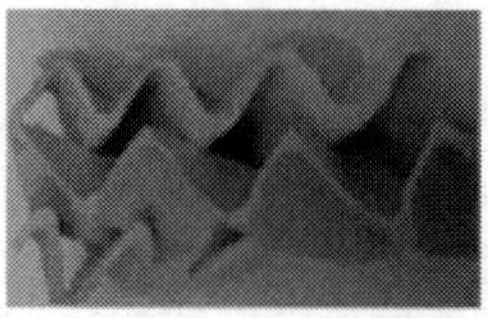

12

## Pressure Drop

- Pressure Drop
  - Issue: High surface area may result in high pressure drop
    - ✓Generally, surface area ⇑ ⇒ pressure drop ⇑
    - ✓High surface area is desired (for NOx efficiency)
    - ✓Low pressure drop is required
  - Mitigation: CFD helps to balance pressure drop vs. surface area

13

## CFD Modeling

- CFD modeling of structured packing for SNCR system
- ✓Three dimensional model was created in SolidWorks
- ✓Software FlowSimulation

- Compare experimental data with modeling data

14

## CFD Modeling

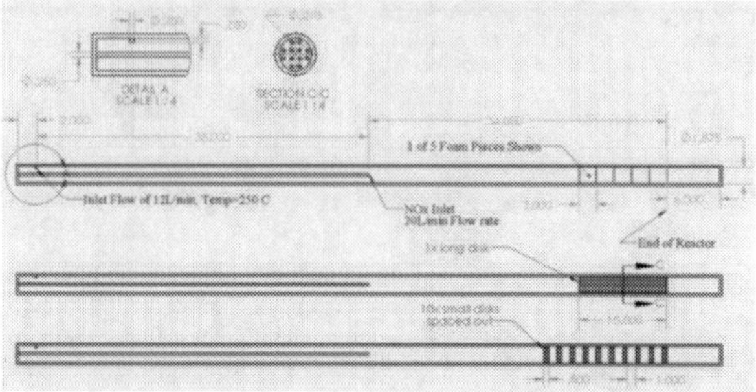


- A - five structured packing
- B - one single insert with channels
- C - 10x small inserts with channels

15

## Mixing

- Air mass fraction contour plot of Configuration A and B

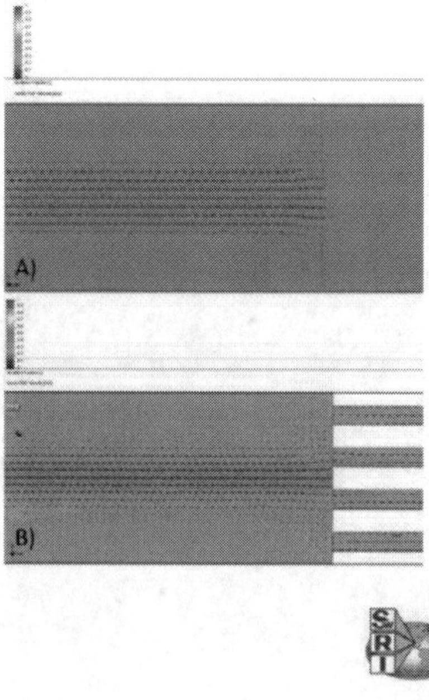


16

## Mixing

- Air mass fraction contour plots with small ranges displayed for Configuration B and C
- Additional vortices created by the second inserted disk in Configuration C

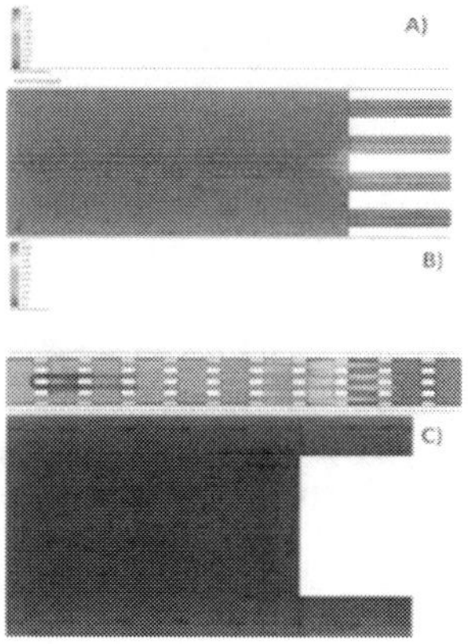


17

## Pressure Drop

- Pressure contour plot of the reactor in Configuration A

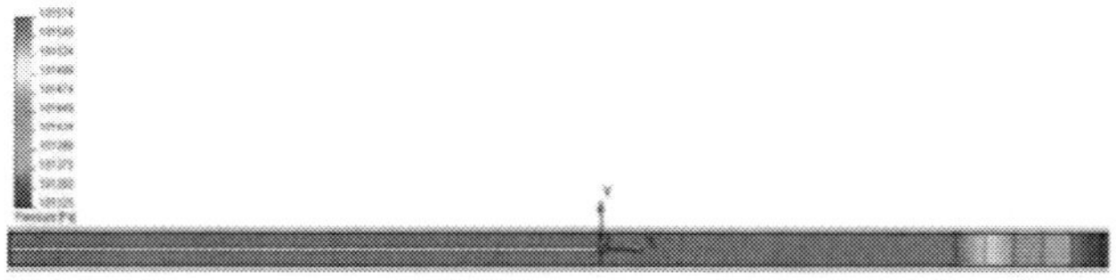

18

# Gas Velocity

- Comparison of velocity contour plot of the three configurations

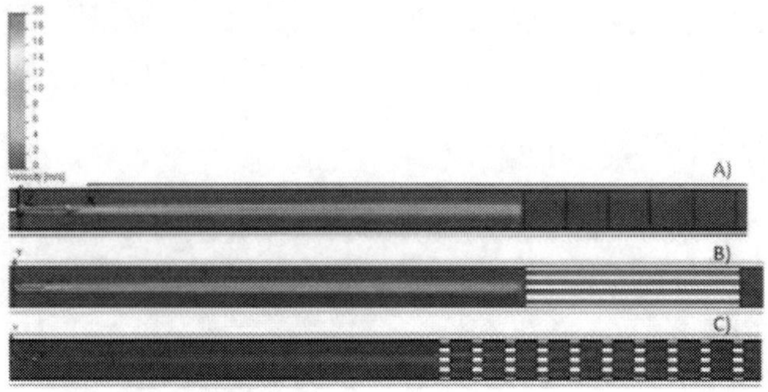


19

# Test Results

- $NO_x$ removal efficiency was increased 10-15% by the structure packing
- CFD modeling predicted the pressure drop of different structure packings
- CFD modeling can be used to simulate the flow and temperature distribution
- Mixing of the reducing agent and NOx is key to SNCR efficiency improvement and preventing reducing agent oxidation

20

## Test Results

- Propane was used as reducing agent for $NO_x$ removal
- A mixture of C1-C4 was used as reducing agent for $NO_x$ removal
- Proprietary oxygenated compounds as reducing agent (US Patent 8,535,627)
- Effect of hydrogen was tested – $NO_2$ reduction

21

## Test Results

- Effect of CO was tested - $NO_2$ reduction
- $SO_2$ effect: Not directly related to NOx removal efficiency; corrosion
- Water effect: slightly decrease the NOx removal efficiency

22

## Summary and Conclusions

- Introducing a well designed "structure" improved SNCR $NO_x$ removal efficiency
- CFD modeling was useful in process intensification
- Oxygenated compounds reduced plugging due to NH3 slip and ammonium sulfate formation
- Less corrosive

23

## Acknowledgement


- This project was funded by the internal research fund of Southwest Research Institute (Project No. 01-R8213).


Thank you!

Questions?

24

# 353086 Flare Emission and Greenhouse Gas Reduction Study during an Ethylene Plant Startup on Recycle through Dynamic Simulation and Process Optimization

**Abstract:**

Currently, emission characterization and reduction have been gaining more attention in the Chemical Process Industry and playing a more prominent part in industrial sustainability. Flare emissions release volatile organic compounds (VOCs) and pollutants into the atmosphere, which react with nitrous oxide in sunlight to generate ozone and smog. Therefore, it is essential to limit this amount to improve local and regional air quality, especially around chemical plant locations. Besides, environmental performance also takes into account material usage, energy consumptions, non-product outputs and pollutant releases. From the economic point of view, reducing burned-off gases saves material, helps avoiding penalty and also creates a more effective way to distribute energy, thus lowering utility cost. Ethylene production process is one of the biggest and most complicated processes in petrochemical industry. A typical start-up of an ethylene plant flares approximately 5.0 million lbs of raw materials and generates at least 15.4 million lbs of $CO_2$, 40.0 Klbs CO, 7.4 Klbs $NO_x$, 15.1 Klbs hydrocarbons, and 100.0 Klbs HRVOCs. Hence, managing flare emission during startup, shutdown and malfunction of an ethylene plant are very important. Since flaring occurs in both normal and abnormal conditions, flare source reduction approach employs simulation and optimization, emphasizes the upstream section of flare system and the whole plant integration, in which more upstream modification and optimization contribute in limiting unwanted products.

In this study, rigorous steady-state and dynamic models of a front-end De-Ethanizer ethylene plant with sulfur recovery unit are constructed to serve as both the generic foundation to further explore causes roots of process abnormal behaviors and subsequently an optimization tool to provide guidance to both design and operational strategies. Rigorous steady-state and dynamic models are constructed and validated before case studies are carried out. Based on these models, possible flaring sources and incidents are identified, which are demonstrated extensively in this paper throughout three case studies of CGC startup, precooling chilling train, and withdrawal from full-reflux columns during an approach of flareless startup on total recycle.

Because charged gas is continuously produced from furnaces and sent to flare to retain compressors' inlet pressure, CGC startup is reportedly the most critical step and has largest amount of vent gas. Several scenario cases of charged gas at different flowrates and compositions are examined to find the least flaring one. Another case study focuses in recycling off-specs from full-reflux C2-/C3-supplier columns to CGC as partial substitute for furnace feed. In the chilling train section, specifications of equipment sizes and heat capacities are considered for simulating pre-cooling process. Optimal procedure is achieved for utilizing methane to cool down cold boxes at a reasonable temperature-dropping rate in order to protect these equipments. In dynamic simulation environment, all of the major steps and options are simulated and evaluated while flaring can also be predicted. These results serve as a basis for emission calculation and greenhouse gas study such as $CO_2$ control. The process can be considered from smaller scale (sectional model) to more comprehensive scale (whole plant model). The best results in these case studies are incorporated and validated in the integrated model of the whole plant in order to stabilize the system in the shortest time while minimizing material and energy loss.

Following this representative, utilizing dynamic simulation as a driving force for operation during abnormal conditions effectively provides an in-depth understanding and provides more detailed planning and expectations for all related personnel before the real action takes place. This procedure contributes a great deal in evaluating new strategies, predicting process dynamic behaviors, and achieving more precise timing to control commissioning and startup strategy.

# 351658 Chemical Plant Startup Simulations for Flare Emission Reduction

**Abstract:**

HRVOC (Highly Reactive Volatile Organic Compounds) emissions from industries are highly responsible for ground-level ozone pollution. Especially, HRVOC emissions from flares of MSS (maintenance, startup and shutdown) & events are the biggest. Thus, reduction of flare emissions could significantly reduce the total HRVOC emissions, and could reduce ozone pollution. On the other hand, reduction of flare emissions could save raw materials and intermediate materials, decrease startup/shutdown time, and thus reduce the total cost of startup and shutdown operations.

In this paper, startup procedures of an ethylene plant were simulated with dynamic simulation. The ethylene plant is a typical front-end demethanizer process. Simulation models have all the major process facilities, including cracking furnaces, quench towers, charge gas compressor (CGC) system, chilling train, demethanizer, deethanizer, depropanizer, debutanizer, C2 converter, C3 converter, C2 splitter and C3 splitter. Dynamic simulation covers all startup steps which involve flare emissions. The simulation results have been validated with real data from ethylene plants.

Three startup scenarios were simulated with dynamic models. The first scenario is a "baseline" scenario, which means no flare reduction methods are used in the scenario. The overall emissions of the baseline scenario are compared with emission data from actual startup events reported to local air quality authorities, and they are within normal range.

The second scenario simulates startup procedure with certain flare reduction methods, including commissioning distillation columns before CGC starting, using gas from columns to help CGC start, and precooling chilling train. An optimized startup schedule is built based on dynamic simulation. Comparing the baseline scenario, the second scenario reduces total startup time by 20%, and reduces total vent gas by 47%.

The third scenario simulates nitrogen startup procedure, which means use nitrogen to start CGC and precooling chilling train. The startup schedule is also optimized based on simulation results. Comparing the baseline scenario, the third scenario reduces total startup time by 36%, and reduces total vent gas by 58%.

# 352687 Dynamic Simulation for Flare Minimization Strategy in an Ethylene Plant Shutdown


**Abstract:**

Ethylene plant shutdowns generate a great amount of flare emissions that inevitably damage air quality of the surrounding environment. During an ethylene plant shutdown, plant generally cut off feed and flared off all off-spec products until the plant was de-inventoried. Here, an optimal shutdown strategy is designed for reducing flare emission and preventing losses of hydrocarbon materials. In this paper, dynamic simulations are employed to model the operations of shutdown for both the general strategy and the optimal strategy.

First, the general shutdown strategy is simulated and dynamic emission profile is presented. The shutdown of an ethylene plant would be executed following three steps: (i) Preparation of shutdown. Feed rates are reduced to minimize the liquid inventories in facilities ahead of the shutting down. (ii)Shutting down. The plant is shut down, beginning with shutting down Cracked Gas Compressor (CGC), followed by gradually cutting off the furnace feed. (iii) Decommissioning including liquid discharge, vapor discharge and $N_2$ purge.

Second, the optimal shutdown strategy is simulated to recover more liquid and vapor hydrocarbon materials in the facilities. Compared with the basic shutdown strategy, the optimal shutdown strategy includes certain flare reduction methods. For example, after shutting down the furnaces, gas from other facilities is used to keep CGC running to help recover leftover hydrocarbon materials in the flash drums and columns, rather than send hydrocarbon materials to flare. In the dynamic model, recycle lines are built and the operations are simulated to monitor the key process variables and study the feasibility. So, through the dynamic simulations, the shutdown strategy can be investigated to help the

the operations of shutdown for both the general strategy and the optimal strategy.

First, the general shutdown strategy is simulated and dynamic emission profile is presented. The shutdown of an ethylene plant would be executed following three steps: (i) Preparation of shutdown. Feed rates are reduced to minimize the liquid inventories in facilities ahead of the shutting down. (ii)Shutting down. The plant is shut down, beginning with shutting down Cracked Gas Compressor (CGC), followed by gradually cutting off the furnace feed. (iii) Decommissioning including liquid discharge, vapor discharge and $N_2$ purge.

Second, the optimal shutdown strategy is simulated to recover more liquid and vapor hydrocarbon materials in the facilities. Compared with the basic shutdown strategy, the optimal shutdown strategy includes certain flare reduction methods. For example, after shutting down the furnaces, gas from other facilities is used to keep CGC running to help recover leftover hydrocarbon materials in the flash drums and columns, rather than send hydrocarbon materials to flare. In the dynamic model, recycle lines are built and the operations are simulated to monitor the key process variables and study the feasibility. So, through the dynamic simulations, the shutdown strategy can be investigated to help the shutdown reduce the risks on safety, reliability, business and environment.

# 352087 Assessing the Use of Ceramic Membrane Bioreactor for Anaerobic Treatment of High-Load Food Wastewater at Bench Scale and Pilot Scale

**Abstract:**

**Abstract:** A novel type of ceramic membrane was tested in two anaerobic membrane bioreactors (AnMBR): a bench scale reactor (12 L) and a semi-industrial pilot plant (2.5 $m^3$). A highly loaded industrial wastewater from corn-processing was successfully treated, reaching COD reductions close to 95%, while up to 99% removal was achieved for a simulated wastewater. The performance of both AnMBRs is still being optimized for the industrial effluent; the digesters show a similar behaviour and the semi-industrial filtration module seems to have an improved permeability. This study confirms many of the benefits attributed to the AnMBR technology, combining the advantages of the anaerobic digestion with the efficiency and robustness of the ceramic filtration.

**Keywords:** Membrane bioreactor; ceramic membranes; anaerobic digestion; corn-processing wastewater

## INTRODUCTION

The Anaerobic Membrane Bioreactor (AnMBR) combines the anaerobic digestion process with a membrane filtration to provide complete solid-liquid separation. AnMBR system makes possible to work with high biomass concentration to treat industrial complex wastewaters heavily loaded with solids, oil and grease, and/or toxic compounds (Futselaar, 2013; Skouteris, 2012). Anaerobic digestion offers several widely known advantages over conventional aerobic processes, since no oxygen is required and the biogas production reduces the associated energy cost of wastewater treatment and it compensates for high-energy requirements related to cross-flow filtration, which is a common drawback of membrane filtration-based technologies. Nevertheless, it presents some limitations that have limited its wide-spread use (lower purification efficiency, poor effluent quality and process instability). The AnMBR technology allows overcoming these limitations as it combines the advantages of the anaerobic process and MBR technology in a robust and compact solution, characterized by the higher biogas production and the excellent effluent quality due to the small pore size of the membranes and the total retention of the suspended solids. As a result, AnMBRs technology has gained popularity in recent years, and more attention has been focussed on the further development of this process for the treatment of high-concentrated wastewater, mainly food and beverage industry wastewaters (Fuchs, 2003) although they have not yet been employed widely at industrial scale (Grant et al., 2008, Christian et al, 2011). However, recent technical advances in membrane and system design are promising great reductions in operation costs, and thus, broader application of AnMBRs (Scott, 2008). As membrane fouling is a limiting issue for implementing AnMBRs, ceramic membranes may have a good opportunity, since the inorganic membrane materials are reported to present less fouling and to possess higher chemical and thermal resistance (Futselaar, 2013; Skouteris, 2012; Sutton, 2006; Wolff, 2013; Kroll, 2010).

In the present work two ceramic AnMBRs have been studied, a bench scale AnMBR fed with both synthetic wastewater and industrial corn-processing wastewater, and a semi-industrial plant fed with industrial corn-processing wastewater. The objective is to evaluate the maximum digestion capacity of the anaerobic reactor, to study the behaviour of the membrane and to optimize the filtration process to get the maximum permeate flow with an optimal transmembrane pressure (TMP).

## MATERIAL AND METHODS

The AnMBRs consist on a CSRT digester coupled to an external ceramic membrane filtration module equipped with backwash and CIP devices. Both, bench and semi-industrial, reactors are operated in mesophilic conditions (35°C) and with average MLSS of 20 g/L. The start-up was carried out with seed sludge from a municipal anaerobic digester.

### Bench scale AnMBR

The bench scale AnMBR is a 12 L reactor, mounted with a single tubular ceramic membrane of 0.02 $m^2$. Cross-flow velocities of 2 and 3 m/s and different backwash sequences have been tested. The bioreactor was initially fed with a simulated wastewater with high starch content, increasing the organic load (OLR) gradually from 0.7 to 7 $kgCOD \cdot m^{-3} \cdot day^{-1}$. Then the substrate was replaced with the industrial effluent at an initial OLR of 6 $kgCOD \cdot m^{-3} \cdot day^{-1}$.

### Semi-industrial pilot plant

The semi-industrial pilot plant with the capacity of 2.5 $m^3$ is equipped with a ceramic ultrafiltration system made up of 4 modules consisting of 7 ceramic membranes with 19 tubular channels (1.7 $m^2$ of filtration area per module). The MBR is operated at a CFV of 2 m/s and the reactor is fed with a highly loaded wastewater from the corn cooking process (COD ≈ 35 g/L). The OLR has been increased from 0.25, in the start-up, to 3.5 $kgCOD \cdot m^{-3} \cdot day^{-1}$.

## RESULTS AND CONCLUSIONS

**In the bench scale AnMBR**, a 95-99% of both COD and TOC was removed from the simulated effluent at a gradually increasing OLR from 0.7 to 7.0 $kgCOD \cdot m^{-3} \cdot d^{-1}$. A high quality permeate with an output COD of 60-300 mg/L and TOC of 20-80 mg/L was achieved, whereby the average specific methane production was 0.34 $m3CH_4 \cdot kgCOD_{removed}^{-1}$.

The treatment of industrial corn-processing wastewater obtained about 90-95% of COD and TOC reduction at an OLR of 6.0 $kgCOD \cdot m^{-3} \cdot d^{-1}$, yielding an average specific methane production of 0.30 $m^3CH_4 \cdot kgCOD_{removed}^{-1}$ and permeate of 300-400 mg/L in COD and 70-150 mg/L in TOC, respectively.

A permeate flux up to 50 LMH could be reached in the bench scale single channel filtration module, according to both the flux-step tests and the long-term studies.

Meanwhile, **in the semi-industrial pilot plant** the OLR was increased from 0.25 to 3.5 $kgCOD \cdot m^{-3} \cdot d^{-1}$ with COD reduction up to 95%, similar to the laboratory studies. The biogas production also increased with the OLR obtaining an average specific methane production of 0.32 $m^3CH_4 \cdot kgCOD_{removed}^{-1}$, close to bench scale results. Ceramic filtration worked satisfactorily, having a permeability over 120 $L \cdot h^{-1} \cdot m^{-2} \cdot bar^{-1}$ with permeate flux up to 45 LMH and average TMP of 0.25 bar.

Many of the AnMBR technology advantages were confirmed. The solids retention by membranes results in a high quality effluent, free of suspended solids, bacteria and reduced COD, which allows to reduce or, even, eliminate a post-treatment. Highly hydrophilic ceramic membranes maximized the performance.

REFERENCES

Fuchs, W., Binder, H., Mavrias, G. and Braun, R. (2003). Anaerobic treatment of wastewater with high organic content using a stirred tank reactor coupled with a membrane filtration unit. Water Research, 37, 902-908.

Futselaar. H., Rosink, R., Smith, G., Koens, L. (2013). The anaerobic MBR for sustainable industrial wastewater management. Desalination and Water Treatment, 51(4-6), 1070-1078.

Grant, S., Page, I., Moro, M., Yamamoto, T. 2008 Full-Scale Applications of the Anaerobic Membrane Bioreactor Process for Treatment of Stillage from Alcohol Production in Japan. Proceedings of the Water Environment Federation. WEFTEC 2008: Session 101 through Session 115: 7556–7570.

Kroll, S., Treccani, L., Rezwan, K., Grathwohl, G. 2010 Development and characterisation of functionalised ceramic microtubes for bacteria filtration. Journal of Membrane Science 365(1-2), 447-455.

Scott, C., Grant, S., McCarthy, P., Wilson, D., Mills, D. (2011). The First Two Years of Full-Scale Anaerobic Membrane Bioreactor (AnMBR) Operation Treating High-Strength Industrial Wastewater. Water Practice & Technology, 6 (2). doi:10.2166/wpt.2011.032.

Skouteris, G., Hermosilla, D., Lopez, P., Negro, C., Blanco, A. (2012). Anaerobic membrane bioreactors for wastewater treatment: A review. Chemical Engineering Journal, 198, 138-148.

Sutton, P. M. (2006), Membrane bioreactors for industrial wastewater treatment: applicability and selection of optimal system configuration, Proceedings WEFTEC 2006.

Wolff, C., Beutel, S., Scheper, T. (2013). Tubular membrane bioreactors for biotechnological processes. Applied Microbiology and Biotechnology, 97(3), 929-937.

PROSYS™ Maximize Your Plant.

# Operator Errors and What Can be Done to Minimize

**Global Congress on Process Safety**

March 31, 2014

10TH GLOBAL CONGRESS ON PROCESS SAFETY

PROSYS™ Maximize Your Plant™

10TH GLOBAL CONGRESS ON PROCESS SAFETY

## Presenter

**Tom Nolan**

- **Graduated from Ohio University with B.S. in Chemical Engineering**
- **24 years experience in chemical process industry in a variety of roles**
- **MAIC and DFSS Six Sigma Black Belt**

PROSYS Maximize Your Plant™

# The Cost of Errors

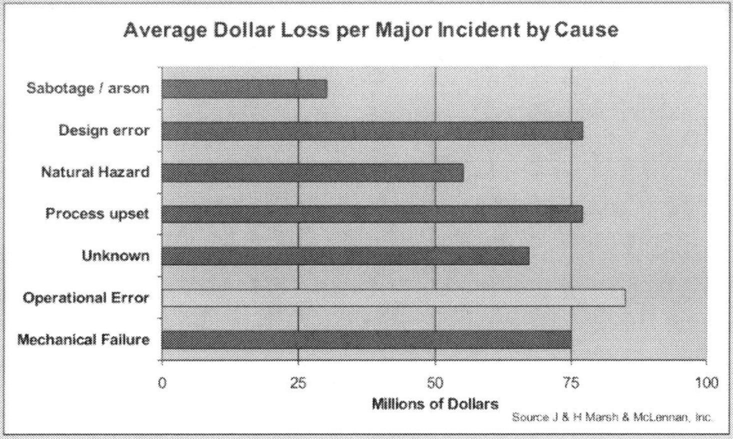


PROSYS Maximize Your Plant™

# The Cost of Operator Errors

- **ASM estimates total loss due to operator error is $8B per year**
- **Chemical Safety Topical Committee - average of one chemical incident per day - cost of over $2 million per incident to comply with requirements**
- **Errors cause 42% of unscheduled shutdowns**
- **70% of process incidents occur during start-up or shutdown**

PROSYS Maximize Your Plant™

# Reasons for Errors

- Lack of Skill
- Lack of Knowledge
- Carelessness
- System Design
  - Operator set up to make errors by inappropriate design – built in errors
  - Errors that are predictable are preventable by better design

PROSYS Maximize Your Plant™

# Focus of this Discussion

- System Design
  - **Alarm Management**
  - **Operator Graphics**
  - Operator set up to make errors by inappropriate design – built in errors
  - Errors that are predictable are preventable by better design

PROSYS Maximize Your Plant™

## Alarm Floods

***In a number of industrial incidents, alarm floods were identified as a significant contributing cause to the incident...***

**As found by EEMUA in 1999 and CSB**

**Alarm Flood defined by ISA 18.2 as -**
***"10 or more annunciated alarms in any 10 minute period per operator"***

PROSYS Maximize Your Plant™

## Why do Alarm Floods Occur?

**One reason is not providing dynamic alarm management**

- **Alarms need to indicate abnormal situations that require operator action**
- **Processes do not operate in one state**
- **What is normal vs abnormal changes with state**
- **Alarms are typically configured for run therefore many alarms are triggered upon a Change of State: Run to Shutdown**
- **Many of the alarms are not applicable or actionable for the new state impleading the operators ability to act quickly on what is important**

PROSYS Maximize Your Plant™

## What is Impacted by Alarm Floods?

- Product quality
- Operability or profitability of the process
- Loss of equipment
- Loss of containment – environmental releases
- Injury and loss of life in plant or community

PROSYS Maximize Your Plant™

## What Makes Alarm Floods so Dangerous?

Can be a problem for three reasons:

- A deluge of alarms can cause critical alarms to be missed
- Floods can be a significant distraction when dealing with process upsets
- Can be an indicator of larger systemic safety issues

PROSYS Maximize Your Plant™

10th GLOBAL CONGRESS ON PROCESS SAFETY

## Impacting Alarm Management Design

- Alarm rationalization is not a process to eliminate alarms - its about quality
- Good rationalization will add alarms when appropriate
- Typical (Static) rationalization is only for run mode
- Dynamic rationalization considers all plant modes

PROSYS Maximize Your Plant™

10th GLOBAL CONGRESS ON PROCESS SAFETY

## Dynamic Alarm Management

- Dynamic rationalization does everything a static rationalization does plus asks "when" for each alarm
- Requires dynamic software to make changes based upon operating mode of the plant
- Eliminates redundant alarms and lowers operator loading during transitions
- Only alarms what is abnormal and actionable for the given state

PROSYS Maximize Your Plant™

10th GLOBAL CONGRESS ON PROCESS SAFETY

# Actual Performance Metrics vs ISA 18.2

| Type | Avg Alarm Rate/hr | Low/Hi of Avg Rate/hr | Peak Alarm Rate/hr | Low/Hi of Peak Rate/hr | % Time in Flood | Standing Alarms >24 hrs |
|---|---|---|---|---|---|---|
| Before Rationalization | 30 | Low=6.2 Hi=61 | 638 | Low=152 Hi=2402 | 17.2% | 9 |
| **ISA 18.2 Target Metrics** | **6** | **---** | **<60** | **---** | **<1%** | **<5** |
| After Dynamic Alarm Management | 2 | Low=0.09 Hi=2.5 | 25 | Low=13 Hi=42 | 0.25% | 4 |

PROSYS Maximize Your Plant™

10th GLOBAL CONGRESS ON PROCESS SAFETY

# Results After Dynamic Alarm Management

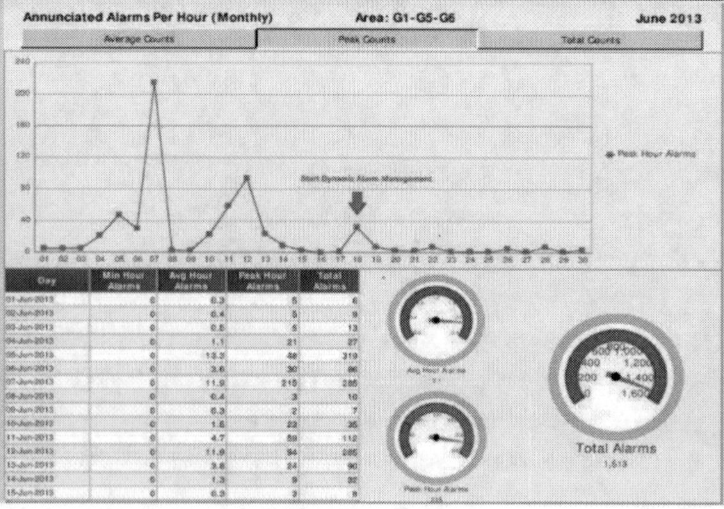

PROSYS Maximize Your Plant™

10th GLOBAL CONGRESS ON PROCESS SAFETY

## Operator Graphics – The Risk of Changing Focus

- Distraction of changing focus from process graphic to faceplate window
- Added workload from managing multiple open windows
- Increased probability of errors when changing values for un-intended tag with multiple faceplates open
- All of above become multiplied and more complex when process is transitioning from one state to another

PROSYS Maximize Your Plant™

10th GLOBAL CONGRESS ON PROCESS SAFETY

## Faceplate Operation

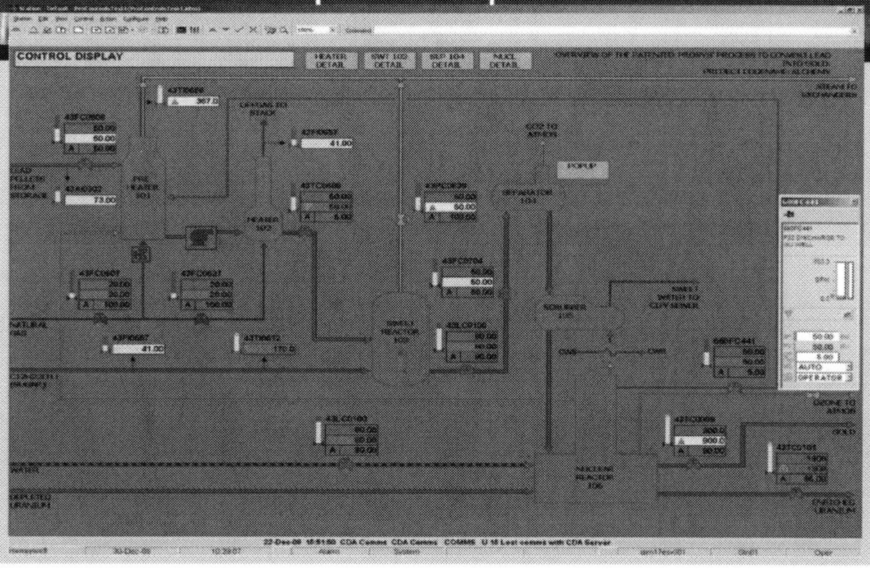

PROSYS Maximize Your Plant™

10th GLOBAL CONGRESS ON PROCESS SAFETY

# Potential for Errors

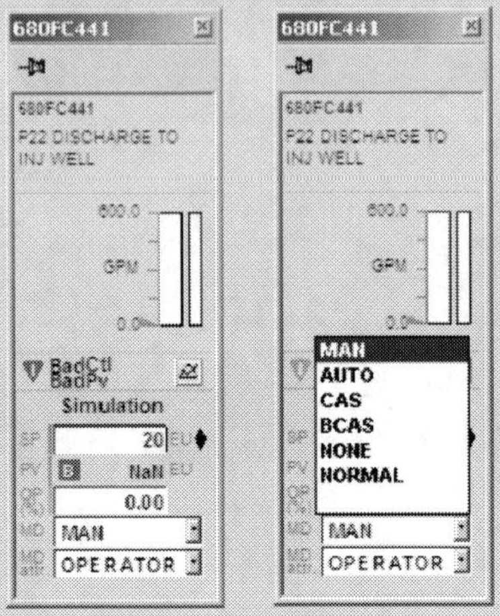


**Faceplate Design -can introduce additional errors**

- Mode drop down list covers SP, PV and OP values
- Operator may select wrong Mode if SP, PV and OP values are hidden from view
- Mode list offers more options than necessary

PROSYS Maximize Your Plant™

10th GLOBAL CONGRESS ON PROCESS SAFETY

# Direct Entry Operation

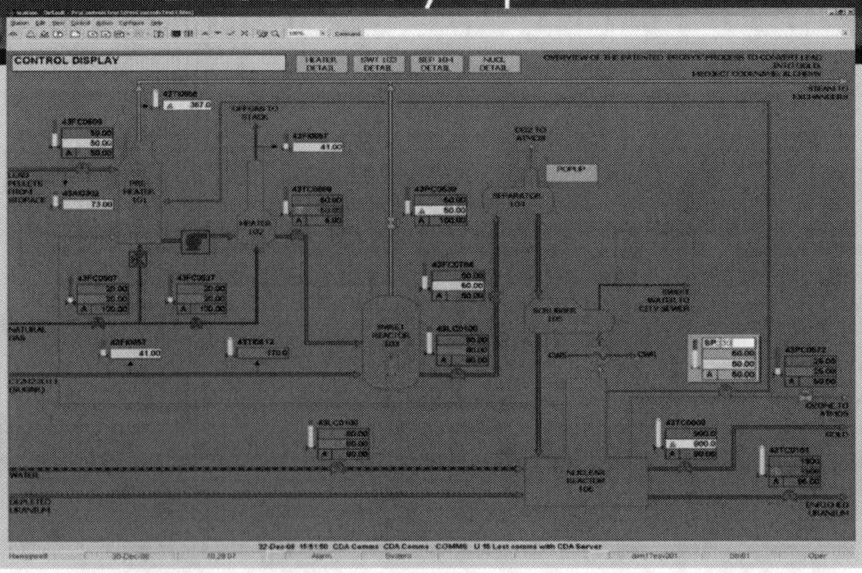

PROSYS Maximize Your Plant™

## Direct Entry Fields

**Values can be entered by:**

- Select point
- Typing value via keypad
- Pressing enter

**Key Factors:**

- By maintaining focus of points after value change, it improves safety for input errors, multiple clicking functions and is more time efficient
- Easily recognizable operator enterable fields

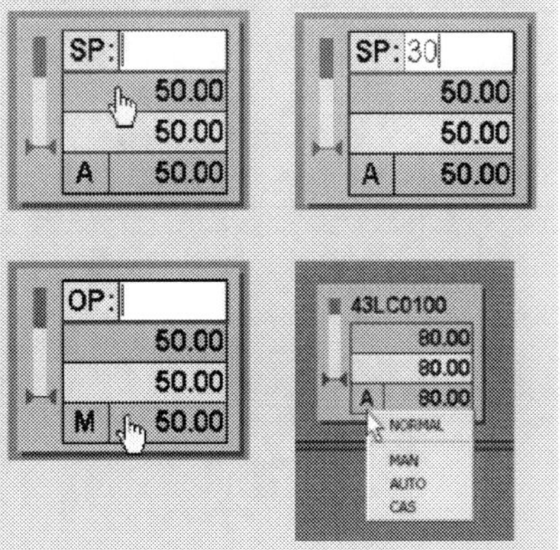


PROSYS Maximize Your Plant™

## Evaluation of Methods

**Comparison of Faceplate to Direct Entry**

- **Keystroke Level Analysis**
- **Operator Loading Analysis**
- **Risk Analysis**

PROSYS Maximize Your Plant™

10th GLOBAL CONGRESS ON PROCESS SAFETY

## Keystroke Level Model

| Symbol | Time (s) | Description |
|---|---|---|
| K | 0.28 | Keystroke |
| P | 1.1 | Point to object |
| BB | 0.2 | Click on object |
| H | 0.4 | Home hands on keyboard or mouse |
| M | 1.2+ | Mental act or routine thinking |

PROSYS Maximize Your Plant™

10th GLOBAL CONGRESS ON PROCESS SAFETY

## Estimate Execution Time Analysis

- **Evaluates execution time by an operator comparing faceplates to direct entry for SP/OP and Mode Changes**

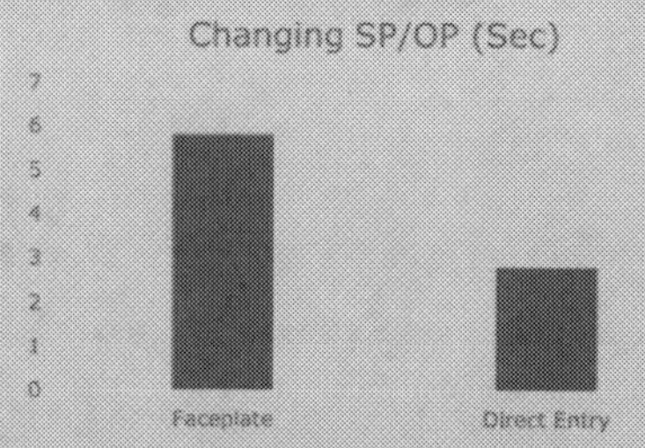


- **50 % Reduction**

PROSYS Maximize Your Plant™

10th GLOBAL CONGRESS ON PROCESS SAFETY

# Estimated Operator Loading Analysis

- **Evaluates Mental and Physical Load on Operator**
  - Tasks that require very little thought are removed – such as (BB), (H)

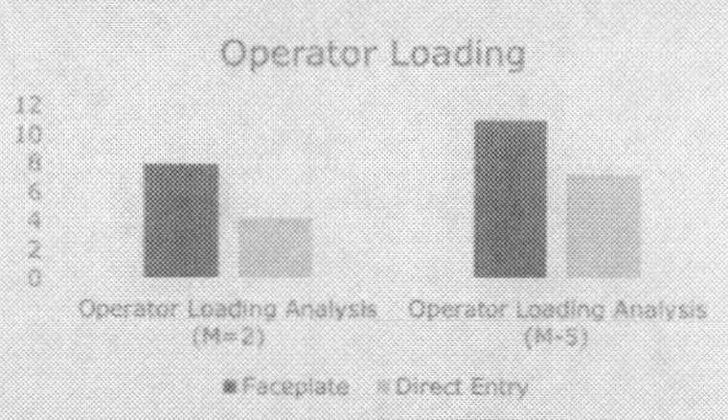


  - Mental and Physical tasks are weighted based on the amount of fatigue they cause – assigned a weight of M=2.0 to 5.0

PROSYS Maximize Your Plant™

10th GLOBAL CONGRESS ON PROCESS SAFETY

# Risk Analysis

- **Identifies actions where operator entry errors can occur**

PROSYS Maximize Your Plant™

10th GLOBAL CONGRESS ON PROCESS SAFETY

# Risk Analysis

| Faceplate operation<br>KLM for changing SP/OP | Potential Error |
|---|---|
| Move to shape (P) | Select wrong point |
| Click on shape (BB) | |
| Move to faceplate (P) | Point in faceplate is previous point |
| Click on SP/OP field (BB) | Change wrong parameter |
| Move hands to keyboard (H) | |
| Type in value and press enter (4K) | Mistype value and press enter |
| Move hands to mouse (H) | |
| Move to close faceplate (P) | |
| Click to close faceplate (BB) | |
| **Potential error count** | **4** |

PROSYS Maximize Your Plant™

10th GLOBAL CONGRESS ON PROCESS SAFETY

# Risk Analysis

| Direct Entry operation<br>KLM for changing SP/OP | Potential Error |
|---|---|
| Move to SP/OP (P) | Select wrong parameter or point |
| Click on SP/OP field (BB) | |
| Move hands to keyboard (H) | |
| Type in value and press enter (4K) | Mistype value and press enter |
| | |
| | |
| | |
| | |
| | |
| **Potential error count** | **2** |

PROSYS Maximize Your Plant™

## Direct Entry Vs Faceplate

**Summary of Benefits**

- **Execution Time SP/OP Changes - 51.5% less**
- **Execution Time for Mode Changes – 50% less**
- **Operator Loading – 33 % to 47% less**
- **Risk of Errors – 50% less**

PROSYS Maximize Your Plant™

## Conclusion

- **Many industrial errors are a result of operators using control systems with flawed designs**
  - Poor design and performance by alarm management distracts operators and/or occludes critical alarms
  - Operator graphics with multiple open faceplates can cause a change intended for one controller to be entered into the faceplate of another
- **The two mechanisms listed above are predictable and therefore preventable through better design available today**

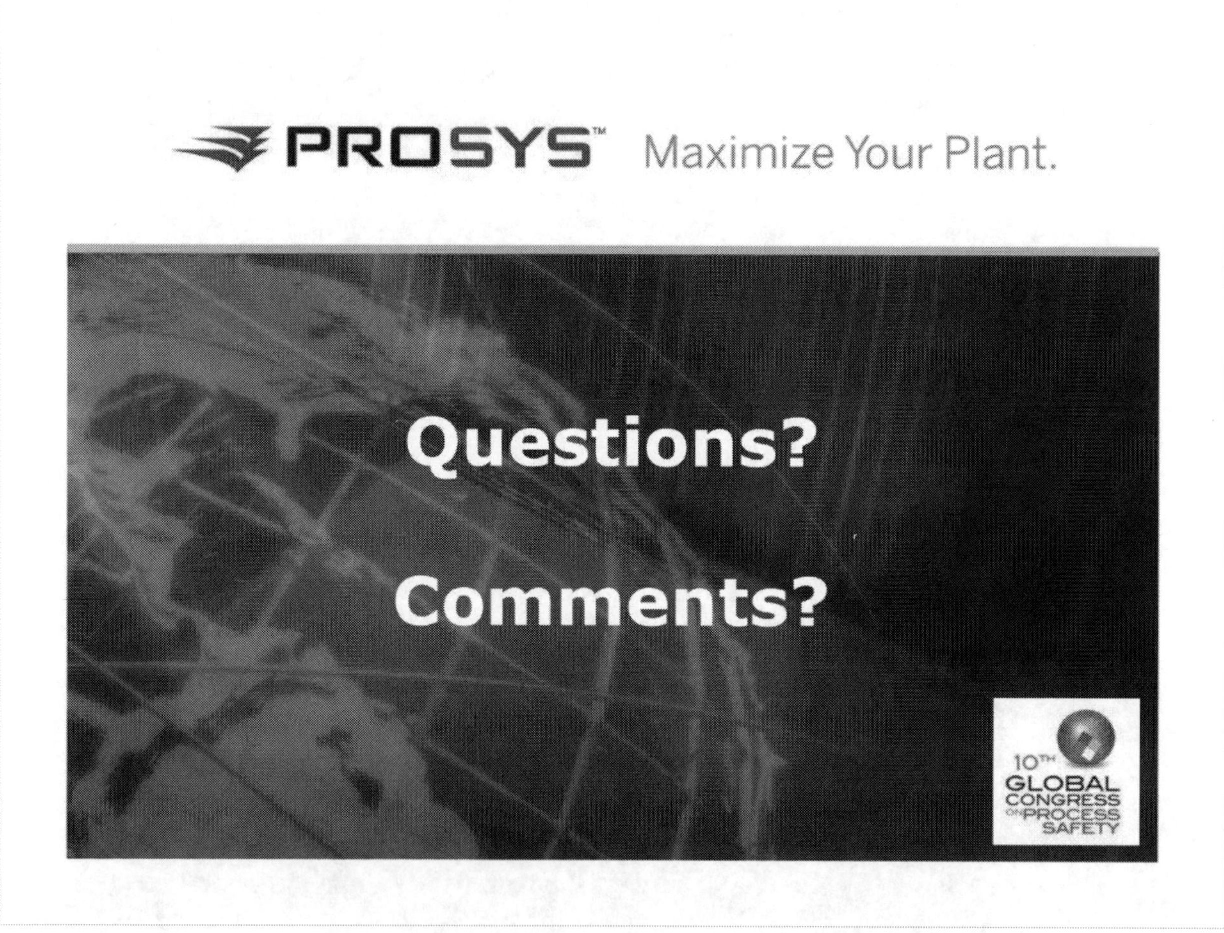
PROSYS™ Maximize Your Plant.
Questions?
Comments?
10TH GLOBAL CONGRESS ON PROCESS SAFETY

## Problem Statement

visium KMS
Process Safety Management Solution

- Any organisation can improve by leveraging knowledge and lessons learned.
- Knowledge can be externally or hidden locally, in a global enterprise, where lessons learned are difficult to share beyond their initial impact.
- There is so much data, information and knowledge (opportunities) coming at your company that can make you stronger or make you look like your not paying attention.
- Not paying attention or managing the available knowledge, can sink your ship... nobody wants that!

Expert Insight Delivered 2



OSyS
Part of the Rolls-Royce Group

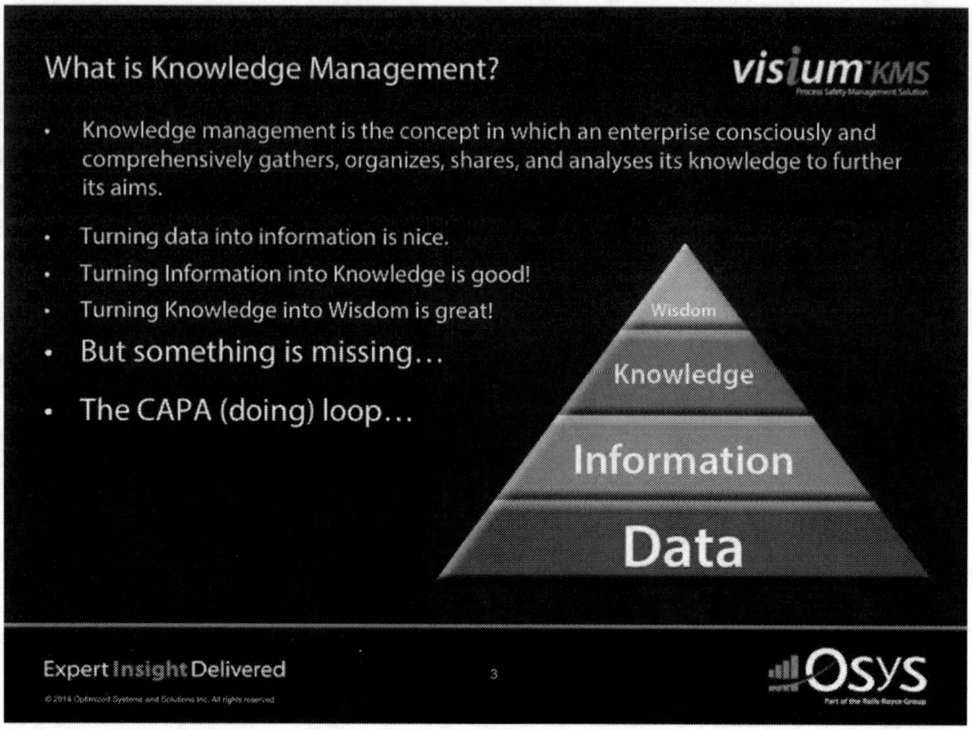
What is Knowledge Management?
visium KMS
Process Safety Management Solution
• Knowledge management is the concept in which an enterprise consciously and comprehensively gathers, organizes, shares, and analyses its knowledge to further its aims.
• Turning data into information is nice.
• Turning Information into Knowledge is good!
• Turning Knowledge into Wisdom is great!
• But something is missing…
• The CAPA (doing) loop…
Wisdom
Knowledge
Information
Data
Expert Insight Delivered
3
Osys
Part of the Rolls-Royce Group

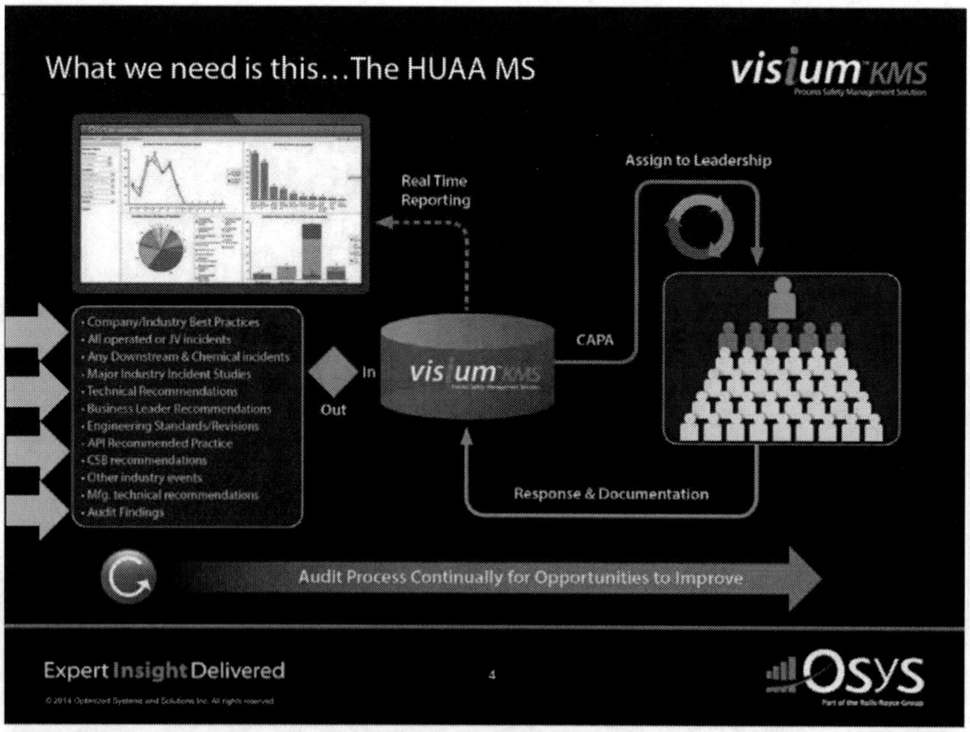
What we need is this…The HUAA MS
visium KMS
Process Safety Management Solution
Real Time Reporting
Assign to Leadership
• Company/Industry Best Practices
• All operated or JV incidents
• Any Downstream & Chemical incidents
• Major Industry Incident Studies
• Technical Recommendations
• Business Leader Recommendations
• Engineering Standards/Revisions
• API Recommended Practice
• CSB recommendations
• Other industry events
• Mfg. technical recommendations
• Audit Findings
In
Out
visium KMS
CAPA
Response & Documentation
Audit Process Continually for Opportunities to Improve
Expert Insight Delivered
4
Osys
Part of the Rolls-Royce Group

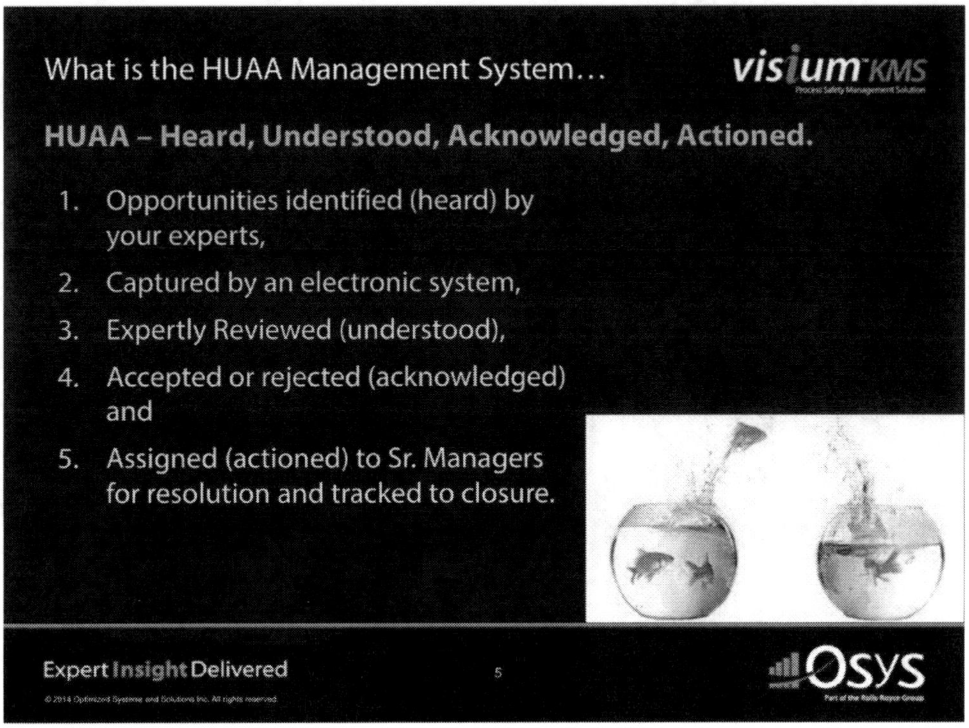
What is the HUAA Management System…
visium KMS
HUAA – Heard, Understood, Acknowledged, Actioned.
1. Opportunities identified (heard) by your experts,
2. Captured by an electronic system,
3. Expertly Reviewed (understood),
4. Accepted or rejected (acknowledged) and
5. Assigned (actioned) to Sr. Managers for resolution and tracked to closure.
Expert Insight Delivered
5
OSYS

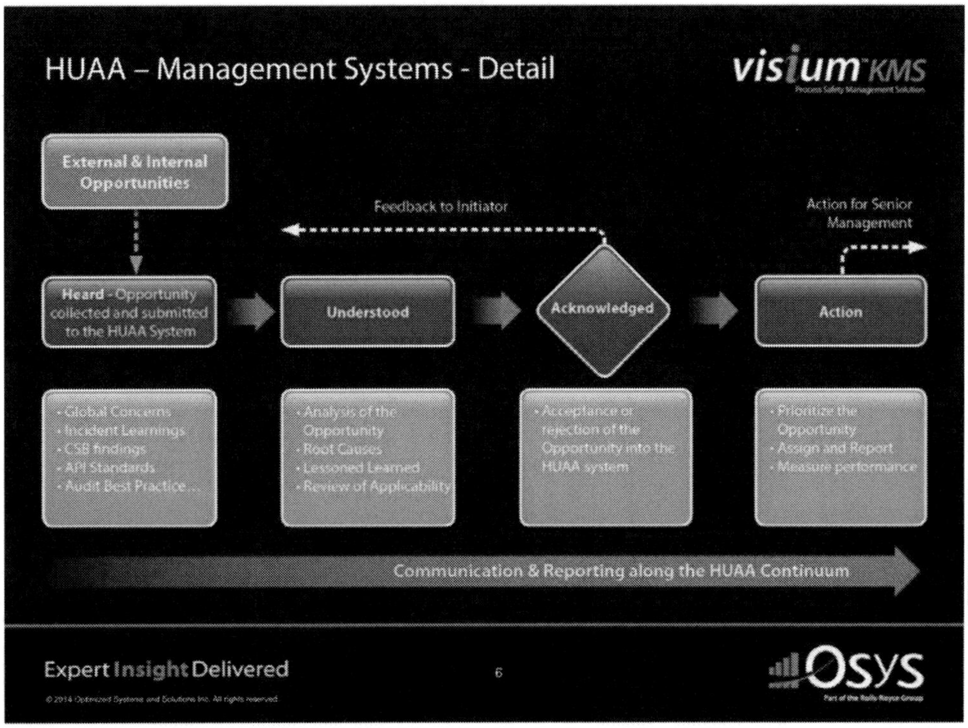
HUAA – Management Systems - Detail
visium KMS
External & Internal Opportunities
Feedback to Initiator
Action for Senior Management
Heard - Opportunity collected and submitted to the HUAA System
Understood
Acknowledged
Action
• Global Concerns
• Incident Learnings
• CSB findings
• API Standards
• Audit Best Practice…
• Analysis of the Opportunity
• Root Causes
• Lessoned Learned
• Review of Applicability
• Acceptance or rejection of the Opportunity into the HUAA system
• Prioritize the Opportunity
• Assign and Report
• Measure performance
Communication & Reporting along the HUAA Continuum
Expert Insight Delivered
6
OSYS

## HUAA Results

visium KMS
Process Safety Management Solution

- Efficient Collection - Important Opportunities are identified and collected from many sources (internal and external), and entered into the HUAA management system.
- Expert Vetting - These opportunities are vetted, parsed and resolved for inclusion or exclusion; documenting & communicating both.
- Accountable Assignment - Included opportunities are systematically assigned to Sr. Leaders and managed across the enterprise at SVP and C levels.
- Action - Improvement will be assured through the accountable actions of these leaders.
- Results - The organization will gain competitive advantage by active listening, learning from their experience, as well as the experience of others, and by taking ACTION!

Expert Insight Delivered



7

OSYS
Part of the Rolls-Royce Group

## Sharing is not enough to keep your ship afloat!

visium KMS
Process Safety Management Solution

Expert Insight Delivered



8

OSYS
Part of the Rolls-Royce Group

# 352631 Autocad - Smart Grid – Agriculture - Architecture and the United States Government

**Abstract:**

The computational methods, process control, process design, process safety, public policy of the smart chip for agriculture industry, and architecture industry in the United States utilizing the computational methods of AutoCAD in the structural development stages of the smart grid for device/component/equipment and geographical development and public policy for grant funding.

# 352747 Production, Characterization and Catalytic Studies of Biobased Carbon Materials

**Abstract:**

Carbon-based nano materials were produced from wood chips, lignin and bio-char. The influence of reaction conditions on the production of nano-carbons was investigated. Carbon-based nanomaterials were characterized using scanning electron microscopy (SEM), transmission electron microscopy (TEM), high resolution transmission electron microscopy (HRTEM), and X-ray diffraction (XRD). The catalytic performance of these nanomaterials as syngas cleaning adsorbents and chemical conversion catalysts for bio-fuel processes has been evaluated.

# 347835 Uncertainty in Sour Gas Viscosities Estimation, What Is the Effect on Your Reservoir Inflow and Tubing Performance

**Abstract:**

In light of the need to find an applicable, consistent and promising correlation to accurately calculating sour gases viscosity under several reservoir conditions, this study aims to compare the existing correlations and other industry standard methods, check their accuracy and range of applicability for sour/acidic gases. Moreover, investigating the impact of uncertainties in estimating sour gas viscosities on reservoir flow performance relationship (IPR), reserves and resources estimations, tubing performance, pipeline and facilities design, the design of compressors and more importantly the need to find such correlation to be used in further advancements of reservoir simulation.

# 354052 Chromonic Nanocarriers for Chemotherapeutics: Size Distribution and Control Release Studies

**Abstract:**

Conventional chemotherapy fails to release the drug in the controlled manner; rather they release the drug instantly due to which re-administration is required as therapeutic index falls. Moreover, they target both the cancer cells and non-cancer cells which make the treatment of the cancer cells highly ineffective due to excessive toxicities. Recently, the nanoparticle mediated drug delivery for chemotherapeutics has gained much attention. The present study reports the use of a novel class of materials – *Chromonics* in cancer drug delivery. Folic acid as a chromonic material has been explored to develop nanocarrier for sustained release of chemotherapeutic drugs *in-vitro*. With the help of a biocompatible polymer as a tool, folate nano-domains were formed from self-assembled liquid crystalline folate solutions. These nano-domains were crosslinked to form a stable nanoparticle system. The effect of different factors like crosslinking agent, relative polymer and folate concentration, on the size distribution was investigated in detail. Based on the observations, we have reported that not only the synthesis but also the size of the developed nanoparticles can be controlled by varying the folate: HPMC weight ratios. The present study then explored the release of drugs from these nanocarriers. We have investigated the release of folic acid as a model drug from these nanoparticles as well as release of methotrexate and cytarabine as an anticancer drug for the period of 60 days. It was observed that release of drug was nanoparticle's size-dependent, release medium dependent, pH of the release medium dependent, crosslinking agent and its concentration dependent. The significant part of the current study is the synthesis of chromonic nanoparticles, how the size can be controlled, how controlled release can be obtained as well as how the release profiles can be modified depending on the dosage forms. Therefore, we can say that folate nanoparticles can be a potent nanocarrier for the sustained release of cancer drugs.

# Double Containment Piping Systems: A Fail-Safe, Leak-Free Solution

IPEX
Committed to Excellence

Darin Johnson, P.Eng.
Applications Engineer
IPEX USA LLC

AIChE – 2014 Spring Meeting
New Orleans, LA

Monday, March 31, 2014

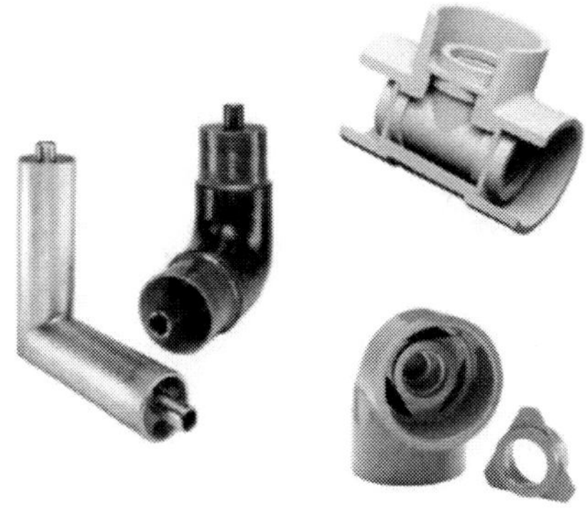

# Why Double Containment Piping Systems?

IPEX
Committed to Excellence

- December 1988 – law change within the ***Resource, Conversation, and Recovery Act*** in the United States regarding underground piping.
- U.S. Environmental Protection Agency (EPA) requirement: Code of Federal Regulations (CFR), Title 40, Part 280.
  - Petroleum Products
  - Hazardous Substances (over 3200, CFR Title 40 Part 302.4)
  - Hazardous Wastes

IPEX
Committed to Excellence

## Why Double Containment Piping Systems?

- Buried tank - all buried piping connected to it must be double contained.
- Above ground tank – associated buried piping consists of 10% or more of the tank volume, it also must be double contained.

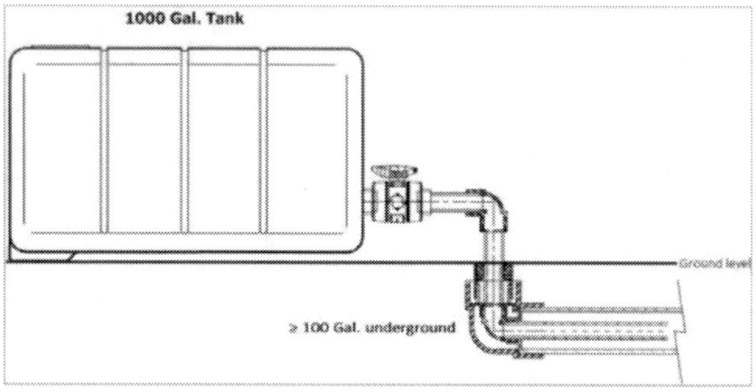


IPEX
Committed to Excellence

## Why Double Containment Piping Systems?

- If the piping is less than 10% of the volume of the aboveground tank, and you choose not to double contain the buried piping, ***<u>responsibility for any clean-up or remediation resulting from a release will still fall on the Client.</u>***

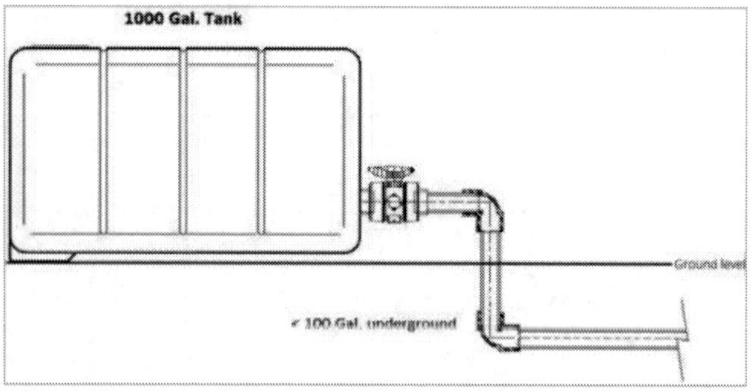

## Why Double Containment Piping Systems?

- **Above Ground Piping Systems**
  - Different jurisdictions have to ability to require chemicals be double contained.

## Nomenclature of Double Containment Piping

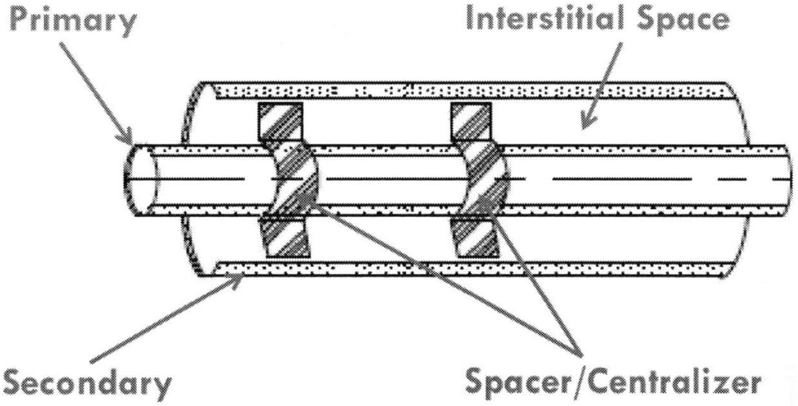


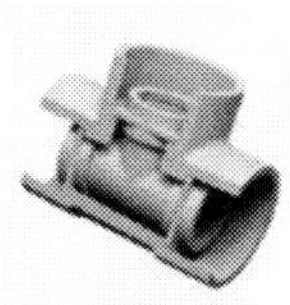

# Double Containment Piping Systems: Material Combinations

| SYSTEM Carrier | Containment | Carrier SIZE inches | SYSTEM Carrier | Containment | Carrier SIZE inches |
|---|---|---|---|---|---|
| PVC 80 | PVC 80 | 1/2 - 12 | S40 A53 BLK SW | FRP | 1/2 - 2 |
| PVC 80 | PVC 40 | 1/2 - 12 | S40 A53 BLK SW | PVC 80 | 1/2 - 2 |
| PVC 40 | PVC 40 | 1/2 - 12 | S80 A53B CS SW | PVC 80 | 1/2 - 2 |
| PVC DWV | PVC DWV | 1-1/2 - 12 | S10 T304L SS BW | PVC 80 | 1/2 - 8 |
| PVC 80 | CS SCH 10 | 1/2 - 12 | S10 T304L SS BW | PVC 40 | 1/2 - 8 |
| CPVC 80 | CPVC 80 | 1/2 - 8 | S10 T0304L SS BW | CPVC 80 | 1/2 - 8 |
| CPVC 80 | PVC 80 | 1/2 - 12 | S10 T304L SS BW | FRP | 1/2 - 12 |
| CPVC 80 | PVC 40 | 1/2 - 12 | S10 T304L SS BW | S10 T304L | 1/2 - 12 |
| CPVC 80 | CS SCH 10 | 1/2 - 12 | S40 T316L SS BW | PVC 80 | 1/2 - 8 |
| NAT PP 80 | PVC 80 | 1/2 - 2 | S40 T316L SS BW | CPVC 80 | 1/2 - 8 |
| NAT PP 80 | PVC 40 | 1/2 - 2 | S40 T316L SS BW | FRP | 1/2 - 12 |
| RED PVDF 80 | CPVC 80 | 1/2 - 2 | S40 T304L SS SW | PVC 80 | 1/2 - 2 |
| RED PVDF 80 | PVC 80 | 1/2 - 2 | S40 T304L SS SW | PVC 40 | 1/2 - 2 |
| NAT PVDF 80 | PVC 80 | 1/2 - 2 | S40 T304L SS SW | CPVC 80 | 1/2 - 2 |
| ENFUSION PP | PVC DWV | 1/2 - 6 | S40 T304L SS SW | FRP | 1/2 - 2 |
| S40 A53B CS BW | PVC 80 | 1/2 - 8 | S10 T316L SS SW | PVC 80 | 1/2 - 2 |
| S40 A106B CS BW | PVC 80 | 1/2 - 8 | S10 T316L SS SW | CPVC 80 | 1/2 - 2 |

# Double Containment Piping Systems: Design Considerations

- **Material Selection**
  - Primary Pipe
    - Chemical Compatibility
    - Pressure Capabilities
    - Thermal Properties
  - Secondary Pipe
    - Chemical Compatibility
    - Loading (Static & Live) for buried applications
    - Hanging requirements for above ground
    - Leak detection requirements
    - UV requirements

## Double Containment Piping Systems: Design Considerations

- **Thermal Expansion & Contraction**
  - Two basic thermal relationships may be present:
    - When only one of the two systems experiences thermal expansion/contraction.
    - When both systems experience different magnitudes of thermal expansion/contraction.
  - The Floating System
  - The Restrained System

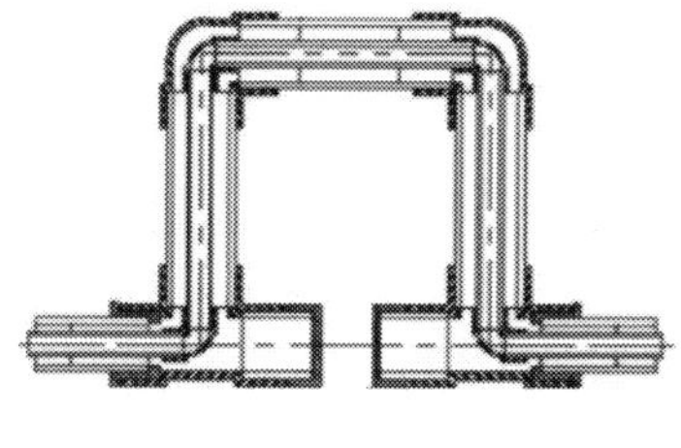

## Double Containment Piping Systems: Leak Detection Requirements

- EPA requirement: (CFR), Title 40, Part 280
  - Pressurized systems require electronic leak detection that must be checked once per year.
  - Drainage systems (gravity flow) require visual leak detection with monthly manual inspection for a media leak.

## Double Containment Piping Systems: Leak Detection Requirements

- The following forms of leak detection are acceptable as per regulations
  - Visual (drainage applications only)
  - Electronic low point probes
  - Continuous cable

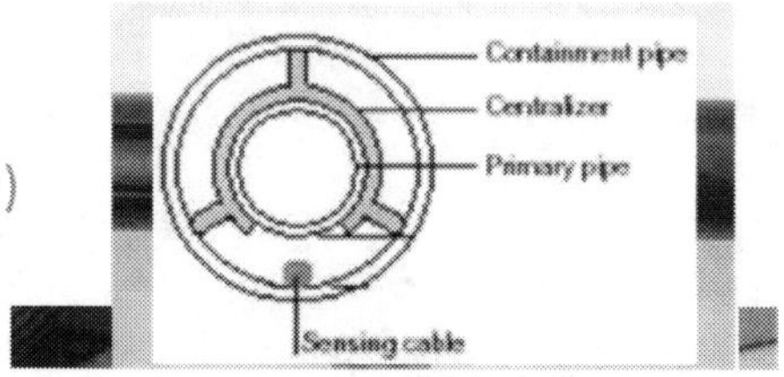


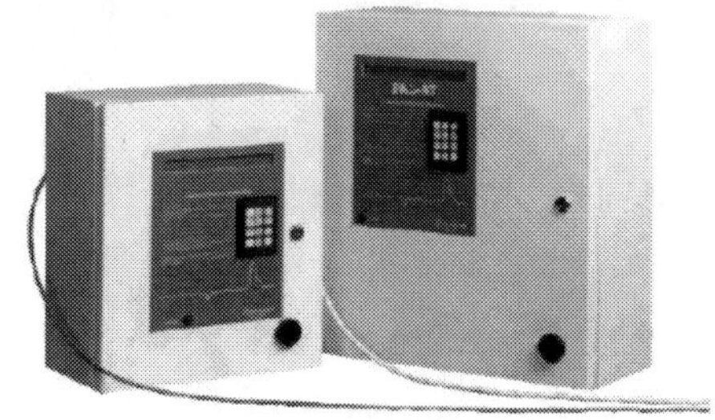

## Double Containment Piping Systems: Applications

- Underground Applications
- WTP & WWTP
- Oil & Gas
- Hospitals
- Museums
- Food and Beverage
- Pharmaceuticals
- Laboratories
- Data Centers

IPEX
Committed to Excellence

## Double Containment Piping Systems: A Fail-Safe, Leak-Free Solution

# Questions?

# Thank you!

# DTP Process

"Now Ready to License"

JGC Corporation
Mitsuo Morita

*2014*



## What is DTP Process ?

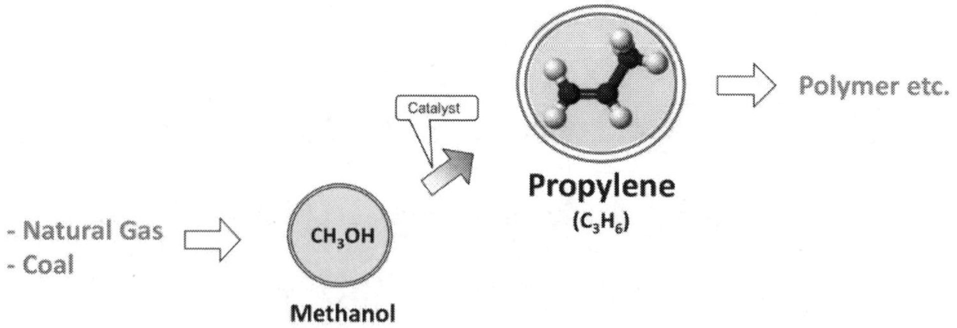


*-On-Purpose Propylene Production from Methanol*

# What is DTP Process ?

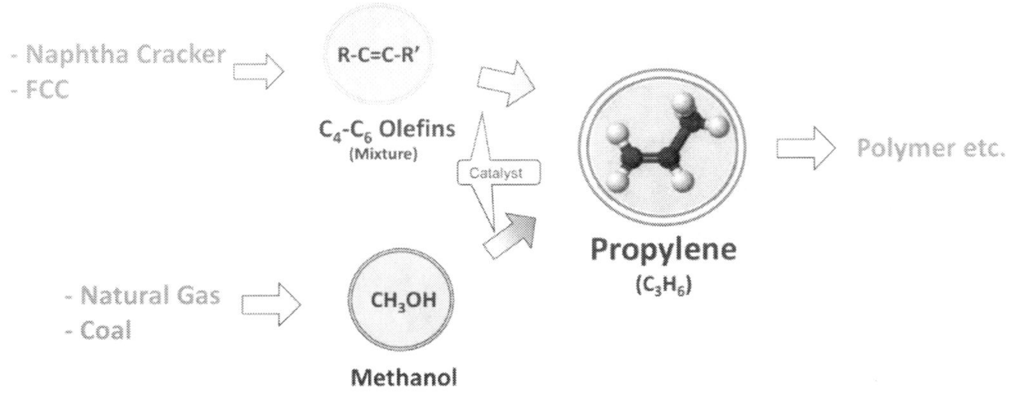


*-On-Purpose Propylene Production from Methanol*

*- Additional Unutilized $C_4$-$C_6$ Olefins as Feedstock*

JGC CORPORATION - 3 - MITSUBISHI CHEMICAL

# Development History and Prospect

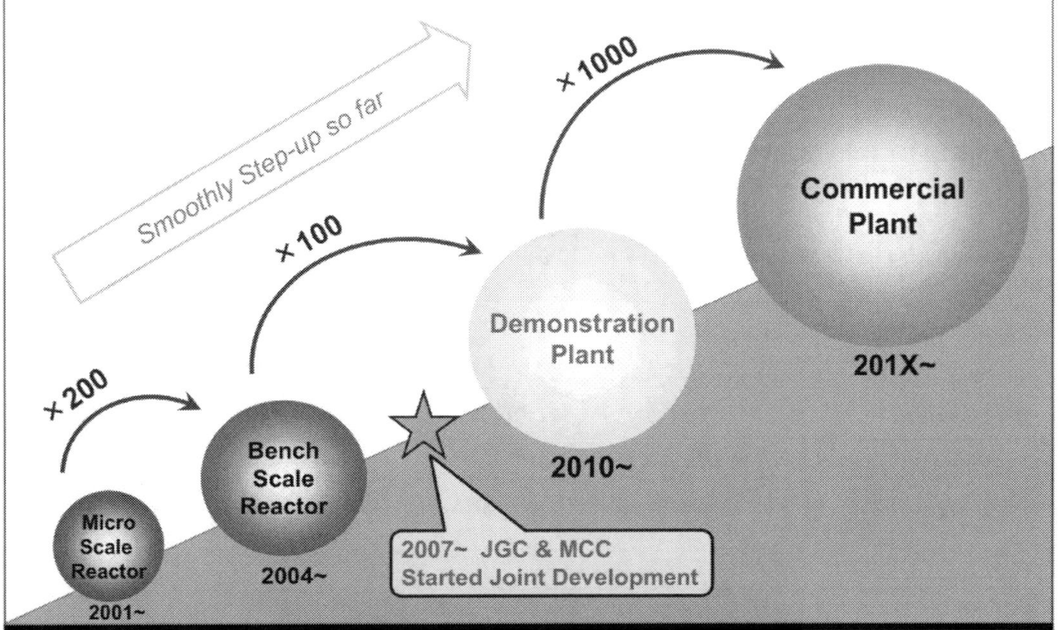


JGC CORPORATION - 4 - MITSUBISHI CHEMICAL

# Demonstration Plant Overview

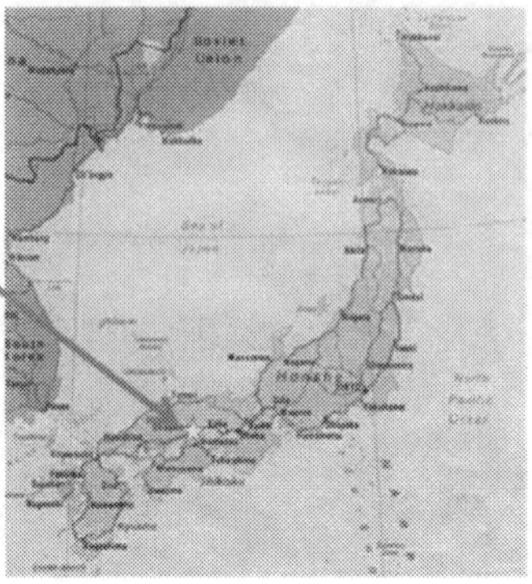

- **Located in Mitsubishi Chemical Mizushima Plant**
- **Successfully Start-up in August 2010**
- **Continuous Operation for Two Years**



# Process Flow (Standalone Case)

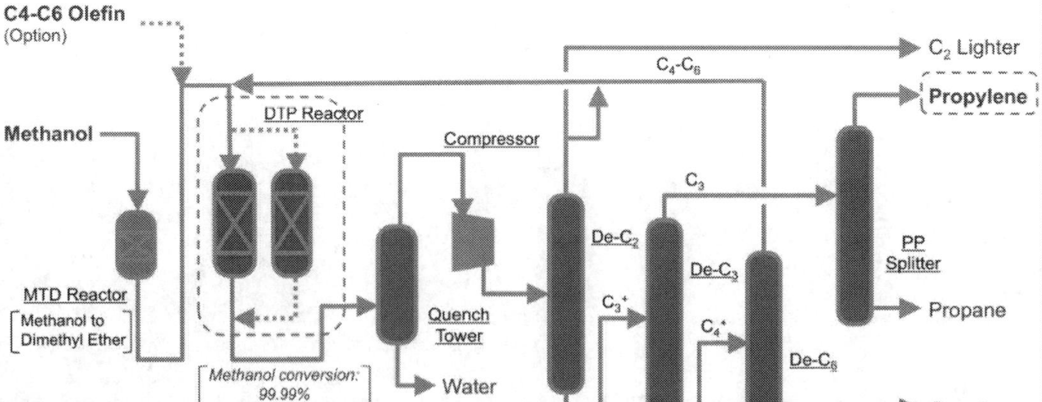


■ **Feature:**

- ***High Propylene Yield: more than 70%***
- ***Applied Existing Proven Technologies***

# Process Flow (Integration Case)

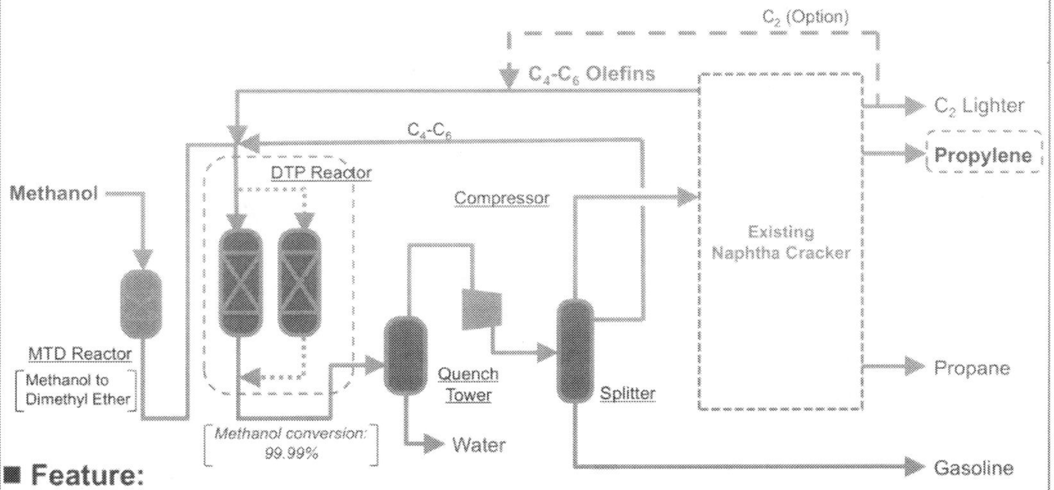


■ **Feature:**

- *High Propylene Yield: more than 70%*
- *Applied Existing Proven Technologies*
- *Low Investment Cost in case of Integration with Naphtha Cracker*

JGC CORPORATION - 7 - MITSUBISHI CHEMICAL

# Application

## (1) Naphtha Cracker Integration Case (Propylene Capacity: 20KTA – 200KTA)

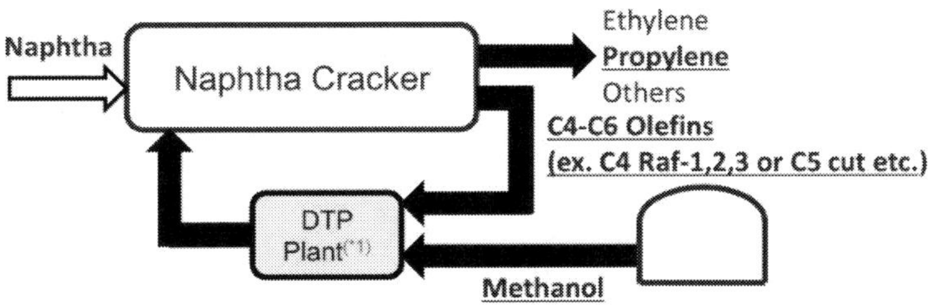


(*1) Utilizing facilities in Naphtha Cracker.

## (2) Standalone Case (Propylene Capacity: 250KTA – 500KTA)

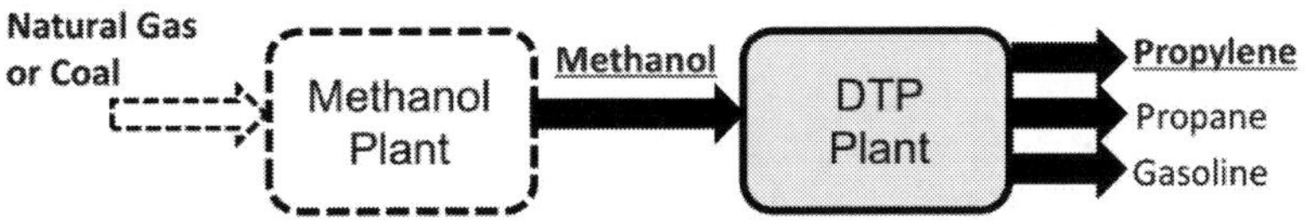


- *Two Cases available depending on Client requirements !!*

JGC CORPORATION - 8 - MITSUBISHI CHEMICAL

# DTP is…

- *High Yield Propylene Production Process*
- *Flexible Feed Methanol to Olefin Process*
- *Process that increase P/E Ratio for Cracker*

*Thank you very much !!*

**Contact:**
Mitsuo Morita
Market Development Dept.
Global Marketing Division
2-3-1, Minato Mirai, Nishi-ku,
Yokohama 220-6001, Japan

Phone: 81-45-682-8371
FAX: 81-45-682-8183
E-Mail: morita.mitsuo@jgc.co.jp

# 352533 Self-Cleaning "Bernoulli" Type Filters Used in Onshore and Offshore Applications

**Abstract:**

Georg Schuenemann GmbH (SAB-BREMEN) has developed a new Filter Type F-480, dubbed the "Compact Giant" for applications in onshore water intake and cooling water systems, sea water desalination plants, and in offshore projects, such as FPSO's, Drillships, and Production Platforms.

This paper will describe the essential design features of the filters, based on the "Bernoulli principle" which result in a fully automated self-cleaning filter system, and choice of materials in a sea water service with high corrosion rates.

A case study will be presented in which stringent project requirements such as substantial flow rates, adaptibility to space limitations and pipe connections had to be observed. Reliability of the system, extended run times without the need for frequent replacement of filter elements, as well as ease of monitoring and operator involvement have been achieved.

# 352653 Effects of Support on Sulfur Tolerance and Regeneration of Pt Catalysts Measured By Ethylene Hydrogenation and EXAFS

**Abstract:**

We investigated the effect of support on sulfur tolerance and regenerability under reducing environments by measuring the rate of ethylene hydrogenation, combined with hydrogen chemisorption, and extended x-ray absorption fine structure (EXAFS). Alumina and titania (P25) supports were used to prepare Pt catalysts at 1 wt.%. The catalysts were tested after treatment in $H_2S/H_2$ at 250 °C followed by regeneration treatments in $H_2$ at 250 °C, 350 °C and 450 °C. The sulfided catalysts exhibited a 20-27 times decrease in the rate of ethylene hydrogenation and a 4-6 times drop in the Pt surface area. Regenerations up to 450 °C were unable to remove all the sulfur, as shown by the presence of Pt-S bonds by EXAFS. However, a partial recovery of the hydrogenation rate per mole of Pt was observed on sulfided $Pt/Al_2O_3$ after reduction at 450 °C, while the induction of strong metal support interactions (SMSI) at reduction temperature above 350 °C was observed on Pt/P25, regardless of the presence of sulfur. For Pt/P25, the reversal of the SMSI state together with sulfur removal by mild oxidation suggests that sequential reduction/oxidation treatments may be more effective in restoring the S-free state of $TiO_2$-supported catalysts.

# 352799 Improved Operational Efficiency and Reliability through Insulation Materials Selection

**Abstract:**

This paper will explore the implications of insulation materials selection on the energy efficiency and reliability for Ethylene production units. Considerations will include insulation properties, structural support, installation methods, installation and maintenance costs.

# BOILER MACT COMPLIANCE: MERCURY IN FUEL GAS

PATRICK LAINE, WES YONGE, & VIC VICKERY
PEI, HOUSTON, TX 77043

Patrick Laine, Ph.D.
Chief Scientist
Houston, TX
850-866-6686
plaine@pei-tx.com

Wes Yonge, PE
Chief Engineer
Houston, TX
713-806-8396
wyonge@pei-tx.com

James 'Vic' Vickery
Chief Technology Officer
Houston, TX
281-536-0899
vvickery@pei-tx.com

## 1 Introduction & Objectives

For most ethylene producers the primary objective of this sampling and analysis effort will be to verify and "establish a mercury fuel specification through which gas-fired boilers that use a fuel other than natural gas or refinery gas may be considered Gas 1 unit" as identified in 40 CFR Part 63.7521.(f).

Measurement and Monitoring Requirements: Decision Rules

❑If Hg concentration ≤ 50% of Gas 1 specification* then no further testing required

❑If Hg concentration > 50% and ≤ 75% of the specification* then semi-annual sampling required

❑If Hg concentration > 75% of the specification* then monthly sampling required

*Specification is **40 µg/Sm³**

The goal will be to execute a sampling and analysis campaign that will produce compliance quality data. The key to producing and validating compliance quality data is a well-designed and executed QA/QC protocol. At a minimum, a performance based QA/QC protocol is recommended that includes the following components:

- Sorbent Trap Breakthrough Bed:
  - ❑ Sorbent traps consisting of a minimum of two sorbent beds in series so that the trailing bed can be analyzed to verify that no breakthrough occurred from the leading bed.
- Duplicate Samples:
  - ❑ The sampling system should be configured with two trains to accommodate the collection of simultaneous duplicate sorbent trap samples, in parallel, providing a means of verifying precision in the sample collection process.
- NIST (National Institute of Standards and Technology) Traceable Spike:
  - ❑ Sorbent traps with a third bed (downstream) that is spiked with a NIST traceable spiked mass of mercury serves as a matrix spike and provides a means of verifying that components in the matrix (the fuel gas) are not affecting the trapping efficiency of the sorbent trap.

## 2 Methods

**Sorbent Trap Mercury Sampling Methods Comparison**

| Method | Description | Method Type: S = Sampling A = Analysis | Evaluation | Sample Collection Phase QA/QC |
|---|---|---|---|---|
| ISO 6978 | Hg in Natural Gas by Thermal Desorption (CVAFS) | S & A | Method calls for low sample volumes can lead to non representative results. | No |
| EPA Method 30B | Determination of Gaseous Hg Emissions from Coal Fired Combustion Sources using Carbon Sorbent Traps | S | Requires modification (Modified M30B) | Yes |
| ASTM D6350 | Hg S&A in Natural Gas (AFS) | S & A | Method calls for short term samples (i.e., low sample volume) which leads to poor accuracy. | No |
| ASTM D5954 | Hg S&A in Natural Gas (AAS) | S & A | AAS is less sensitive than AFS. | No |
| EPA Method 1631 | Measurement of Hg in Water | A | No | Yes |

## 3 EPA Method 30B QA/QC Performance Criteria

| QA/QC Test or Specification | Acceptance Criteria | Frequency |
|---|---|---|
| Analyzer calibration | Correlation Coefficient: 0.999 | Daily |
| Transfer efficiency of analyzer | %Transfer Efficiency: 90%–110% | Once per sorbent trap |
| Initial calibration blank | < 0.5 ng | Once before each calibration |
| Sorbent Trap Section B breakthrough | ≤ 10% of Section A Hg mass, for Hg concentrations > 1µg/dscm<br>≤ 20% of Section A Hg mass, for Hg concentrations ≤ 1µg/dscm | At each sample point |
| Spike Recovery Test | Average recovery between 75% and 125 % for $Hg_0$ | Once per sample point |
| Paired Sorbent Traps Agreement | ≤ 10% RD for Hg concentration > 1µg/dscm<br>≤ 20% RD or 0.2 µg/dscm absolute difference for Hg concentrations ≤ 1µg/dscm | At each sample point |

## 4 Equipment

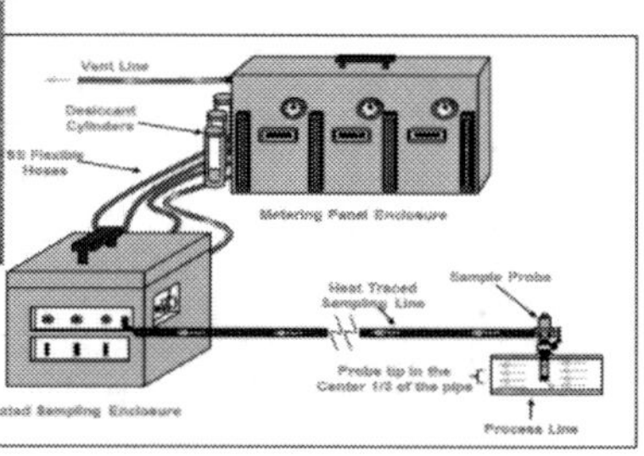


| Flow | Primary Bed | Breakthrough Bed | Spiked Bed |
|---|---|---|---|
| QA/QC Acceptance Criteria: | 99% of Total Collected Hg | ≤ 10% of Primary Mass | Recovery between 75% and 125% |

Gold Trap:

## 5 Conclusions/PEI Recomendations

❑ There are five methods accepted by the Boiler MACT rule. PEI has extensive experience and expertise employing **all five** of these methods.

❑ Operators can achieve compliance without submitting a Fuel Analysis Plan by using ASTM D6350, however, we have noticed variable results with this method.

❑ The **PEI recommendation** for obtaining the most accurate compliance quality data is to use Modified EPA Method 30B in conjunction with EPA 1631 where traps should be analyzed in a certified ISO 17025 laboratory. The recommended pathway includes a rigorous, performance based QA/QC protocol that validates accuracy and precision of all sample collection procedures.

# 351376 Young Professional Simulation Tutorial

**Abstract:**

## 18C00 Young Professional Simulation Tutorial

Authors: Naomi Hua and Mike Donahue, Invensys



Process simulation is a complex skill to master. Even with years of experience, a typical user is constantly facing new problems that need to be addressed. There are many issues that can lead to unexpected and unnecessary errors with the use of process simulation tools. Regardless of what simulation program you are using, these issues are common from one program to another. In this presentation, targeted for the young professional, we will discuss the "Most Common Pitfalls in Process Simulation". For each topic, we will discuss various techniques and methods these troublesome problems and prevent their occurrence.

Topics to be discussed include:

1. Erroneous data entry

2. Simulation defaults

3. Bad plant data

4. Thermodynamic method selection

5. Simulation tolerances and convergence
6. Recycle identification and handling
7. Model complexity
8. Difficult or impossible specifications
9. Assay pseudo-component development
10. Assay property blending

1. Erroneous data entry is the most common mistake made when building a process simulation model. This presentation will cover methods that can help ensure proper data entry in your process simulation model, such as checking the data you entered, particularly the units of measure, using "alternate forms" of data input such as copy/paste and data import, and reviewing all warning and error messages.

2. Defaults are methods, values, and conditions that are automatically selected for you by the program when you begin a new simulation model. These defaults will affect the results calculated by your model and every simulation program has them. Most defaults within simulation programs fall under categories such as convergence criteria, thermodynamic methods, tolerance and units of measure. The problem with defaults is unintended results from not fully understanding the effects a particular default has on the model. Take the time to understand the defaults of the program and how these defaults will affect the results. If you do not agree with the defaults selected by the program, most simulation programs will allow the user to change the default.

3. Plant data changes over time. The feed information and process conditions of the unit operations may be different. Not only does plant data change over time, but measurement information is often very inaccurate. This could be due to faulty or poorly calibrated measurement equipment, leaks in lines, sensor failures, and other issues. Because of these changes and inaccuracies over time, you always have to validate the information that is entered in your model to ensure that you are obtaining the proper results. To do so, you can utilize programs that can help with data validation and reconciliation like SimSci's ROMeo online optimization program. With it, you can connect your model to your data historian to constantly import current data measurements into your model and then reconcile these measurements to determine when problems arise within your plant and where those problems are occurring. You will also need to use your engineering judgment to determine if the data is valid. More details will be provided during the presentation.

4. The proper selection of thermodynamic methods for your simulation is the most important decision that you make when modeling a process. The method you select could have a drastic effect upon your results. Often, any number of thermodynamic methods is suitable for a given application. It is up to the user to determine the best method to use based upon knowledge, personal experience, and best judgment. Some important factors that need to be considered are the components being used and the operating conditions.

Oftentimes, the selected thermodynamic method still needs to have additional inputs such as binary interaction parameters to simulate the process accurately.

5. Many of the calculations in process simulation are iterative. When these calculations iterate, the program compares the results calculated from one iteration to the results calculated in the next iteration in order to determine

whether it is converged to a solution or needs to continue calculating. This method of comparing results is where tolerances come into play and simulation programs use tolerances to determine when a solution is reached. One of the major mistakes when modeling a process is not understanding or using the correct tolerance values. A looser tolerance may make the file easier and quicker to converge, but it also reduces the accuracy of the model. A tighter tolerance may be more accurate, but might make it impossible for the model to solve or increase the convergence time exponentially. A common practice is that when you have reached a converged solution with your model, tighten the tolerances within the model and solve it again. If you notice that your answers have changed significantly, you need to continue tightening the tolerance values until your results change little from one iteration to another. You will then know that you have reached an accurate solution for your model.

6. Recycles in process simulation occur when you return to a previously calculated unit operation to recalculate the unit due to a change made elsewhere in the model. They are very common and can often be difficult to converge. Some suggestions on troubleshooting recycle convergence include initially solving your model without the recycles to see how it behaves, providing initial estimates for recycle streams, using tighter tolerances on unit operations within recycle loops than the total recycle tolerances themselves, and ensuring that all components have a way in and out of a recycle by performing mass balances and using purge or makeup streams where necessary.

7. The general rule of thumb when modeling is to start simple and add complexity as you go. Each time you add a few streams or unit operations converge and save the model before moving on. Make sure you justify any additions that you make to the model. You can most likely model your system without including every unit from your PFD in the simulation. When modeling, think about the goal of your simulation and not about literally translating every real-world piece of equipment into the simulation model. Make the judgment call on what is most important and what the purpose is of every unit you add to your simulation. For example, if your real-life process shows a heat exchanger, a pump and a splitter, but all that is necessary is that the temperature, pressure, and separation are achieved properly, consider using a flash drum to accomplish all three.

8. Specifications within process simulation are everywhere and every stream and unit operation you place in your model contains specifications. In its simplest form, a specification is a constraint that you place on the model that the model needs to meet in order to successfully converge. For streams, these are usually temperatures, pressures, flowrates and compositions. For unit operations, they range from process conditions, like temperatures and pressures, to product conditions, like an overhead composition in a column. There are some pitfalls, however, that you can run into with specifications such as azeotropes in columns, high purity specifications, and the possibility of multiple solutions.

9. Petroleum assays are sets of distillation data that are used to characterize all of the components that reside in a given crude sample. When you enter distillation data into a process simulator, the program converts it into components that it can use when modeling your process. The components that are generated are called pseudo-components, and each pseudo-component represents all of the unknown components that would be within a given boiling point range in the crude. Some of the things you will want to ensure when properly developing your assay components are did you use the correct distillation curve basis, did you enter the data correctly, did you use enough cuts and are you using the correct characterization methods .

10. When you have more than 1 assay in a simulation model, you need to be concerned with assay blending and how it will affect the results of your file. Assay blending is the mixing of 2 or more groups of components (i.e. "C sets of cuts/pseudo-components) in order to calculate averaged component properties for cuts that are in the same boiling point range as one another. Once these new averaged component properties are determined, the program will use these averages from this point on in the simulation model any time it needs to access component

properties for a given pseudo-component. All component properties are blended, including normal boiling points, specific gravity, molecular weight and refinery inspection properties, and usually the blended component properties are calculated using a weighted average

These topics are based on experience and feedback from end-users to technical support representatives for process simulators. They often represent the most commonly asked questions or areas that most users find to be problematic. While the specific examples highlighted in each topic come from SimSci software, the intent of this tutorial presentation is to be broad-based and is particularly applicable to young professionals seeking to develop their experience with simulation.

# Polymeric Sulfonic Acid Composites for Fuel Cell Applications: Effects of Casting Solvent and Sonication Time on Membrane Performance

*Jimoh K. Adewole*[1], *Abdullah S. Sultan*[1,*], *Amir Al-Ahmed*[2], *and S. M. Javid Zaidi*[3]

[1]*Center for Petroleum & Minerals, King Fahd University of Petroleum & Minerals, Dhahran 31261, Saudi Arabia*

[2]*Center of Research Excellence in Renewable Energy, King Fahd University of Petroleum & Minerals, Dhahran 31261, Saudi Arabia*

[3]*School of Chemical Engineering, The University of Queensland, Australia*

## Abstract

Polymer blends comprising sulfonated poly (ether ether ketone) (SPEEK) and 3wt% (2-acrylamido-2-methyl-1-propanesulfonic acid) (PAMPS) were prepared by solution casting method. Two solvents (N,N-Dimethylacetamide (DMAc) and N-Methyl-2-pyrrolidone (NMP)) were uesd separately. Residual solvents in the membrane samples after drying were quantified using thermo gravimetric analysis. Effects of periods of sonication (0 to 90 minutes) and types of solvent on the membrane performance were evaluated interms of morphology and water uptake. An attempt was made to correlate membrane performance with the time of sonication in order to select the most suitable solvent and membrane casting conditions for subsequent membrane preparations. For both solvents, the water uptake was observed to first decrease at 10 minutes of sonication and then increased as the sonication time was increased. The minimum water uptake recorded for DMAc samples was 19.23% at 20 minutes sonication time while that of NMP sample was 24.24% at 10 minutes. However membrane morphology was found to deteriorate with increase in sonication time. No correlation was obtained between the membrane properties and the sonication time. In general, membrane performance was better for membranes samples using DMAc than NMP solvent. Therefore, DMAc was selected as the casting solvent for membrane preparation on which further studies will be carried out.

**Keywords:** morphology, water uptake, residual solvent, thermogravimetric analysis

## Introduction

Water soluble polymers have received academic and industrial interests due to their potential applications in fuel cells and other electrochemical devices (Ye et al. 2012). One of the water soluble polymers that have attracted publication interest is poly (2-acrylamido-2-methyl-1-propanesulfonic acid) (PAMPS) (Florjańczyk et al. 2003). Generally, some of the major performance criteria for fuel cell materials include proton conductivity, water and methanol uptake, good mechanical properties, chemical stability, thermal stability and methanol permeability (Gil et al. 2004; Swier et al. 2005; Ye et al. 2012). PAMPS was reported to posses higher proton conductivity than partially hydrated Nafion. In addition, it has been found to be more tolerant to water fluctuation in high temperature operations. Some of the major drawbacks in the use of PAMPS are its poor mechanical properties, thermal stability, water and methanol uptake and solubility in many solvents (Florjańczyk et al. 2003). One of the approaches that have been proposed to solve these problems is development of polymer composites.

This research paper investigates the synergistic effect of combining water soluble PAMPS with sulfonated poly (ether ether ketone) (SPEEK) which is non - water soluble polymer to prepare composite membranes for application in fuel cells. Specifically, preliminary results on the performnance of the composite were presented. Performance were evaluated based on the composite morphology and water uptake. Effects of sonication time and types of casting solvent on membrane properties were also evaluated.

## Material and Method

### Material

Sulfonated poly (ether ether ketone) (SPEEK) purcahsed from FuMA-Tech, Germany, and poly (2-acrylamido-2-methyl-1-propanesulfonic acid) (PAMPS ) purchased from Aldrich Chemical Co., are the two major polymers used in this research. The SPEEK has a DS of 0.533 and solubility of 99% in DMAc. SPEEK is derived by sulfonation of Poly (ether ether ketone) (PEEK). PEEK is a commercially successful polymer due to its excellent thermal stability and solvent resistance (Mohr et al. 1991). For sometimes, further application of PEEK have been stalled due this solubility issues. By sulfonation, PEEK can be made soluble in most commercial solvents. Importantly, the solubility of SPEEK also depends on its degree of sulfonation (DS). Since, this work involves the use of two polymers, a solvent which will disolve both the PAMPS and SPEEK must be selected. Literature review showed that PAMPS dissolves in N,N-Dimethylformamide (DMF), N,N-Dimethylacetamide (DMAc), propylene carbonate (PC), and water (Florjańczyk et al. 2003). Initial solubility studies that was carried out in our laboratory was done for N,N-Dimethylacetamide (DMAc) and N-Methyl-2-pyrrolidone (NMP). The choice of NMP was due its unique solvent properties and its structtural distiction from DMF and DMAc (Figure 1). Moreover, no literature was found where NMP was used as casting solvent for PAMPS. Thus, DMAc and NMP were used as solvents in this work. All materials were used as received from the manufacturers.

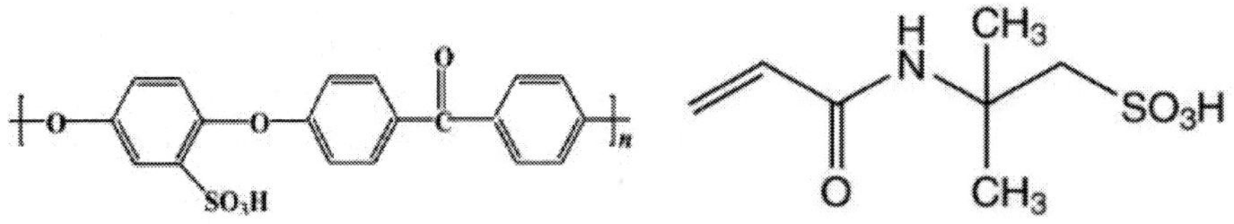

**Figure 1** Structure of SPEEK and PAMPS

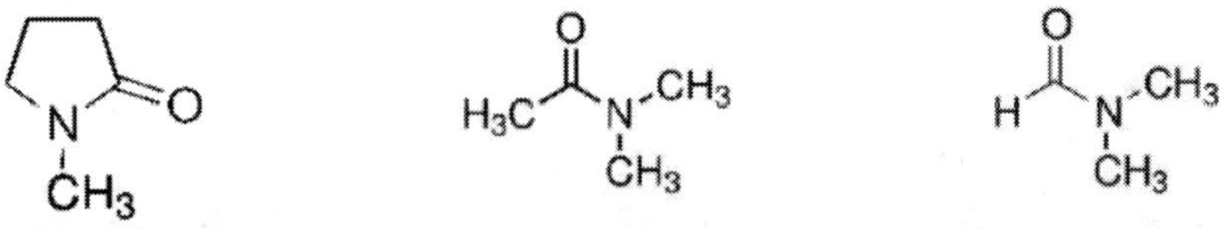

N-Methyl-2-pyrrolidone N,N-Dimethylacetamide N,N-Dimethylformamide

**Figure 2** Structure of Solvents

## Experimental Procedure

### Membrane Preparation

SPEEK was first dissolved in solvents (DMAc and NMP) and stirred for 2hours at 70°C. PAMPS was then added and the blend was stirred for 24hours. The solution was sonicated for various periods of time ranging from 10 to 90minutes using an Ultrasonic bath. In untralsonic bath, the final temperature achieved for a given time of sonication depends on the tempearutre of the laboratory (summer or winter). It is therefoe genrally advisable that maximum temperature of the bath is determined and maintains (as the so - called equilibrium tempearture), when operating continuously under ambient conditions (Santos et al. 2009). During the sonication, equilibrium temperature was maitained by simply filling the bath with water that was preheated to sample temperature.

**Membrane Characterization**

Water uptake and morphological tests were performed. Water uptake was carried out by first vacuum drying all membrane samples at 100°C until constant weights were obtained. Dried samples were then immersed in deionized water for 24h at room temperature (25°C). Samples were taken out of the water after 24h, wipped to remove surface water and then weighed immediately. Water uptake was calculated using Eq. 1.

$$water - uptake = \frac{W_{wet} - W_{dry}}{W_{dry}} \qquad \text{Eq. 1}$$

Surface morphology of the membrane was done using scanning electron microscopy (SEM). Thermogravimetric analysis (TGA) was done from room temperature to 350°C using TA Q600 analyzer at heating rate of 20°C/min under nitrogen atmosphere with flow rate of 20mL/min using ~ 8 -10mg membranes samples.

**Results and Discussion**

**Membrane Morphology by Scanning Electron Microspcopy (SEM)**

Results of surface morphological analysis performed on membrane were shown in Figure 3 and 4. The overall pictures of this figure depicts a dense membrane with some surface defects observed for most of the sonicated samples. For example, Figure 3A showed some kind of layered arrangement with no visible pores at the surface. Also, the PAMPS seems to be well distributed throughout the SPEEK. The micrograph in Figure 3B can be observed to be rough and possesses some uneven ramdom distribution of PAMPS agglomerates indicating that the distribution of PAMPS have been disturbed by the 10minutes sonication. An increase in sonication time to 20 minutes was observed to improve the distribution of PAMPS. This was however accompanied by some cracks believed to be due to the slight increase in the viscousity of the blend solution. The vicousity increase was brought about by the excape of some fraction of the solvent during sonication. As the sonication time was increased, PAMPS was observed to become more evenly distributed and the cracks become more visible and larger in size (Figure 3D to F). SEM images of samples at sonication time of 10 and 30 minutes were reproduced in higher magnification of 10μm (Figure 4) in order to further revealed the effect of sonication on the morphology of the somposites. Similar morphological behaviour has been reported in literature, though for inorganic/organic clusters of nanoclay/epoxy composites membranes (Lam et al. 2005).

**Water Uptake**

Water uptake have been reported to be affected by the degree of sulfonation, membrane pretreatment, hydration, ambient relative humidity and water temperature (Xing et al. 2004; Wu et al. 2006). Figure 5 showed the effect of sonication time and solvent type on water uptake that was measured at room temperature (25°C). For both solvents, water uptake is same at zero sonication time. There was however a further decrease in the water uptake for NMP at 20 minutes sonication time before it beging to increase. Thus, the best sonication time for both solvent lied between 10 and 20 minutes with respect to water uptake. Interestingly, this result correspond to that reported by Lam (2005) on the effect of ultrasonication on microhardness of composites.

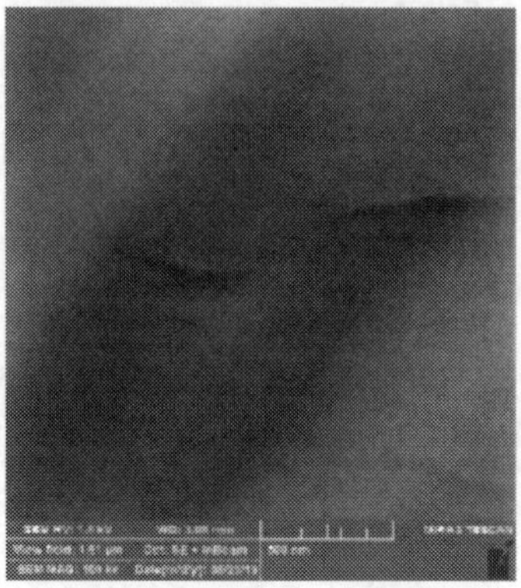

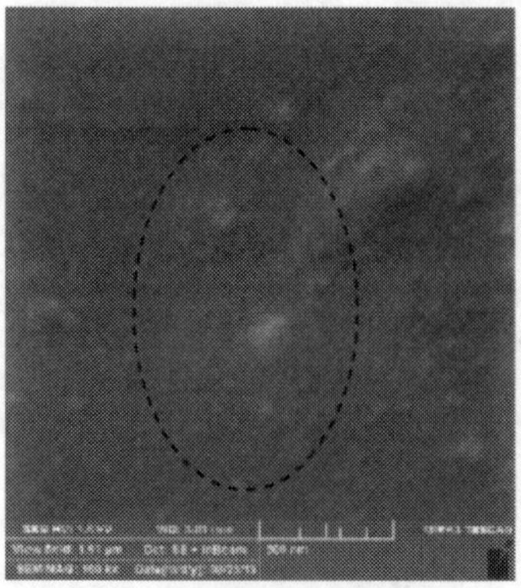

A) No sonication

B) 10 minutes sonication

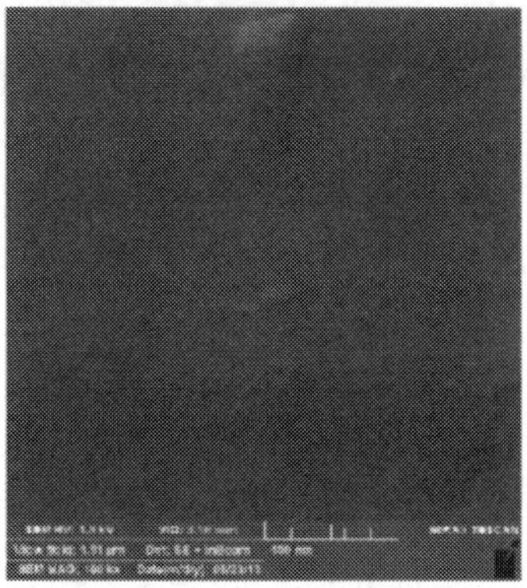

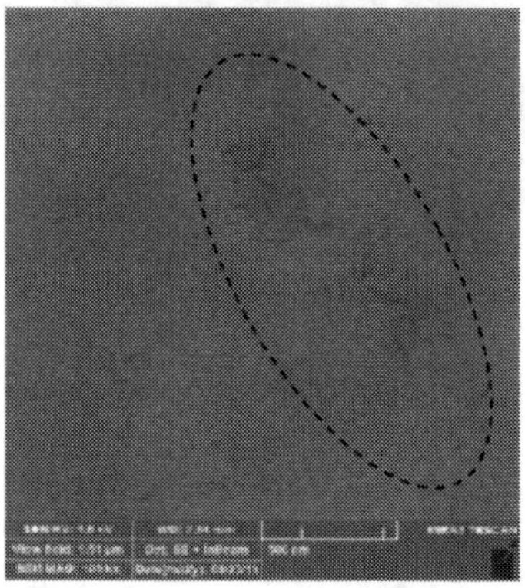

C) 20 minutes sonication

D) 30 minutes sonication

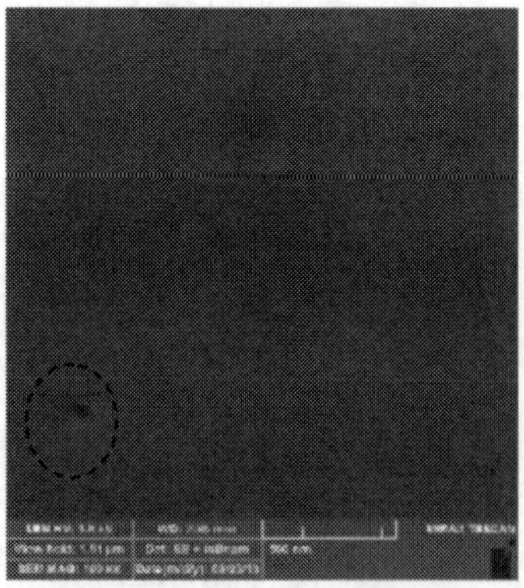

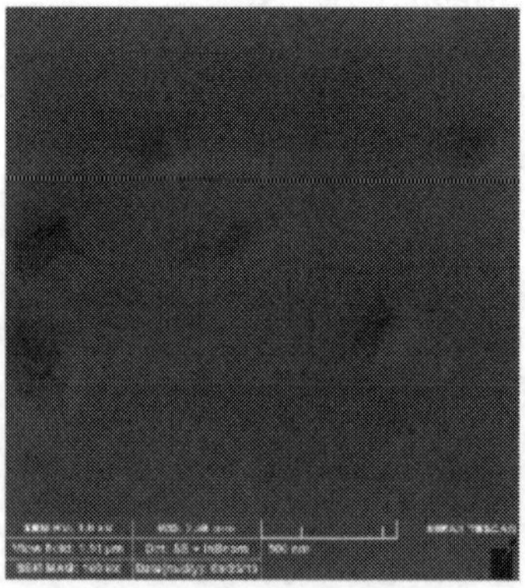

E) 60 minutes sonication

F) 90 minutes sonication

Figure 3 SEM micrograph of SPEEK-PAMS composite with various times of sonication with magnification of 500nm

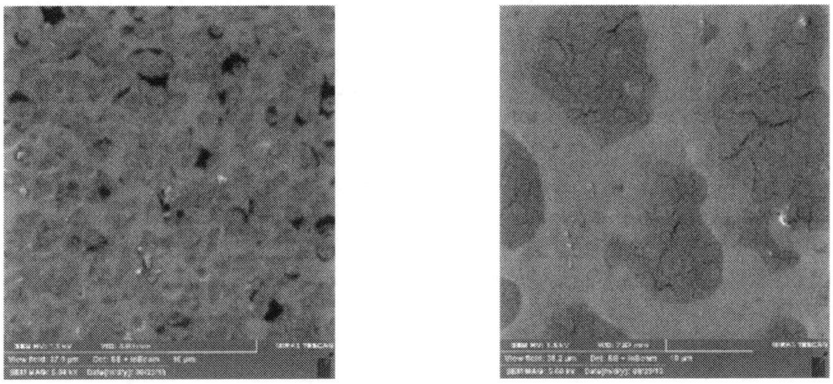

A) 10 minutes sonication B) 30 minutes sonication

Figure 4 SEM micrograph of SPEEK-PAMS composite with various times of sonication with magnification of 10μm

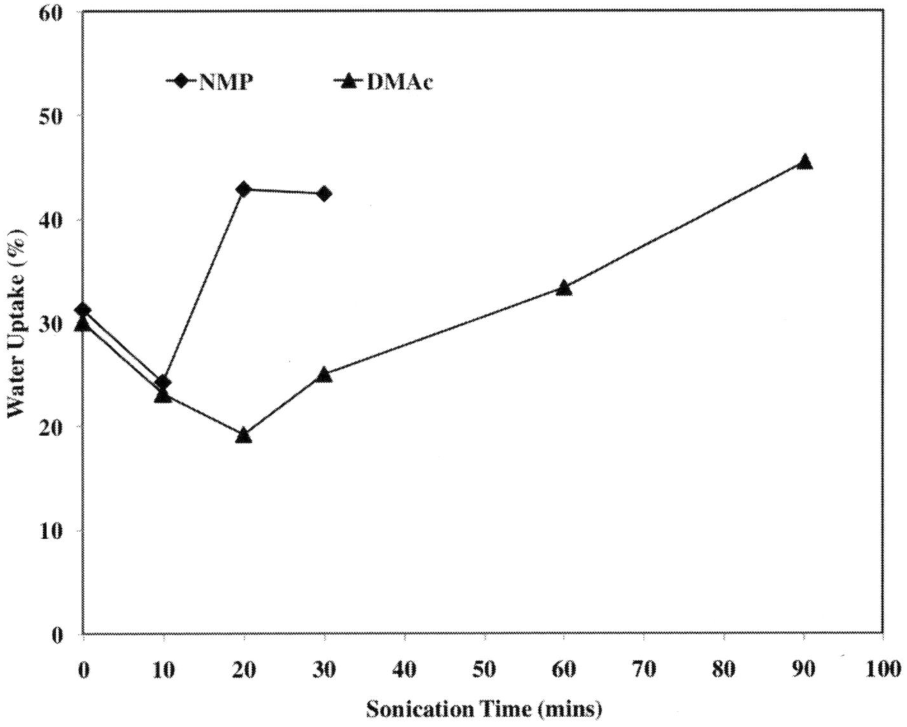


Figure 5 Water uptake of SPEEK-PAMS (3wt%) composite with variuos sonication times

**Thermogravimetric analysis (TGA)**

The difference in membrane performance resulting from the use different casting solvents have been attributed to the presence of residual solvents in the membrane samples after evaporation (Joly et al.

1999). TGA was thus used to quantify the residual solvent and evaluate its effect on the membrane performance as has been previously demonstrated in lieterature (Recio et al. 2007; Iqbal et al. 2008). Figure 6 shows the TGA curve for samples prepared from NMP and DMAc with different post vacuum heat treatment conditions. From this figure, weight loss can be attributed to evaporation of variety of substances, including solvents, from the samples (Rich et al. 1997). Table 2 contains the solvent and its boiling points, and weigth loss at the boiling point range that were extracted from TGA results. For sample heat treated for 24hrs, the overall percentage weight loss up to 204°C was about 8wt% of membrane sample for NMP while it is about 6.3wt% for DMAc. These will definitely include compnenents such as water, solvents and other unidentified substances. To quantify residual solvents, weight loss was estimated between 192 -214°C for SPEEK-PAMPS-NMP and between 154 and 176 for SPEEK-PAMPS-DMAc. The results obtained showed that the amount of the residual solvent in NMP is larger than that of DMAc by 1.25wt% (Table 1). To further investigate if this difference is responsible for higher value of water uptake for the NMP sample, membrane samples made from both solvents were further heat treated under vacuum for 72hrs. Residual solvent after this treatment was slightly higher in DMAc than NMP. Despite this, the value of water uptake of the NMP membrane is still higher than that of DMAc by approximately the same value as before the heat treatment (1.71wt%).

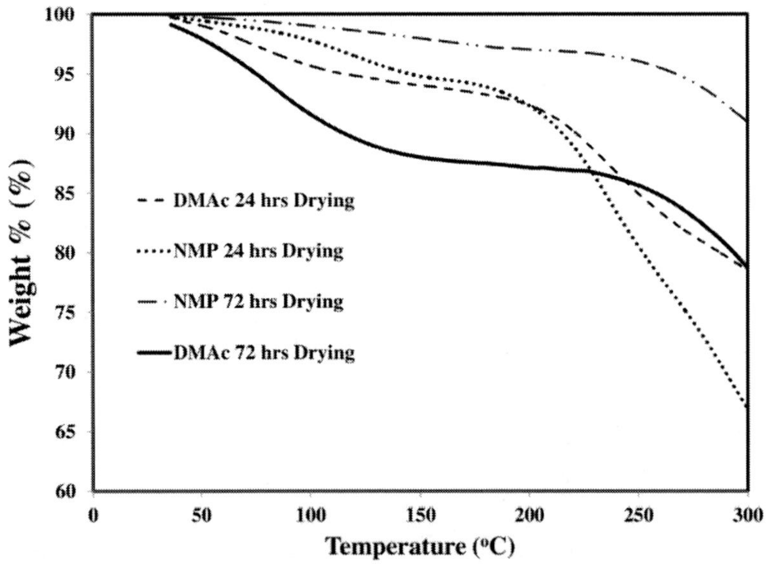


Figure 6 TGA of SPEEK-PAMPS-NMP and SPEEK-PAMPS-DMAc

Table 1 Residual solvent in SPEEK-PAMPS composites membrane samples

| Membrane Sample | Solvent | B. Pt. (°C)* | Residual Solvent (wt%) | Water Uptake |
|---|---|---|---|---|
| DMAc 24 hrs Drying | DMAc | 164.5 – 166 | 0.57 | 30.00 |
| NMP 24 hrs Drying | NMP | 202 – 204 | 2.85 | 31.25 |
| DMAc 72 hrs Drying | DMAc | 164.5 – 166 | 0.367 | 21.21 |
| NMP 72 hrs Drying | NMP | 202 – 204 | 0.217 | 22.92 |

*Boiling Points of Solvents

On a general note, results from Figire 1 – 5 showed that the best perfomance can be obtained for variety of sonication time depending the property of preference. Also, performance of DMAc solvent was observed to be better than that of NMP on the basis of water uptake. For this reason, DMAc was selected as the casting solvent for membrane preparation on which further studies will be carried out. Water uptake of DMAc membrane, prepared without sonication, was comapred with those reported in previuos literature. Table 2 illustrates the comparison between SPEEK-PAMPS and other relevant membranes previuosly reported in literature. In addition, this table serve to provide insight into the possible potentials of the composites. From this table, it was observed that the water uptake for SPEEK-PAMPS is within the range of values of other membranes on the table. Its water uptake is less than that of Nafion 112 which is a commercial membrane. The differences observed between the water uptake values of SPEEK – PAMPS and other membranes from the table can be explained based on three important factors - $SO_3H$ group, the degree of sulfonation and experimental temperature (Gil et al. 2004; Shan et al. 2006; Wu et al. 2006). Water uptake of SPEEK – PAMPS is higher due to the presence of $SO_3H$ group in the PAMPS molecules (Figure 2). $SO_3H$ group have been reported to enhance membrane acidity and hydrophilic properties (Shan et al. 2006). Infact, the higher the $SO_3H$ group the larger the water uptake. The degree of sulfonation has been reported to affect the water uptake of membranes. The higher the degree of sulfonation, the larger the value of water uptake (Gil et al. 2004). Thus, water uptake of SPEEK – PAMPS is lower than SPEEK -3 due their different degree of sulfonation, 53.3 and 72% respectively and the experimental temperature 24 and 60°C, respectively. The experimental temperature of SPEEK – 2 is close to that of the composite. However, the results cannot be compared in this case since the DS of SPEEK – 2 was not known. Above explanation is also true for the values of water uptake for SPEEK – 1.

Table 2 Comparison of water uptake values obtained with those from Literature (Xing et al. 2005)

| Membrane | Water Uptake (wt%) | Experiment Temperature (°C) | **D S (%) | Reference |
|---|---|---|---|---|
| Nafion 112 | 22.85 | N/A | - | (Xing et al. 2005) |
| SPEEK – 1* | 16.00 | 30 | 45 | (Ren et al. 2005) |
| SPEEK – 2* | 44.90 | 25 | N/A | (Zhong et al. 2008) |
| SPEEK – 3* | 85.00 | 60 | 72 | (Wu et al. 2006) |
| SPEEK– 4* | 15.00 | N/A | N/A | (Handayani and Dewi 2011) |
| SPEEK/PAMPS (3wt%) | 21.21 | | | This work |

*The numbering have been included for explanation purposes.

**DS means Degree of Sulfonation

**Conclusion and Recommendations**

Blend of SPEEK and PAMPS was prepared by solvent casting method and its performance was evaluated with respect to its morphology and water uptake. Effect of sonication time (10 – 90minutes) and solvent types (DMAc and NMP) were also investigated. Performance of membrane prepared using DMAc was observed to be better than that of NMP. There was no correlation between the sonication time and membrane properties, but it was observed that the sonication deteriorate the membrane morphology. Membrane prepared without sonication shows better surface morphology, whereas the best water uptake was observed between 10 and 20 minutes. Thus, other properties need to be investigated in order decide the best sonication time. Therefore, further investigation such as thermal stability, methanol permeability, ion exchange capacity and gas ($O_2$ and $H_2$) permeability of the membrane are recommended for a complete evaluation of the membrane performance for direct methanol fuel cell (DMFC) application. All these investigations are in progress in our laboratory.

**Acknowledgement**

Authors would like to thank the National Science, Technology and Innovation Program (NSTIP) of KACST for funding this research under project number 10-ENE1374-04. Also, authors extend their appreciation to Center of Petroleum & Minearls, Center of Excellence in Renewable Energy, Mr. Dan Jerry D.Cortes of the Nanotechnology Laboratory, Chemical Engineering Department, King Fahd University of Petroeleum & Minerals, Saudi Arabis for their supports.

**REFERENCE**

Florjańczyk, Z., et al. (2003). "Proton conducting electrolytes based on poly(2-acrylamido-2-methyl-1-propanesulfonic acid)." Electrochimica Acta **48**(14–16): 2201-2206.

Gil, M., et al. (2004). "Direct synthesis of sulfonated aromatic poly(ether ether ketone) proton exchange membranes for fuel cell applications." Journal of Membrane Science **234**(1–2): 75-81.

Handayani, S. and E. L. Dewi (2011). "Influence of Silica/Sulfonated Polyether-Ether Ketone as Polymer Electrolyte Membrane for Hydrogen Fueled Proton Exchange Membrane Fuel Cells." Internat. J. of Sci. and Eng. **2**(2): 26-30.

Iqbal, M., et al. (2008). "Solvent effect on morphology and CO2/CH4 separation performance of asymmetric polycarbonate membranes." J. Membr. Sci. **318**(1–2): 167-175.

Joly, C., et al. (1999). "Residual solvent effect on the permeation properties of fluorinated polyimide films." Separation and Purification Technology **16**(1): 47-54.

Lam, C.-k., et al. (2005). "Effect of ultrasound sonication in nanoclay clusters of nanoclay/epoxy composites." Materials Letters **59**(11): 1369-1372.

Mohr, J. M., et al. (1991). "Gas transport properties of a series of poly(ether ketone) polymers." Polymer **32, 13**: 2387-2394

Recio, R., et al. (2007). "Gas separation of 6FDA–6FpDA membranes: Effect of the solvent on polymer surfaces and permselectivity." J. Membr. Sci. **293**(1–2): 22-28.

Ren, S., et al. (2005). "Surface modification of sulfonated poly(ether ether ketone) membranes using Nafion solution for direct methanol fuel cells." Journal of Membrane Science **247**(1–2): 59-63.

Rich, D. C., et al. (1997). "Effect of cure conditions on Probimide 32 polyamide-imide." Journal of Applied Polymer Science **63**(9): 1113-1126.

Santos, H. M., et al. (2009). The Power of Ultrasound. Ultrasound in Chemistry: Analytical Applications. J.-L. Capelo-Martinez. Weinheim, Wiley-VCH.

Shan, J., et al. (2006). "Sulfonated polyether ether ketone (PEEK-WC)/phosphotungstic acid composite: Preparation and characterization of the fuel cell membranes." Pure Appl. Chem **78**(9): 1781 - 1791.

Swier, S., et al. (2005). "Sulfonated poly(ether ketone ketone) ionomers as proton exchange membranes." Polymer Engineering & Science **45**(8): 1081-1091.

Wu, H.-L., et al. (2006). "Preparation and characterization of poly(ether sulfone)/sulfonated poly(ether ether ketone) blend membranes." European Polymer Journal **42**(7): 1688-1695.

Xing, D. M., et al. (2005). "Characterization of Sulfonated Poly(Ether Ether Ketone)/Polytetrafluoroethylene Composite Membranes for Fuel Cell Applications." Fuel Cells **5**(3): 406-411.

Xing, P., et al. (2004). "Synthesis and characterization of sulfonated poly(ether ether ketone) for proton exchange membranes." Journal of Membrane Science **229**(1–2): 95-106.

Ye, Y.-S., et al. (2012). "Water Soluble Polymers as Proton Exchange Membranes for Fuel Cells." Polymers **4**(2): 913-963.

Zhong, S., et al. (2008). "Modification of sulfonated poly(ether ether ketone) proton exchange membrane for reducing methanol crossover." Journal of Power Sources **180**(1): 23-28.

10TH GLOBAL CONGRESS ON PROCESS SAFETY

# Fixed Plant Facility Siting – A Mega Project Lifecycle Framework

**M.F. Haider**
**mech_haider@yahoo.com**
**mhaider@suncor.com**

Prepared for Presentation at
American Institute of Chemical Engineers
2014 Spring Meeting
10th Global Congress on Process Safety
New Orleans, LA
March 30 – April 2, 2014

# Fixed Plant Facility Siting – A Mega Project Lifecycle Framework

**M.F. Haider**
**mhaider@suncor.com**
**mech_haider@yahoo.com**

**Keywords:** Facility Siting, Project Management, Mega Project, Consequence Analysis (CA), Quantitative Risk Assessment (QRA), Adaptive, Change driven, Agile methodology

## Abstract

Facility Siting based on tools such as Consequence Analysis (CA) and Quantitative Risk Assessment (QRA) is often times successfully applied for Brownfield operating unit projects, projects with relatively smaller footprints and for projects lasting relatively short time-periods. Challenge remains to apply and realize the full potential and value of Facility Siting for Mega-Projects. This is mostly due to project complexity and length – this paper will introduce a Management Framework to successfully apply CA and QRA as part of Facility Siting for Mega-Projects. The paper will also outline key strategies to ensure successful (i.e. value adding) implementation of facility siting, suggest a list of activities, associated sequence and pitfalls to avoid. Project Management team (e.g. Project Sponsor, Project Engineers, Managers) of both owner/operator and contractor/vendor may find this paper more useful as it provides a broad overview (i.e. when to initiate, how to plan, execute and close such assignment to maximize value) including real life examples of facility siting process over the project lifecycle.

## 1. Introduction

The driving force for process safety has been primarily based on catastrophic events [1]. With an increasing number of incidents, the process industry and governments started to take measures such as Facility Siting initiatives with hope to minimize loss of life and property. In the 1970s and 1980s, some of the world's most tragic industrial accidents took place. The Flixborough explosion in 1974 was reinforced by that of the Seveso tragedy in 1976 [2]. The Macondo blowout in the Gulf of Mexico on April 20, 2010, killed 11 employees and led to an uncontrolled oil spill lasting 87 days. This blowout was the most significant offshore incident in the US to-date. Finally, the Fukushima Daiichi nuclear plant incident in Japan in March 2011 drew the attention of the global process and power industries and encouraged them to incorporate natural disaster risks in a hazard analysis study [3]. As part of inherently safer design, many companies now-a-days apply CA and QRA as part of facility siting for their new projects. Due to the scarcity of information at the beginning, the studies in many cases start later in the project lifecycle for them to be effective and value adding. This paper will focus more on how to manage such initiatives for new Mega-Projects over the technical intricacies of the CA/QRA processes. The intention of the paper is to enhance the Body of Knowledge publicly available for management team in as much non technical languaga as possible to make their greenfiled Mega-Project initatives a success i.e. safer for people, environment and property.

So, what is different for industrial Mega-Projects?

- Industrial Megaprojects, as the name implies, are very large. To provide a simple and simply applied definition, we are defining a megaproject as any project with a total capital cost of more than $1 billion (U.S. dollars) as measured on January 1, 2003 [4].
- Most often engineering and construction activities run in parallel and not in sequence
- Multiple sub-projects under the 'Mega-Project umbrella' run in parallel
- Complex – design is almost 'never' complete in early stages – goes through many iterations
- Progressive elaboration is evident
- Scarcity of critical workforce with mega-project experience
- Scarcity of information – requires project team to make assumptions for CA/QRA
- Timeline: 2-5 years

## 2. Key Strategies

### *2.1 Facility Siting Charter*

It is not only important or good to do but critical that the stakeholders need to identify, understand and agree on the purpose and value to conduct facility siting for their projects. The Author's personal experience over the years has been in order to maximize value of Facility Siting the project team has to start 'early' to develop project Charter and start with the end in mind. As can be seen from the following generalized Figure 1 that as the company starts to appraise opportunities, it is

Figure 1: Stages and Milestones to apply CA/QRA in a Mega-Project

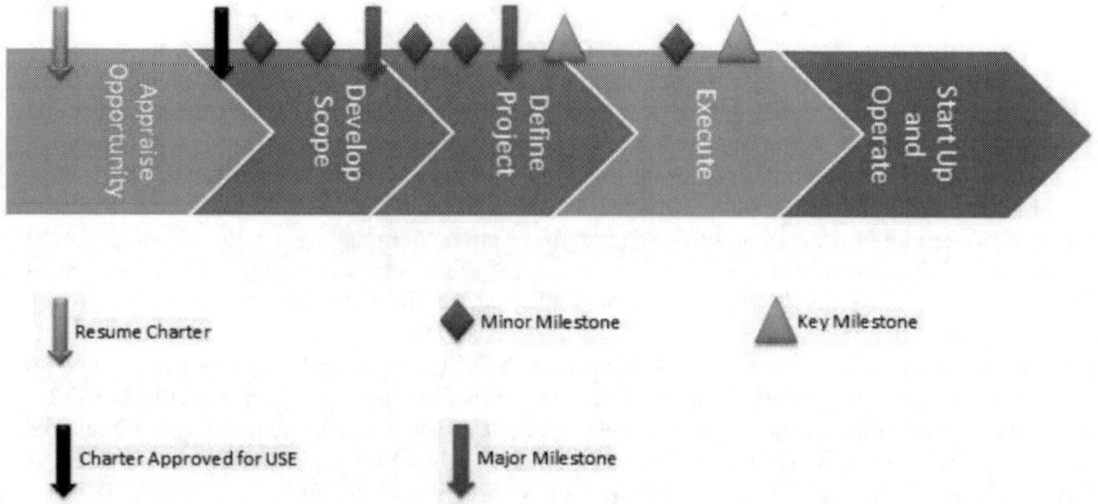


prudent to resume developing the Facility Siting Charter by bringing in the stakeholders together. Figure 1 shows a generalized project lifecycle for a typical Petro-Chemical Mega-Project. The

Figure also shows that the Key Deliverable is developed through a successive series of repeated cycles over the project life-cycle. The Milestones will be defined later in the paper. Part of starting with the end in mind is to clearly outlining ultimate goal/s, e.g. vision is to design a safer facility and utilize Mitigated QRA Report to formally document and communicate that facility is within the Risk Threshold or justification for accepting risk profile as is by Senior Management. The Vision in the Charter also provides opportunity to strike a balance between business goals and climbing the Risk Mountain – at each stage of increasing safety (decreasing risk). As part of striking a balance, the Project Engineer has to Identify options up-front if and/or when it becomes apparent that aggregate Residual Risk will not be within the Risk Threshold or acceptable so that team can put more effort in developing justification for either accepting or rejecting the risk profile by Management

Figure 2: Strike a balance between Business goal and Climbing the *expensive* Risk Mountain

### *2.2 Management Strategy*

In order to successfully manage Facility Siting, CA and QRA for a Mega-Project, the stakeholders have to align on a management strategy. There are two primary management approaches for the purpose of managing such work:

1. Predictive or Plan driven
2. Adaptive or Change driven

The Author's past experience has been if the organization undertook very 'similar' projects in the near past (<5 years), critical experienced workforce is retained, Lessons Learned is readily available, and most importantly, the past project is still considered to be a success, it is advisable to use Plan driven approach. On the other hand, if the organization is introducing new technology, has limited past experience with similar project, involves relatively high number of stakeholders, capital cost > $1B, project life spans >2 years, it is recommended to utilize Change driven approach to maximize value. It is to be noted that it is not essential to meet all the suggested criteria mentioned above to pick one management strategy over the other.

2.2.1 Benefits of Adaptive or Change Driven Approach

In Adaptive/Change Driven Planning, project team intentionally repeats one or more project activities as the project team's understanding of the product increases ..........Also known as ...agile methods .....intended to respond to high levels of change and ongoing stakeholder involvement [5]. The heart of Facility Siting involves making timely but informed decision. Some of the characteristics of this framework:

1. To add small incremental improvements that will deliver value to stakeholders
2. Milestones develop the product (e.g. Mitigated QRA Report) through a series of repeated cycles, while increments successively add to the functionality of the product
3. For each Milestone, a deliverable (read 'model') or set of deliverables will be completed. Future iterations may enhance those deliverables or create new ones. Each iteration incrementally builds the deliverables until the exit criteria for the phase are met, allowing the project team to incorporate feedback.

The biggest challenge with Mega-Project is scarcity of key information needed as inputs to develop scenarios and models. Some of the critical benefits, a project may gain by applying Change Driven approach:

1. The approach acts as Leading Indicator on Process Safety for Plant Layout - allows for multiple opportunities to make critical decisions on plant layout over the project life-cycle capitalizing on what is known at a given time. This lets the project team to use "what they know", to ensure "what they don't know" falls into the right place
2. Opportunity to make decisions over the project lifecycle not only works to save life but also helps with striking a balance between business needs and climbing the risk mountain
3. Aides design-development with progressive elaboration while producing final CA/QRA deliverable for the project

### *2.3 Execution Strategy*

The main idea of Change Driven process starts with decomposing the known preliminary layout and process information into a set of requirements, work to be performed and their priority sequence, sometimes referred to as a *Product Backlog*.

It is advisable that the owner/operator forms a multi-use agreement (contract) with the chosen vendor. This way the plant areas under the Mega-Project umbrella can get their specific work done while the CA/QRA Project Team provides overall oversight and guidance.

At the beginning of an iteration, the team will work to determine how many of the highest priority items on the backlog list can be delivered within the next iteration. The time-length of each iteration should be less than 4 weeks. As shown in Figure 1, during "Develop Scope" and "Define Project", the length of each iteration could be even less than 2 weeks. This iterations (otherwise known as "Minor Milestone" as shown on Figure 1) will develop scenarios, assumptions and preliminary CA/QRA models. Assumptions are required due to lack of available quality information in the early stages of a project.

At the end of each iteration, the Minor Milestone/deliverable should be ready for review by the customers/stakeholders. This does not mean that the customer is required to accept delivery, just that the product should not include unfinished and/or incomplete features.

The sponsor and customer representatives should be continuously engaged with the other project stakeholders to provide feedback to Contractor on deliverables as they are created and to ensure that the *Product Backlog* reflects Sponsor/stakeholders' current needs. Below, you will find two examples using two different management strategies. For confidentiality purpose, the specific name/location is kept generic and few items were modified to suit.

#### 2.3.1 Project X Facility Siting Using Change Driven Approach

A CA/QRA team was given the assignment to support design development during "Develop Scope" phase (Figure 1). The Sponsors already selected the site for the project. But the fixed plants were yet to be located on the site. The Vision for the initiative was for the QRA result to come under the organization Risk Threshold. Multiple EPC's were given to engineer different plant areas. The CA/QRA team selected a vendor to support them with defining the *Product Backlog* based on the preliminary Process Design Basis, intended product the facility is expected to sell, the feedstock necessary etc. The CA/QRA team had prioritized the Product Backlog based on hazard level of feedstock, preliminary PHA schedule and project schedule. This also involved obtaining feedback from different Process Engineering and Operations teams to collect necessary inputs to put together the assumptions and scenarios to be modeled within the 1$^{st}$ iteration.

Out of 25+ scenarios, only 7 perceived to be the most hazardous were modeled for CA that gave ample input (e.g. explosion, toxic and fire contours) for the CA/QRA team to work with the stakeholders to recommend location for specific process units on the given site. Note that, CA/QRA team integrated the potential consequence to understand local incident and global consequence events. At this time QRA was not performed. This was used as input for minor modification of preliminary plant layout. It is to be noted that the site plan was still preliminary after the first iteration but assisted different EPC accountable for their specific area to recognize where to locate the specific plants i.e. finalized key plans for the overall site plan.

With further design-development, more refined information became available from each EPC. This new information led to 130+ scenarios and was used for further CA/QRA iterations that created either new model and/or embellished the past ones. Each EPC was accountable for facility siting within their boundary while the CA/QRA team provided technical oversight and integration support for the overall project. The Sponsor/stakeholders were engaged on a regular basis to ensure that the Product Backlog reflects future modeling requirements to properly site facilities e.g. Fire Water Pump house, Tank Farms, site Medical Centre etc. Early engagement of CA/QRA team and higher management willingness to incorporate their feedback with the progressive elaboration of the project led to QRA result meeting the organization risk threshold and ultimately safer facility. Changes suggested based on CA/QRA work in earlier stage (i.e. Develop Scope, Define Project) were easier to incorporate into the design.

2.3.2 Project Y Facility Siting Using Plan Driven Approach

The sponsors for project Y selected the site for the project and decided to hold on to not perform CA/QRA work with the progressive elaboration of the project activities. This was mostly decided based on the fact that not all information was available to properly conduct the 'studies'. Fundamentally, the sponsors looked CA/QRA as deliverables requiring a more 'complete' plan. Once the Develop Scope (as referred in Figure 1) is almost complete, the sponsors decided to engage CA/QRA team and a vendor. By this time, the project had significantly high number of stakeholders. Many of the stakeholders were not very familiar with CA/QRA and the value it could bring to them. There was not much time to schedule and train the stakeholders as well to bring them up to speed. As a result, the stakeholders were unable to align on a Vision. The Vendor was asked to provide explosion, toxic and fire contour maps for specific scenarios and locations that were almost 'fixed' on the site plan at that point in time. The resultant contour maps showed significant consequence on nearby facility with over 700+ people. It was also noted that the plant area consuming the most hazardous substance would impact the nearby facilities significantly from consequence perspective. It should be noted that the result was only based on a selected few scenarios. By the time the project reached the "Execution" phase, more scenarios were added as part of the scope. But it was already too late to incorporate most of the recommended changes resulting from CA/QRA work. The project went through several area by area risk reduction workshops only to accept (by the company executives per organization risk matrix) the aggregate residual risk level which was also found to be not within the organization risk threshold.

2.3.3 Importance of Starting Early

Figure 3: Impact of variable based on project time [5]

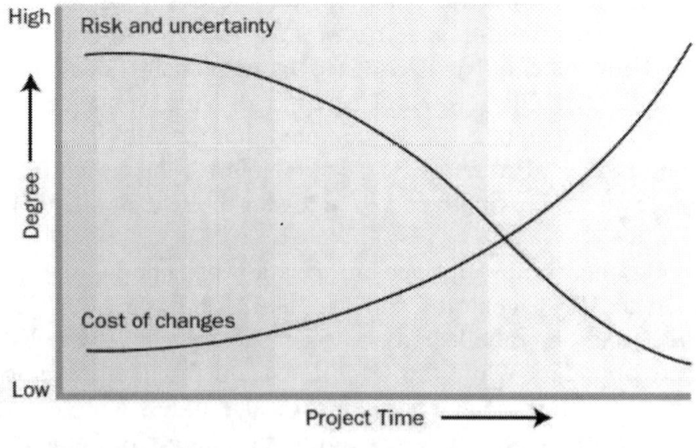


The above two examples highlight the importance of starting CA/QRA work early enough to realize and maximize the value CA/QRA brings to a Mega-Project. The high-level view provides common frame of reference for comparing projects for CA/QRA initiative, even if the projects were dissimilar. Change driven methodology allows for applying the known information at early stage in a way that supports laying out plant facilities to reduce risk and thus maximize potential value. While these characteristics remain present to some extent in almost all project life cycles, they are not always present to the same degree [5]. Change Driven framework, in particular, keeps stakeholder influences higher and the costs of changes lower especially in the early stages of the life cycle than in predictive/plan life cycles. The Figure 3 highlights a well-known fact of project management that even though risk and uncertainty remains high in the early stages of a project, cost of incorporating changes remains much lower. This translates into a key opportunity for project stakeholders to align on laying out the key areas on a site plan intelligently to reduce risk and maximize value by applying tools such as CA and QRA.

### *2.4 Key Decision Points*

Facility Siting, from a management point of view, is all about bringing the stakeholders closer and find the 'right' spot for the facilities. Finding the 'right' spots requires making timely decisions. The challenge has always been to make timely decisions as the stakeholders more often find themselves forced to make those decisions based on information that are frequently of poor quality and at times uncertain, especially in the early stages of a project. At times, due to the scarcity of information, stakeholders rely on their past experience to make those decisions on plant layout. As a result, depending on the individual experience, two different teams for a project area may come up with two very different plant layout. CA/QRA work, if started early enough, will also depend on input that most often requires embellishment through assumptions to create needed scenarios and models. Even with this limitation, CA/QRA provides a key 'input' in allocating real estate for key facilities on a plant site at early stage of a project lifecycle. With each iteration and deliverable 'Change driven' methodology allows for multiple opportunities to make critical decisions on plant layout over the project life-cycle. It allows the stakeholders to decide when to 'sharpen the pencil' (e.g. PHAST to Computational Fluid Dynamics model), relocate facility or accept the consequence and/or risk, 'pull the plug' on further modeling/refinement and start developing justification, if required, to recommend management to accept/reject the aggregate risk profile.

### *2.5 Alignment with Process Hazard Analysis (PHA) Planning*

CA/QRA results are a useful input to PHA (e.g. HAZOP, LOPA) activities. This is particularly true as projects move forward after establishing the plant layout on a high level and starts to conduct "Define Project" (Figure 1). PHA activities may also result in recommendations (i.e. new scenarios) that goes into the *Product Backlog* and requires project team to develop new models or refine old ones. Mega-projects have multiple sub-projects running in parallel most of the time, having an Integrated Schedule for all the reviews is a good strategy so that the right stakeholders are booked for the sessions. The integrated PHA schedule also supports prioritization of *Product Backlog* for CA/QRA activities over the lifecycle of the project.

### *2.6 Integration Strategy*

It is critical for a Mega-Project involving multiple plant areas/silos to understand the BIG picture so that site risk can be dealt with on an aggregate basis. There is strong possibility of one area to impact other plants in the vicinity. It is recommended to review integration aspect and aggregated results on a regular basis. The following Figure 4 shows that the origin of an incident may reside in one particular plant area, while it may impact other areas in the vicinity.

Figure 4: Consequence of an incident in one area impacting multiple areas

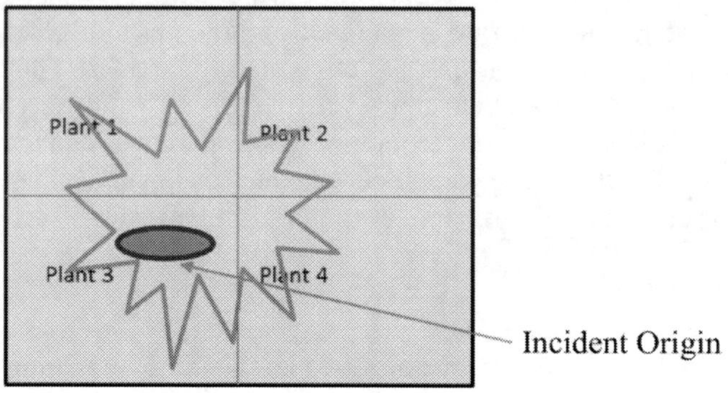


Recommended Strategy:

- Setup Strong Center - Task Force Single Point of Contact (SPOC)
- Collaborate – SPOC will promote Collaboration among Plants/Areas

## 3. List of Activities (suggested)

### *3.1 Overall Sequence*

The CA/QRA work typically should start with the approved Charter as shown on Figure 1. A project Task Force i.e. SPOC is formed comprising of Project Engineer, Operations rep, Process rep and CA/QRA Technical Expert. The SPOC is ultimately accountable to produce the CA/QRA deliverable studies but also acts to support the design development and site plan finalization efforts in the early stages of the project.

The team collects 'input' such as Preliminary site Layout and design information, works with area process and operations team to identify scenarios, develop assumptions for known-unknowns and unknown-unknowns, develop modeling 'Product backlog', prioritize backlog based on PHA schedule and produce modeling output (i.e. deliverable) for stakeholder use. The stakeholder usually has the option of accepting or rejecting consequence. If the result of an iteration is acceptable within area causing the incident – communicate to outside stakeholders to ensure there is no cross boundary impact. If the result of the iteration is NOT acceptable to outside of the area – use the result as input to PHA for the areas to mitigate. Collaboration recommended between origin area and impacted area/s to identify appropriate mitigations. There is also a possibility that the consequence resulting from the iteration is not acceptable to the area causing the incident, in that case the area will initiate a PHA to mitigate the consequence. As part of developing mitigations, consider the following:

- Develop QRA if appropriate
- Perform Building Assessments for Critical Buildings
- Apply Cost-Benefit analysis to optimize risks/opportunities

- Determine Safe Haven locations and their characteristics

If Mitigation is not possible from Business perspective – perform As Low As Reasonably Practicable (ALARP) justification for Senior Management to formally accept the posed risk

- To apply mitigations on a site wide basis, review Milestone deliverables by taking on a BIG picture perspective
- Future iterations may enhance those deliverables or create new ones as new information is available. Each iteration incrementally builds the deliverables to produce the Mitigated CA/QRA report to support Facility Siting for the stage

The overall effort is expected to produce a maximum of two single reports for the overall site:

1. Mitigated version of QRA Report
2. and Consolidated CA Report

It is recommended to develop as-built, pre-start up CA/QRA reports prior to start-up. Table 1 and 2 outlines typical activities performed by SPOC and plant areas respectively once key process areas are established/finalized on Site Plan. Please note that the activities are project dependent and may warrant modifications to suit specific situation.

**Table 1. SPOC Sequence of Activities to conduct Facility Siting**

| Initiative | Key Steps |
|---|---|
| Setup Facility Siting Team | Sponsor to support to setup a team (i.e. Project Engineer, Technical Expert, Operation, Process representative) to act as SPOC |
| Integration | Provide functional (technical and management) guidance to Areas |
| | Assess cross boundary impact/opportunity |
| | Communicate incremental outputs after each Area PHA/Risk Reduction |
| IFA versions (prior to "Define Project" stage) | Consolidated CA Report |
| | Mitigated and Integrated QRA Report |
| IFD CA/QRA (prior to "Execution" stage) | Re-validate IFA CA/QRA post PHA and vendor PHA |
| As-built versions | Post-oil version |

### *3.2 Typical Area Activities*

**Table 2. Area (Typical) Sequence of Activities to conduct Facility Siting**

| Initiative | Key Steps |
|---|---|
| Plant 1 (Typical) | Perform Modeling and Calculations |
| | Conduct Risk Reduction Workshop (if appropriate) – Apply mitigations |
| | Incremental output for use based on reduced risk level |
| | Assess Buildings/facilities |
| | Identify Safe Haven/shelter Locations (if required) |
| | Estimate Area Project Scope impact |
| | Perform area scope updates/changes e.g. passive fire protection requirement |
| | Carry out QRA (if appropriate) |
| | Update Product Backlog |
| | Develop/update contract to support area specific activities |

## 4. Pitfalls to Avoid

The following observations were noted by the Author during his time to manage Facility Siting scope:

1. Facility Siting does not replace good and common sense engineering practice e.g. locating a plant that uses high volume of hazardous material will significantly increase risks to other plants/areas (e.g. camp) within the vicinity
2. Evaluate effectiveness of 'Cost Reduction' philosophies as opposed to added risk e.g. different modularization techniques may decrease the plant footprint by reducing pipe lengths but makes the plant more compact and likely to increase severity of explosion consequence
3. Avoid developing models with hope to realize risk reduction
4. Sharpening the pencil for models most likely may not reduce risk
5. Avoid jumping into CFD or other similar methods without evaluating cost-benefit for undertaking such effort
6. Utilization of QRA results is much more controversial than the methodology. Watson [6] has suggested that QRA should be considered as an argument, rather than a declaration of truth. In this view – focus on accuracy over precision
7. CA/QRA Task Force should be part of Management of Change (MOC) process
8. Evaluate carefully before locating facilities e.g. tank farms close to main entrance/egress as part of Facility Siting

9. Keep the focus on VISION for the overall effort to avoid scope creep
10. Note the difference between Facility Siting as a Deliverable and tool for design-development activities

## 5. Conclusion

This paper presented Adaptive/Change driven framework to manage initiatives like facility siting/CA/QRA as part of Mega-Projects. Change driven framework does not mean "lack of planning or no planning". As demonstrated in the paper, this methodology starts with a vision and then develops the ultimate product both iteratively and incrementally when it is possible to define small incremental improvements that will deliver value to stakeholders. This allows making key decisions on laying out the facilities in early stages of a project rather than waiting for design to develop to a significant extent. Often times, project team wait too long for design to mature enough before applying CA/QRA to validate consequence impact and/or risk level. Most often, especially for Mega Projects, this leads to delays in executing the CA/QRA work properly at the right time to influence site layout. This delay in execution leads to scenarios when scope changes cannot be incorporated anymore. This translates into value erosion as project was unable to reap the full benefit of Facility Siting. The management framework recommended in this paper i.e. Change driven/adaptive framework allows for the stakeholders to start early in the project lifecycle to apply CA/QRA in laying out the plant facilities. The key is to start EARLY but START with the END in mind. Alignment among stakeholder on a VISION up-front is also important as this is expected to save life while striking a balance between business needs and climbing the Risk Mountain. Vision also allows avoiding scope creep while applying Adaptive/Change driven approach. It is highly recommended that stakeholders focus more on accuracy over precision. In the early stages of project lifecycle, quality information is scarce. Change driven/Adaptive framework gives you as the project stakeholder the opportunity to use "what you know" intelligently to ensure "what you don't know" falls into the right place.

## 6. References

[1] M. Mannan, A. Chowdhury and O. Reyes-Valdes, "A portrait of process safety: From its start to present day," College Station, 2012.

[2] M. Mannan, Lees' Loss Prevention in the Process Industries, Volumes 1–3 (3rd Edition), , 2005, Elsevier, 2005.

[3] K. McAndrews, "Consequences of Macondo: A Summary of Recently Proposed and Enacted Changes to US Offshore Drilling Safety and Environmental Regulation," in *Society of Petroleum Engineers, Americas E&P Health, Safety, Security and Environmental Conference*, 2012.

[4] E. W. Merrow, Industrial Megaprojects Concepts, Strategies and Practices for Success, New Jersey.: John Wiley & Sons, Inc., 2011.

[5] A Guide to the Project Management Body of Knowledge (PMBOK® Guide) – Fifth Edition, Newtown Square: Project Management Institute, Inc., 2013.

[6] S. Watson, "The Meaning of Probability in Probabilistic Risk Assessment," *Reliability Engineering and System Safety,* vol. 45, pp. 261-269, 1994.

# 348480 Global Energy and Transportation and Mobil Oil


**Abstract:**

Biofuels, Computational Methods, Engineering Economics, Polymers, Thermodynamics. Polymers-plastic, Thermodynamics- oil, Engineering Economics – the money it takes to finish the method of Global Engineering Research, Computational Methods – numerical input of analytical data, Biofuels – the transport fluids that utilized in more than one method from industrial usage to everyday living which enters the atmosphere. Exxon has completed a $600 million dollar deal for production of algae biofuel, amongst Mobil Oil has performed major research Global Energy and the effects of oil. Reading a most recent article on Mobil website.

# 348748 Characterization of Iron Phthalocyanine As the Active Material for Lithium Batteries

**Abstract:**

Results on the determination of the lithium-ion diffusion coefficient in the iron phthalocyanine (FePc) active material as a function of intercalation level, using a modified electrochemical impedance spectroscopy technique, will be presented. Also, developed thermodynamic functions such as the Gibbs free energy, entropy, and enthalpy changes for the lithium insertion reaction with FePc, derived from cell open-circuit voltage measurements as a function of intercalation level, are presented. Also presented are estimates of the maximum amount of lithium that can be intercalated into the FePc based on constant volume and temperature conditions.

# 349086 Design of a Free-Fall Reactor for Fast Pyrolysis of Waste Plastics

**Abstract:**

Thermal pyrolysis has earned significant reputation in the recovery of valuable chemicals, especially petrochemical based fuels from various materials, including biomass and plastics. Of many available technologies, *fast* pyrolysis was preferred in his study owing to its minimal heat and mass transfer limitations and better control on process conditions. The focus of this study is towards efficient conversion of waste plastics into gaseous products consisting primarily of alkanes, alkenes and aromatics. Lab scale experiments were performed using a micro reactor to determine favorable conditions, including reaction temperature and particle size that would result in maximum yield of ethylene. Using these results, a simple industrial size radiatively heated free-fall reactor was designed for the purpose of pyrolyzing waste plastics. The goal is to design a *fast* pyrolysis reactor that improves upon inefficiencies in traditional reactors while resulting in high yield of ethylene.

# 349732 Biosorptive Dehydration of Ethanol/Water Azeotropes Using Compound Starch-Based Adsorbent

**Abstract:**

**Abstract**: Recently, the interest in the production of bioethanol has greatly increased as a result of the increasing fossil fuel prices and associated environmental issues. Bioethanol is made from either grains or biomass and the broth contains 6 to 12 percent ethanol. In addition, water removal represents a serious problem as ethanol form azeotrope with water at atmospheric pressure. Distillation to water-free ethanol consumes 50 to 80 percent of energy used in the fermentation ethanol manufacturing process. Various techniques have been developed to break the azeotropic point, such as salting-out method, azeotropic distillation, extractive distillation, reactive distillation, adsorptive distillation and pervaporation. Among the methods, the adsorptive option is the most attractive, especially the biosorption, which has distinct advantages over conventional methods as it is environmentally friendly, highly selective, easily available, easy to operate, cost effective and reusable in repeated cycles in the treatment of aqueous organics.

In this work, a specially formulated compound starch-based adsorbent (CSA), which consists of corn, sweet potato and foaming agent, was developed. As an exploratory study, this research was focused on the investigation of CSA for biosorptive dehydration of aqueous ethanol and optimization of conditions for its maximum adsorption under various conditions of vapor feed flow rate, bed temperature and bed height through an orthogonal design method. Furthermore, to investigate the adsorption mechanism, dynamic saturated adsorbance of CSA was studied and characterization of the adsorbent before and after adsorption, as well as after regeneration, was performed.

In the first step, the net retention time and separation factor of ethanol and water were obtained using inverse gas chromatography (IGC), the results indicated that the net retention time of water on CSA is significantly longer than that of ethanol which attested the feasibility of this adsorptive separation purpose and low temperature could be propitious to the process which was consistent with the traditional theory of adsorption.

Then, through a batch adsorption experiment, the optimum adsorption condition was determined under vapor feed flow rate, bed temperature and bed height by orthogonal design method, which indicated that the optimization condition for ethanol was at vapor feed flow rate of 1.1g/min, bed temperature of 83 and bed height of 25cm. The packing density was 445kg/$m^3$. The residence time is determined by vapor feed flow rate and bed height. The residence time is more influenced than bed temperature. The longer the residence time, the more the output is. But it cannot be ignored that the adsorption selectivity may decreases. Low bed temperature is benefited to the adsorption process which agrees with the result of IGC. Five cycles experiments of adsorption-regeneration were carried out at the optimum adsorption and regeneration condition, the results showed that there is practically no change in the regeneration efficiency of CSA after the cyclic process adsorption–regeneration. It is due to the properly chosen regeneration conditions assuring complete desorption of the water adsorbed without changes in the CSA structure. This unchanged adsorption capacity after a number of cycles demonstrates reusability of the adsorbent.

To study the adsorption mechanism, experiment was conducted to study the dynamic saturated adsorbance of CSA. The water adsorption capacity of CSA has been found to be similar to some molecular sieves, which is about 0.15g/g. Besides, field emission scanning electron microscopy (FESEM) and mercury porosimetry (MP) were applied to investigate the change of surface morphology and microstructure of CSA before and after the adsorption, as well as after regeneration, as shown in Fig.1 and Fig.2. The FESEM images showed that compared with the particles before the adsorption, they became plump and the void spaces became smaller after adsorption, which indicated that the CSA was saturated with water. However, the CSA particles after regeneration became wizened like the nature particles, which demonstrated that almost all the water had been desorbed from CSA. The pore size distribution (PSD) indicated that compared with the PSD curve before adsorption, the peak at 6000nm disappeared and the peak at 17000nm became remarkably smaller after adsorption, which manifested that CSA is mainly composed of macropores which play a major role in the adsorption. However, the peaks at 6000 and 17000nm reappeared after regeneration, and the curve looked exactly the same with that before adsorption, which further demonstrated that CSA after regeneration can be totally regenerated and recovered to the original structure.

These results indicated that the starch-based adsorbent can be considered as an effective and low-cost bio-sorbent to dehydrate alcohol.

# 349781 Purification of 2-Amino-1-Phenylethanol Enantiomers By a New Technique Combining Distillation and Crystallization

**Abstract:**

Stripping crystallization (SC) is a new separation technique which combines distillation and crystallization. Basically, SC is operated at a triple-point condition, in which the liquid is simultaneously vaporized and crystallized due to the three-phase equilibrium. In essence, the process is continued until the liquid phase is completely eliminated and only the pure solid crystals remain in the feed. Thus, crystal washing is not required since only pure solid crystals remain in the feed and no impurities are adhered on the crystal surfaces at the end of SC. In the previous research, SC has been successfully applied to separate p-xylene from the mixed xylenes by Shiau et al. (2005, 2008).

Phenylethanolamine, also known as 2-amino-1-phenylethanol (2-APE), is an important basic skeleton in many adrenergic drugs and is often used as ingredients of cold medicines and anorectics. As the biological activity and pharmacokinetics of the enantiomers of a chiral drug can be different, the objective of this research is to apply SC to purify R-2-APE from the chiral 2-APE mixture. The experiments show that no crystallization occurs during SC without seeding due to subcooling of the liquid enantiomers. On the other hand, seeding can eliminate subcooling during SC, leading to the R-2-APE crystals gradually formed preferentially on the seeds. Thus, seeding with magnetic stirring can be successfully employed to purify R-2-APE from the 2-APE enantiomers.

1. Shiau, L. D., Wen, C. C. and Lin, B. S. Separation and purification of p-xylene from the mixture of m-xylene and p-xylene by distillative freezing, *Ind. Eng. Chem. Res*, **44(7)**, 2258-2265, 2005.

2. Shiau, L. D., Wen, C. C. and Lin, B. S. Separation of p-xylene from the multicomponent xylene system by stripping

# 350841 Fabrication of Low Cost Insulating Material from Kaolin Clay for Construction Purposes

**Abstract:**

A new low cost, thermal insulator, construction mineral polymeric material was synthesized from local Saudi Arabia resources using Inorganic Mineral Polymerization technique (MiP). In this technique, the mineral polymer is produced from the following components: clay mineral rich in koalinite, sand, sodium hydroxide solution, and water. Inorganic polymerization experiments are conducted by mixing known amounts of clay minerals and sand in a variable speed stainless-steel mixer for some period of times until a homogenous solid mixture is obtained. Sodium hydroxide solution is added to the mixture and the polymerization reaction takes place at different mixing speeds and times. The produced mixture is divided into a number of samples of known weights. Each sample is moulded in a stainless-steel cylinder at a medium pressure. Then the moulded samples are cured in an oven for 24 hours. The curing temperature was 80°C. The mineral polymerization conditions such as the ratio of clay to sand filler, particle size of the clay, particle size of the sand, concentration and amount of sodium hydroxide, speed and time of mixing were optimized to obtain stable and hard mineral polymer. Compressive strength for different specimens of fabricated mineral polymers was tested under dry and wet conditions. Values of compressive strength with values ranging from 11.5 MPa to 27.9 MPa were obtained.

# 351141 Thermodynamic Study of Binary PAH (Anthracene + Phenanthrene) Solid Mixtures

**Abstract:**

Polycyclic aromatic hydrocarbons (PAHs) are common components of many materials, such as petroleum and various types of tars. They are generally present in mixtures, occurring both naturally and as byproducts of fuel processing operations. It is important to understand the thermodynamic properties of such mixtures in order to better understand and predict their behavior (i.e., fate and transport) in the environment and in industrial operations. The Knudsen effusion technique was combined with gas chromatography and differential scanning calorimetry to measure the vapor pressure, composition, and phase behavior of binary anthracene and phenanthrene mixtures at near ambient temperatures. Mixtures of anthracene and phenanthrene exhibited non-ideal mixture behavior: In mixtures with less than 80% phenanthrene, this more volatile component preferentially sublimed and the vapor pressures decreased as the mixtures became enriched in anthracene. The vapor pressure data do not follow Raoult's law (that is, ideal mixture) predictions. Mixtures with high phenanthrene content (> 80%) showed azeotrope like behavior, in that anthracene and phenanthrene sublimed at nearly constant vapor pressure and molecular ratio. These results confirm that the behavior of few-component PAH mixtures does not follow ideal mixture models. Hence, as tars and petroleum (and potentially other organic compound-containing mixtures) weather to few component mixtures over time, complicated, non-ideal phase behavior might be observed.

# 351160 How Confident Are You That Your Major Accident Risks Are Under Control?


**Abstract:**

Major accidents in the process industries including those at Texas City , Buncefield UK and in the Gulf of Mexico have highlighted the critical role played by senior managers and executives. Effective leadership is essential if companies are to develop a positive safety culture that remains constantly vigilant towards the risk of a major process safety incident. In the UK the HSE has made leadership a key priority for the high hazard industries and there is a cross-industry initiative to develop and deliver training focussed towards board members responsible for process safety performance.

This paper explores the leadership failings that contributed to recent major accidents, and the cost to both the company reputation and the individuals involved. Key principles are discussed that can help improve leadership in the major hazard industries. These include; attitudes and decisions of senior manager affecting the investment strategy and the safety culture of the organisation, reinforcing the importance of safety by personal example, thorough understanding of major accident hazards and key risk control systems, investigating process safety incident and near misses to find the underlying causes, developing world class safety management systems, and identifying weaknesses in these systems using targeted performance indicators.

The author bases this paper on the development of a training standard for 'Process Safety Leadership' involving a cross-industry 'expert panel'. Training of senior executives in compliance with this standard has been carried out from late 2011 and feedback from these courses and the key requirements for effective leadership will be discussed.

# 351236 Sil Determination: Shortcomings with the Use of LOPA


**Abstract:**

Proposed Topic Area: Managing Risks

For a number of years, industry has been trying to apply the principles outlined in IEC 61508. In the process sector, this has resulted in a focus on the requirements of IEC 61511 (2003) and ANSI/ISA 84.00.01 (2004). These are risk based standards. They involve the identification of potential hazardous events, the causes that could lead to each event and the necessary risk reduction for each event. Where the risk reduction involves an instrumented function (e.g. a trip or an interlock), the required performance of the instrumented function can described in terms of the average probability of failure on demand. In the standards, the performance for such instrumented functions is assigned to one of four Safety Integrity Levels. The assessment of what performance level is needed in respect to a specific hazardous event is referred to as SIL Determination. In the field of SIL Determination, IEC 61511 provides examples of a number of different approaches – these are shown in IEC 61511 Part 3. These include Event Trees, Risk Graphs, Safety Matrices, Fault Tree Analysis and Layer of Protection Analysis (LOPA).

In the first few years, following the publication of IEC 61508 Edition 1 in 2000, industry generally embraced the use of Risk Graphs as the method of choice. However, over the subsequent few years, many companies using this method found that it was (a) giving higher safety integrity level targets than really needed and (b) was less flexible than other methods and consequently more difficult to use in many real situations. This has resulted in a general move to other methods and in particular to the use of Layer of Protection Analysis. Indeed, LOPA may now be the predominant "Method of Choice" for SIL Determination. As a method, LOPA gives the impression of simplicity and appears quite straightforward. It almost invites anyone with a minimum of understanding of the method to be a practitioner.

However, there are a number of shortcomings in the use of the LOPA method that have become evident through its use across the globe.

This paper will review the application of LOPA and provide insights into what may be seen as the proper use of the method and its core principles. It will also highlight various shortcomings in use that have been found during expert reviews carried out of a number of existing studies. The paper will be of benefit both to those who are new to the LOPA method, but also of value to those who have been using the method for some time.

# 353790 A Recipe to Excel: Beyond Technical Competency

**Abstract:**

The presentation will highlight the fundamental soft skill sets needed to be successful in managing diverse work forces in an organization. The skill sets are invariant of the type of organizational environment. This could span R&D driven organizations to manufacturing & production to engineering, procurement and construction driven organizations and to other service industries.

To be successful, it is taken for granted for any technically driven project or organization, that to carry out his/her assigned responsibility, the manager needs to have technical skills and understanding. The factor that differentiates a truly successful manager, however, is the manager's extraordinary strength in soft skills, sometimes referred to as people skills. The necessity of these skills is becoming more important and pertinent than ever before. The reason behind this special requirement is the diversity of groups involved in any work environment. Such diversity is due to rapid globalization. Challenges associated with diversity, when managed properly, bring strength rather than weakness.

In this presentation the importance of effective communication will be emphasized. Effective communication is the most difficult skill to acquire and practice but the most rewarding when a manager/leader acquires and proficiently applies it. Through effective communication the leader brings team spirit and trust. Trust in turn brings success.

# Management Careers in Consulting and Engineering

**R. Benson Pair, Jr., PE, BCEE**
*Chief Technology Engineer - Environmental*
*KBR*

KBR



## Career Opportunities in Consulting and Engineering

Consulting companies and Engineering firms offer engineers a wide range of career opportunities in management

KBR

## Management Careers in Consulting and Engineering

- Technical and Technology
- Engineering
- Project Management
- Engineering Management
- Project Controls
- Offices
- Business Units
- Corporate
- Sales
- IT/IM
- Procurement
- Construction

KBR



## Consulting

- De-centralized, serving regional markets
- Smaller offices and smaller projects compared to EPC firms
- Historically has been engineering only, but that is changing
- May be focused on one market sector: Industrial or Government (Federal/State/Local)

KBR

## Engineering, Procurement, Construction

- Fewer, larger offices
- Larger projects
  - Contract value
  - Schedule
  - Task force can be >100
- Often serves Industry and Government

KBR



## Technical Skills

- Engineering Disciplines
- Accounting
- Cost and Scheduling
- Contracts
- HSE

KBR

## People Skills

- Leadership
- Communication: Listening and talking
- Negotiating
- Respect for others
- Patience
- Humility in yourself / Pride in others
- Sense of Humor
- Delegation
- Mentoring

KBR 

## Career Development - The "Y" Approach

- Recognizes and values the technical specialist
- Shows transition from generalist to technology specific roles
- Identifies three broad bands of skill levels
- Continues to recognize and value the technical leaders who combine technology and management skills
- Shows transfer routes to Project Engineering / Project Management and Sales Coordinator / Sales

KBR 

## The "Y" Model for Career Development

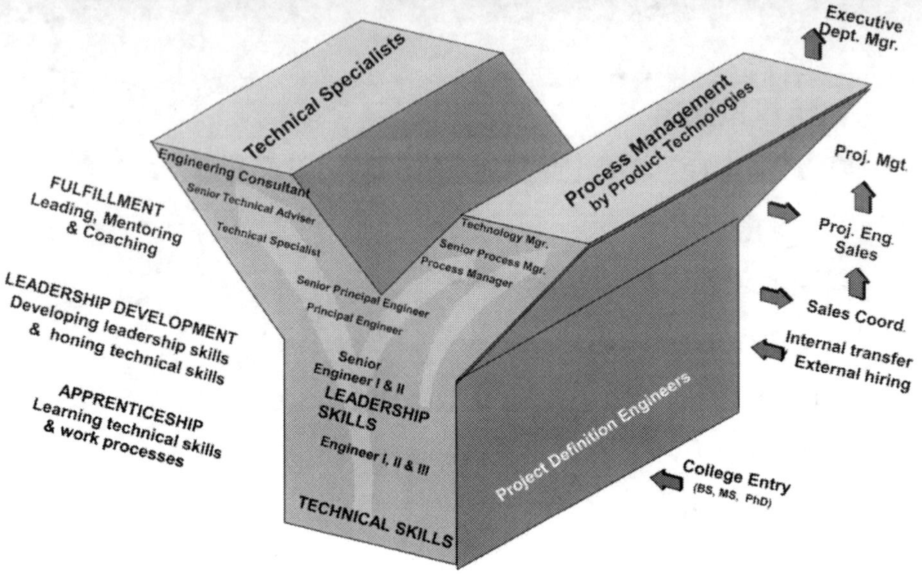


KBR



## Management Is…

… a set of well-known processes, like planning, budgeting, structuring jobs, staffing jobs, measuring performance and problem-solving, which help an organization to predictably do what it knows how to do well. Management helps you to produce products and services as you have promised, of consistent quality, on budget day after day week after week.

KBR

## …(Still) not Leadership

Leadership is associated with taking an organization into the future, finding opportunities that are coming at it faster and faster and successfully exploiting those opportunities. Leadership is about vision, about people buying in and about empowerment and most of all about producing useful change.

Source: "Management Is (Still) Not Leadership", John Kotter, Konosuke Matsushita Professor of Leadership, Harvard University, *Harvard Business Review*, January 9, 2013

KBR



## Management and Leadership

Managers

- administer
- maintain
- focus on systems and structures
- rely on control

Leaders

- innovate
- develop
- focus on people
- inspire trust

Source: "The Wall Street Journal Guide to Management," Alan Murray

KBR

## Honesty

"Your business and its employees are a reflection of yourself and if you make honest and ethical behavior a key value, your team will follow suit."

Source: "Top 10 Qualities that make a Great Leader", Tanya Prive, Forbes, 12/19/2012

KBR



## Engineering Ethics

**Engineer's Creed**

As a professional engineer, I dedicate my professional knowledge and skill to the advancement and betterment of human welfare

I pledge:

- To give the utmost of performance;
- To participate in none but honest enterprise;
- To live and work according to the laws of man and the highest standards of professional conduct;
- To place service before profit, the honor and standing of the profession before personal advantage, and the public welfare above all other considerations.

In humility and with need for Divine Guidance, I make this pledge.

Adopted by the NSPE, June 1954

KBR

# 353824 Management Careers in an Industrial Setting

**Abstract:**

Management careers in an industrial setting offer a broad range of opportunities and challenges.

It certainly is important to gain a clear understanding of your personal goals and objectives before embarking on a career in management.

What are your motivations? Do you have the skills and aptitude? Are you willing the do the things that need done?

Learning and development are essential. Time commitments are significant.

Rewards can be amazing when you see people & operations develop beyond expectations.

The key becomes tapping into the capability of the team and help move the organization forward.

# 353793 Managing Through Extreme Change

**Abstract:**

The Department of Energy Weapons Complex, a nuclear chemical enterprise employs hundreds of thousands of scientists and engineers. Through its birth during the Manhattan project in the 1940's to the end of the Cold War, the DOE Weapons Complex was a rapidly expanding enterprise that provided career growth and service towards a national mission. After the Cold War ended, the weapons complex has been slowly decreasing in size through a series of layoffs, furloughs, reductions in benefits and contract changes. While the DOE Complex was originally built by world class industrial companies, the current contract practices and potential legal liabilities have changed the field of companies that now operate the complex for the Department of Energy. These new types of companies formed as limited liability corporations, are typically partnerships of engineering and construction firms, aerospace companies or not for profit research companies. Managing businesses in a declining industry and opportunities within the nuclear chemical industry will be presented.

# 353905 You Can Have it All, But Not at the Same Time

**Abstract:**

Women (and men) in science and engineering have struggled with the Career Life Balance. With the rise of dual income couples and more female breadwinners, the working world has a goal of becoming better at balancing work and family life for both men and women. As managers women may face additional challenges not experienced by our male counterparts. Challenges occur in all sectors of the workplace: private industry, academia and government. Numerous articles and books have been written on the subject. Recently the Huffington Post featured and article titled "Do Women Make Better Bosses?".

# Managing Volunteers Is A Little Like Herding Cats

1

Joseph J. Cramer, PhD, PE

Director Emeritus, Program Development

Careers in Management Session

AIChE Spring Meeting

April 1, 2014

New Orleans, LA

## How did I select this title?

2

- Clearly chosen to attract interest
- Analogy has some validity
- Both cats and volunteer Ch.E.s are:
  - Intelligent
  - Fiercely Independent
  - Can scratch when irritated
- Strict military style of management won't work
- More than 60 million volunteers in USA
- Volunteers (and their management) are very important

## Background Information

3

- Have been managing volunteers for >30 years – almost entirely involving AIChE
- Three degrees in Chemical Engineering
- Didn't like life in a lab or plant
- Didn't like just doing calculations all day
- Gravitated towards project management and later department management
- Enjoyed interaction with people, like dealing with crisis situation but also liked doing planning to avoid the crisis in the first place
- Slowly transitioned from managing 'direct reports' to 'volunteers'

## Types of Authority

4

- Line Authority (Traditional)
- Staff (Expertise) Authority
- Coercive/Reward Authority (penalties and rewards)
- Functional Authority (specific tasks for designated time)
- Referent Authority (based on charisma, charm, personality)

## 5 Developing Your Style of Managing Volunteers

- Expertise (spoke the language, been there/done that)
- Rewards (important to volunteers even if they won't admit it)
- Granting Functional Authority
- Referent Authority (personality based-honesty, credibility, trust, knowledge)
- Motivating (nurturing enthusiasm)

## 6 Volunteers vs Employees

- Even disinterested workers can be motivated by compensation
- Disinterested volunteers disappear (or worse, can be a negative force)
- Poor workers can be fired (not always easy)
- Poor volunteers are difficult to fire
- Must word harder to really engage volunteers
- Must match tasks with skills - carefully

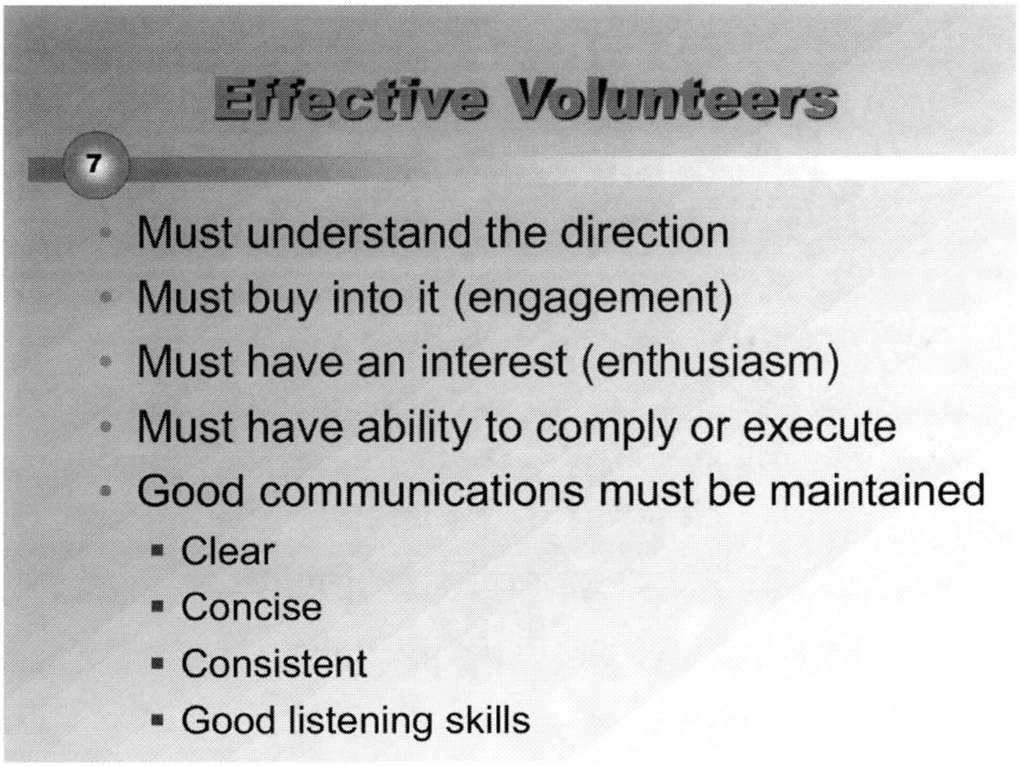

Effective Volunteers
7
Must understand the direction
Must buy into it (engagement)
Must have an interest (enthusiasm)
Must have ability to comply or execute
Good communications must be maintained
Clear
Concise
Consistent
Good listening skills


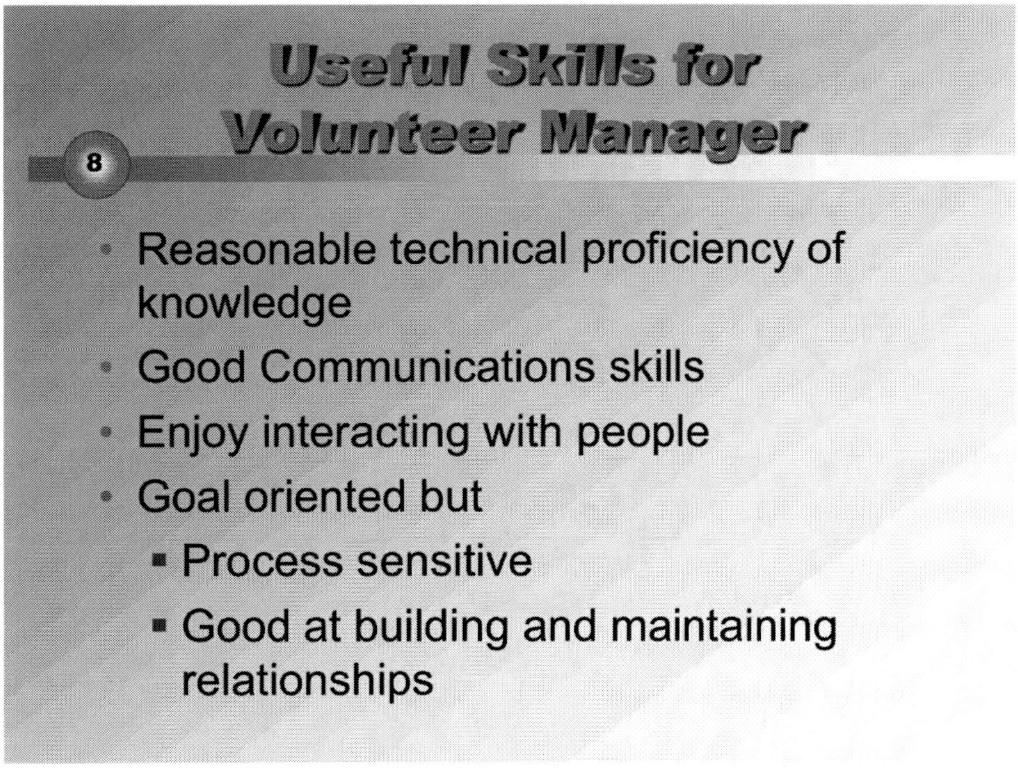

Useful Skills for Volunteer Manager
8
Reasonable technical proficiency of knowledge
Good Communications skills
Enjoy interacting with people
Goal oriented but
Process sensitive
Good at building and maintaining relationships

## The Benefits

9

- The organization can accomplish so much more
- The volunteer learns about 'managing'
- Volunteers acquire management experience more quickly
- The experience is transferrable
- The skills are good for a lifetime
- Relationships can last for life

10

## Thank you for your attention.

Any questions?

Comments?

Anecdotes?

# 353795 Panel of Careers in Management Participants

**Abstract:**

An open panel segment will allow for an open Q&A discussion involving chairs, speakers and the auduence.

# Oil in Water

William E Moore
Westech Engineering

14 AIChE Spring Meeting, New Orleans, LA

WESTECH Process Equipment. Process Driven.

## Oil in Water

- Despite the popular saying
  - oil and water do mix
- Thereby lies the fun and employment.

14 AIChE Spring Meeting, New Orleans, LA

# Oil in Water

# Oil in Water

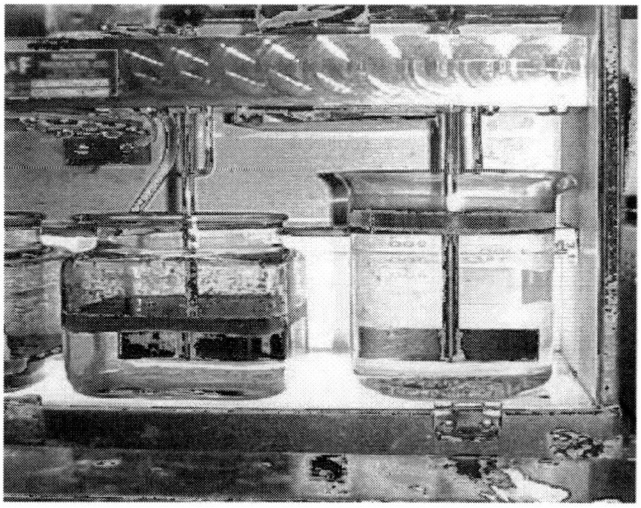

## Oil in Water

- Oil (and grease) are a problem in a lot of industries
- Oil tends to coat everything it comes in contact with.  This makes it a great lubricant and a killer of many organisms.
- Oil can act as a glue in process equipment causing fouling and media clumping
- Can lead to foaming

## Oil in Water

There are four main types of oil

- Free or floating
- Dissolved
- Emulsified
- Grease

## Free Oil

- Most common type of oil
- Usually floats on top of the water
- Relatively easy to remove using an API separator
- There are many patented equipment for free oil removal. Most are limited to removal of small quantities of oil.

## Dissolved oil

- Water is a universal solvent
- Polar VS Non-polar
- Difficult to remove
- Fortunately concentrations are usually low

## Emulsified Oil

- Emulsified oil is a colloidal dispersion of oil drops in water
- Can be stable for months or even years
- Made by chemicals and/or mixing
- To remove must break emulsion with chemicals and/or heat
- Then float the oil out on gas bubbles (DGF VS IGF)

## Grease

- Grease is just thick oil
- Can have all the forms of oil in water
- Typical test outside of petrochemical industry is Total Oil and Grease
- Don't forget cooking grease

## Forces acting on a Particle in Water

- Gravity
- Flow
- Friction
- Brownian motion
- Electrostatic
- Van der Waals

## API 421

- The "Industrial Bible" for free oil water removal
- Based on Stoke's Law which also is the basis for clarification and many other precipitation systems

The oil-globule rise rate ($V_t$) can be calculated by Equation 1 or 2 shown below. Equation 1 should be used when the target diameter of the oil globules to be removed is known to be other than 0.015 cm and represents a typical design approach. Equation 2 assumes an oil globule size of 0.015 cm.

$$V_t = \frac{g}{18\mu}(\rho_w - \rho_o)D^2 \qquad (1)$$

$$V_t = 0.0123\left(\frac{S_w - S_o}{\mu}\right) \qquad \text{(where } D = 0.015 \text{ cm)} \qquad (2)$$

# API 421

$V_t$ = vertical velocity, or rise rate, of the design oil globule, in cm/s.

$g$ = acceleration due to gravity (981 cm/s$^2$).

$\mu$ = absolute viscosity of wastewater at the design temperature, in poise. Note: 1 P = 1 g/cm.s; 10 P = 1 Pa.s.

$\rho_w$ = density of water at the design temperature, in g/cm$^3$. Note: 1 g/cm$^3$ = 1 kg/litre.

$\rho_o$ = density of oil at the design temperature, in g/cm$^3$.

$D$ = diameter of the oil globule to be removed, in cm.

$S_w$ = specific gravity of the wastewater at the design temperature (dimensionless).

$S_o$ = specific gravity of the oil present in the wastewater (dimensionless, not degrees API)

# API 421

Restrictions

- Horizontal velocity (vH) through the separator should be less than or equal to 1.5 cm/s (0.015 m/s) or equal to 15 times the rise rate of the oil globules (*V*t), whichever is smaller.
- Separator water depth (*d)* should not be less than 1 m, to minimize turbulence caused by oil/sludge flight scrapers and high flows. Additional depth may be necessary for installations equipped with flight scrapers. It is usually not common practice to exceed a water depth of 2.4 m.
- The ratio of separator depth to separator width (*d/B*) typically ranges from 0.3 to 0.5 in refinery services.
- Separator width (*B*) is typically between 1.8 and 6 m in refinery services.
- Effluent has < 5 ppm free oil.

# API 421

- The amount of freeboard specified should be based on consideration of the type of cover to be installed and the maximum hydraulic surge used for design.
- A length-to-width ratio (*L/B*) of at least 5 is suggested to provide more uniform flow distribution and to minimize the effects of inlet and outlet turbulence on the main separator channel.

# Picture of an Operating API

# API Design

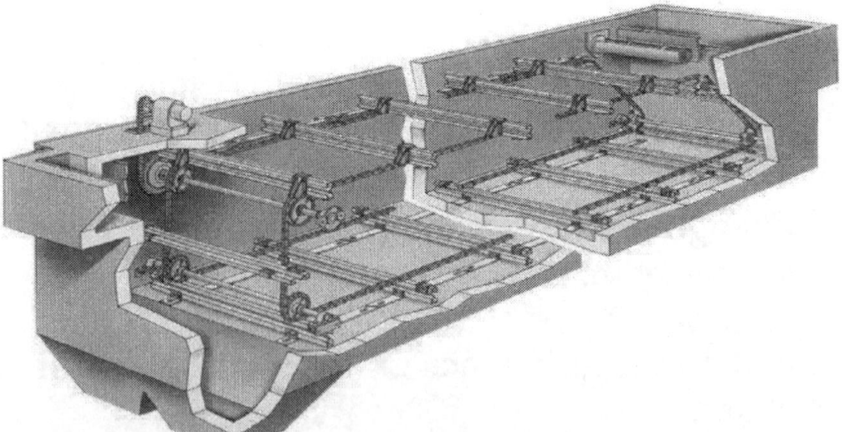

## Free Oil Removal problems

- How do you get the oil out?
- How do you get out the settled solids?
- What about the “rag”?
- Usually covered

## Corrugated Plate Interceptor

- Sometimes called a “Lamella Separator”
- An API with slanted media installed to increase surface area
- Will save capital and plot space
- Much easier to plug

# Plate Pack

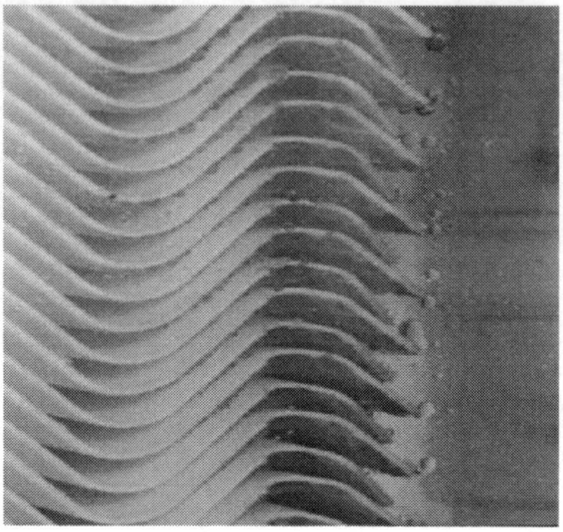

# How to Break Emulsified Oil

- Lower pH
- Heat
- Add emulsion "breakers" – can be inorganic but usually organic. Must be specific to the emulsion.
- Add air bubbles

# DAF VS IAF

- Also known as DGF VS IGF
- Both have very strong proponents and most refineries only have one type.
- Most industrial systems are considered explosive and have covers
- A major problem is solids removal from the bottom of the tank.

# Design Flexibility

*Using customer specifications, standard design methods, computer software, and exper sizing, WesTech customizes each project to suit your specific application. Cost-com and proven results ensure that your process needs are fully met in the design of each*

### Typical DAF Configurations

*Circular Column Supported DAF*

- Larger diameter tanks (50 to 85 feet in diameter)
- Heavy duty cage drive
- Typical bottom fed influent with top fed recycle

*Circular Bridge Supported DAF*

- Smaller diameter tanks (less than 50 feet in diameter)
- Heavy duty shaft drive
- Typical side or top fed influent with concentric recycle

*Rectangular DA*

- Available up to 20
- Ships fully shop-
- Non-metallic or st
- Optional ZICKEF float removal me

# DAF VS IAF

- In a DAF, liquid from the DAF vessel is pumped to a separate pressurized tank. Compressed air is added and the water flows back to the DAF tank which is at atmospheric pressure. The dissolved gas is released and small bubbles form.

# DAF VS IAF

- In an IAF, liquid from the vessel is pumped through a venturi which sucks in air. The liquid then goes back into the tank and air bubbles rise.

## DAF VS IAF

+ DAF produces smaller bubbles and a better dispersion.

+ IAF does not require a compressor or a pressurized tank. Thus it is cheaper and has slightly less operating costs (power).

## Dissolved Oil Removal

Good Luck

- May try activated carbon, ion exchange, or proprietary media
- Usually requires chemistry specific to what is removed

# Final Oil Removal

- Walnut shell filter < 1 ppm
- Activated carbon
- Other adsorbents

# Walnut shell filter

# Activated Carbon

- Can be made from any carbon source – charcoal, coal, coconut shell, etc.
- Made by pyrolysis, steam or chemicals
- Creates lots of pores with high surface area
- Iodine, molasses and permanganate #

# Activated Carbon Uses

Organics removal

- Better removal for non-polar than polar
- Water interferes
- Does not like oil
- Works by surface adsorption – micro and macro pores
- Regenerated by heat

## Activated Carbon Uses

Chlorine Removal

- Works by reacting with the chlorine
- Works on other oxidizers- speed is determined by strength of the oxidizer
- Chloramines

## Other Activated Carbon Uses

- Air purifier
- Mercury adsorption
- Catalysis
- Medical
- Warning- Don't forget the fines

# Measuring Oil in Water

- Standard lab test is a solvent extraction
- Field equipment looks for the sheen or uses UV light
- XRF
- Stripper with sensors
- No instrumentation will give total oil in water

**WorleyParsons**

resources & energy

# Synergies Between Refinery & A Waste Dump

Peter A.Nick, PE

Sr. Supv. Engr Worleyparsons WCO

Peter.nick@worleyparsons.com

Principal - Renewable Energy Technologies

www.retphreg.com

EcoNomics

## Two Sides of a Coin

- Our appetite for energy to power our lives has run up against the problems of finite and unstable sources, economic generation and environmental sensitivity. New Tight Oil & Gas procedures that reduce foreign dependence, match existing efficiency and provide sustainable alternatives are the goals of new technologies, but accomplishing these ends in a timely, cost-effective manner is not easy.
- A REFINERY is thus considered an ENVIRONMENTAL INSULT that could readily be replaced by renewably generated electric cars and power generation units
- As landfills reach capacity in the US , and restrictions on the location and environmental impact of such facilities become problematic, a crisis looms for states and municipalities as to how best dispose of MSW. With over a half million tons per day of waste generated in this country, local and state governments are confronting spiraling costs for transporting and disposing of this trash as well as political pressure to find lower-impact disposal methods.
- A LANDFILL is also considered an ENVIRONMENTAL INSULT that could be avoided if only we just recycled enough.

## SIMILARITIES

- SO THEY BOTH ARE ENVIRONMENTAL *INSULTS*
- BOTH HAVE PUBLIC MYTHOLOGY DESCRIBING HOW TO GET RID OF THEM FOREVER
- BOTH HAVE A HIGH FEED & STOCK MATERIAL CONCENTRATION OF COMBUSTIBLE ORGANIC MATTER

  1. Refinery – primarily C-H petroleum hydrocarbons
  2. Landfill – residential and commercial waste of organic and non organic origin

## What is an Organic

- Refinery feeds and products are CxHy types of hydrocarbons with only a small percentage of heteroatoms such as O, N, S and an even smaller amounts of heavy metals such as Ni, V, Fe, U, Hg, Se
- Municipal Waste primarily has lots of C5-C6 sugar derived C-H-O polymers (cellulose, plant matter/food waste, med waste) but with a substantial amount of petroleum-like polymeric material(polyolefins and other polymers, latex paints, composite plastics, etc. Lots of non-organic salts, glass, metal components..some recoverable some not
- Most of these wastes have a considerable fuel potential

  (MSW HeatOfComb = 4000 -12000 BTU/lb .....Oil is ~ 18000 BTU/lb).

## Dichotomy -> Singularity

- Municipal and industrial waste has been a growing societal pollution problem for years.
- With both long term energy needs and waste disposal costs continuing to rise and be subject to regulatory pressures (many not justified), one finds opportunity to apply basic chemical engineering principles to take advantage of such a situation if a refinery is nearby.
- Is there a commonality say between what the trash could produce and the refinery needs ?

**Yes there is !**

## So How Is This Accomplished ?

- Refineries need Power, hydrogen, fuel gas……all of which they gladly for a reasonable economical price rather than buy OSBL.
- Several MSW conversion processes can make the above utilities from MSW and other waster material.
- None of which would yet make economic sense but for one fact:

*Cities and their trash haulers increasingly would like to pay you to take the fuel (trash)* ***and keep it*** *rather than dump it in the ground. Not easy to do that anymore in most urban areas.*

## So How Is This Accomplished ?

- High temperature Anaerobic Pyrolysis/Gasification or advanced-control incineration with Cogen units are the only two technology groups that could do this economically and reliably.

- Every other process marketed today leaves too much toxic residue and/or air emissions that overwhelm the process cost needs. Many of these processes are " Rube Goldberg" nightmares and essentially end up having to burn significant amounts of raw trash just to be able to maintain heat rates for sustainable operations. Scalability to better economic design is limited at best.

## Why Pyrolytic Gasification ?

- Incineration is pretty much limited to Combined Cycle CoGen units. PHREG product gas can be used in a number of ways.
- Incinerators generate large amounts of ash which is difficult to handle and typically must be treated and disposed of as a hazardous waste.
- Greater flue gas and ash cleanup
- Inherently higher cost of a steam power system
- Net cost per ton of equivalent feed is 20-25% greater than pyrolysis
- Grate efficiency of burn is at best 30% - more likely below 25%.
- PHREG net efficiency is 22% of feed LHV
- Capital cost 20% below that of "Trash burn " plant
- Eliminates ash disposal headaches.

## So How Is This Accomplished ?

- *Even most gasifier/pyrolytic units have levels of non-feasibility.*
- *Incinerators are still burn units, regardless of how efficient they are now, and will always have air emissions problems unless a lot of money is spent on the unit.*
- *Plasma-Arc processes are self-limited in unit size and have costly maintenance problems*
- *Coal Gasification knockoffs run at high pressure and temperatures which means very high capital investment that can only be amortized by processing 5-10 K+ ton/day rates if at all*

## Why Pyrolysis & Not Incineration ?

Incineration inherently emits oxidized hazardous pollutants (NOx, SOx, ClOx, FlOx, dioxins, furans, aldehydes, ketones, others) promoted by direct combustion of waste material

Pyrolysis in PHREG unit occurs in a reducing atmosphere.

Hazardous components of off-gas are $NH_3$, $H_2S$, HCl, HF, HCN, amines, etc.

easily and economically scrubbed with existing technology.

Cleanup equipment capital costs are 20-40% less for PHREG than for incineration processes.

PHREG process avoids conditions favorable to the production of dioxins via the deNovo synthesis reaction.

Low nitrogen in Syngas Product reduces NOx formation tendencies during combustion. Air combustion of waste makes NOx formation a recurring problem for incinerators. Impossible to make a true low NOx solids burner.

SOME PROPOSED GASIFIERS ARE IN FACT INVOLVE PARTIAL OXYGEN COMBUSTION OR RAW OR PARTIALLY PYROLYZED WASTE MATRIX

# So Here is Our Candidate !

- The PHREG reactor aka the Pyrolytic Heat Recovery Enhanced with Gasification process
- Patent (US # 7452392B2) is a property of Renewable Energy Technologies, a California corporation.
- PHREG Is an extension of and improvement to the old Union Carbide Purox MSW process that was a technical success back in the 1970's but was discontinued (1977) due to the low energy prices it was attempting to provide an alternative to. Trash dumping was still cheap back then.

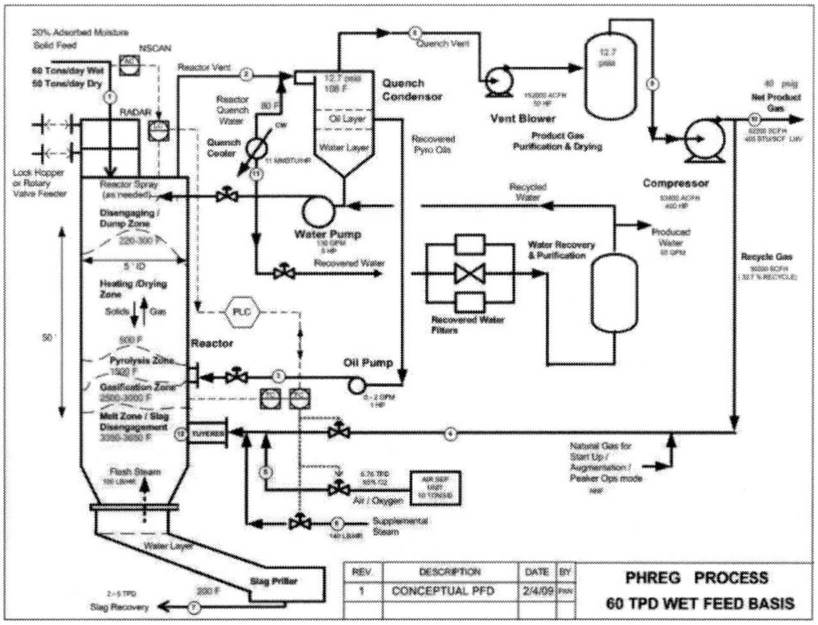

# Value Recovery Solutions for Organic Waste Disposal

Pyrolytic Gasification yields a high quality Synthesis Gas from Waste Materials such as:

- Municipal Solid Waste (Landfill after resource recovery at $25 -$150/ton)
- Biomass and Agricultural Wastes (Burn on Site – emissions problems)
- Sewer Plant Sludge (Limited landfill availability - Landfill at $50 -$150/ton)
- Medical & Hospital Waste (Hazardous Landfill at $250 - $500/ton)
- Refinery & Oil Processing Wastes (Hazardous Landfill at $250- $1500/ton)
- Used Tires (Hazardous Landfill, Local Nuisance, Fire Danger – equiv fuel to coal)
- Food, Beverage, & Cosmetic Manufacturing Waste (Very High BOD to Sewer, Landfill Disposal at $25 - $150/ton)
- Construction & Demolition Wastes (Landfill after resource recovery at $25 - $150/ton)

If it contains organics…it's a fuel to us and a disposal problem solved for you !

# SynGas Utilization

## Several commercially viable options available

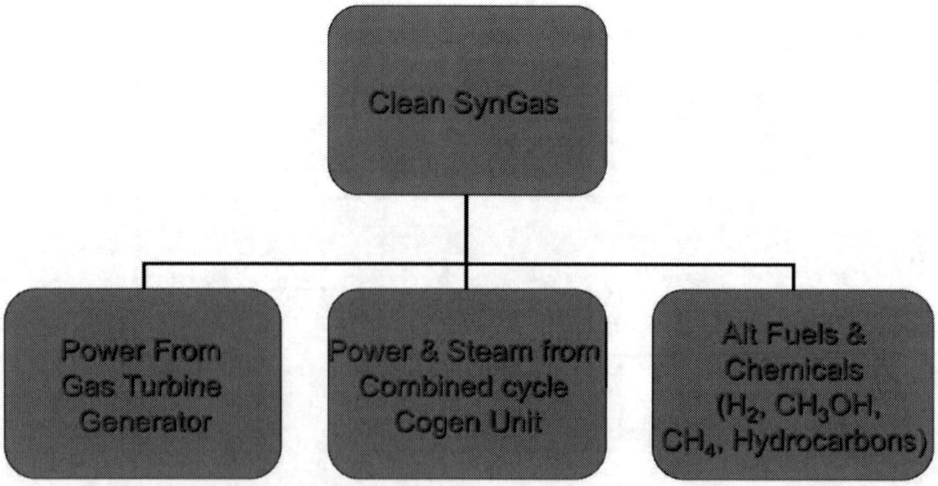

# FEED SYSTEM

- Upon arrival at the plant facility, the wastes will be placed in a surge pile on an apron with double containment to prevent runoff which would otherwise contaminate soils or groundwater.
- Material is pulled from the surge pile and run through a shredder with 4 counter rotating shearing shafts. Shafts and screens are designed for rapid replacement, and can be completely replaced in 2 hours or less as needed.
- Shredded feed is placed on an eddy-current sorting table to sort out and recover any metallic materials, sorted by composition.
- The separated shredded waste is placed on a pan conveyor, which carries it through the Fast Neutron Chemical Analyzer to the surge hopper for the airlock charging valve on the gasification reactor.

# REACTOR

- The gasification reactor is a vertical shaft furnace operated as an updraft falling bed reactor to pyrolyze the feed and produce SynGas and slag. The refractory lined shell is approximately 50’ high. Diameter is a function primarily of capacity.
- Slag is tapped continuously into a water quench tank and the solids are recovered by a drag-chain conveyor.
- Steam rises from the quench tank and enters the reactor via the slag notch. A hot (3500 – 3800 ° F), sub-stoichiometric combusted driver gas is injected through tuyeres and provides the primary heat to drive the reactions and gas flow to buoy (lift) the reactor solid charge.

## GAS CLEANUP

- Crude syngas exits the top of the reactor via the draft flue at approximately 300 °F and is carried through the connecting ductwork (residence time to be adequate for sterilization but less than 5 min to avoid dioxins) and then is quenched with water prior to entering the gas cleaning system.

- The raw gas exiting the reactor is hot enough to kill any pathogens, but too cool for the de novo synthesis of dioxins and furans. The gas is reducing in nature, which also tends to block the possibility of forming these undesired compounds. The water quench puts a final closure to any such possibility of significant formation.

- The Quencher saturates the raw product gas, and brings the temperature to around 170 °F, condensing out any higher hydrocarbons (C 6 through C10) which are separated by gravity in a settling tank The quench water is kept at alkaline pH in order to strip out any acid gases ($H_2S$, HCl, etc.)

## GAS CLEANUP

- The quenched and stripped process gas now enters a multi-stage scrubber which removes fine particulates and residual acid gases.

- Amounts of HX (mostly CL, F, Br, S, N) gases will occur and they will come out in the net water condensed from the overhead.

- The stripped and scrubbed gas is cooled in a dessicator (or possibly a glycol contactor) to remove excess water vapor and produce a SynGas product well suited for:
  - 1. Firing a combined cycle gas turbine to produce electricity.
  - 2. Conversion to any of several syngas-derived chemicals such as methanol, hydrogen, acetic acid, alcohols......

## What does it cost to generate ELECTRICAL power from PHREG versus competing processes ?

- Operation of a 500-2500 ton/day RET commercial plant will allow for electricity at a competitive 10-15 cents/kilowatt-hr given the bases of Table I below.
- This would take care of the MSW taken from a base of 200000-1000000 generators.
- This cost of generation is comparable to that of new power plant construction using existing generation methods such as gas, oil, hydroelectric or coal.

## Cost of Energy from Various Sources

| Energy Source | Cents/Kilowatt-Hour |
|---|---|
| RET Municipal Solid Waste Process | 10-15 |
| Coal (hypothetical for CA) & Fuel Oil | 8-12 |
| Municipal Solid Waste Combustion (MWC) | 14-22 |
| Natural Gas | 6-12 |
| Hydroelectric | 6-20 |
| Wind | 35-60 |
| Nuclear | 18-22 |
| Solar | 45-60 |

Table 1

## SynGas Utilization Option 1

### Power from Gas Turbine-Generator Set

| FEED LHV BTU/LB | FEED RATE TONS/DAY | NET POWER GEN MW | CAPITAL COST APPROX $MM | CAPITAL RECOVERY APPROX YRS | REMARKS |
|---|---|---|---|---|---|
| 4500 | 500 | 0 | 50 | 6 | MSW MINIMUM SELF SUSTAINING |
| 8500 | 500 | 20 | 75 | 7 | BASELOAD MSW |
| 8500 | 2500 | 100 | 250 | 5 | BASELOAD MSW |
| 12500 | 50 | 4 | 15 | 2 | MED WASTE TYP |
| 15000 | 100 | 11 | 28 | 24 | OLD TIRES TYP. |
| 15000 | 500 | 55 | 85 | 5 | OLD TIRES TYP. |

BASIS: GIVEN TONS/DAY OF TRASH @ LHV 7500 HRS/YR UTILIZATION
ANNUAL OP COSTS @ 10% OF CAPITAL
BASELOAD POWER REVENUE @ $200/MWH
FEED TIPPING FEE AT $ 100 /TON MSW
AT $ 500 /TON MED WASTE
AT $ 100 /TON TIRES

# The assumptions for generating these numbers are:

1. New plant construction in California, 2nd quarter 2014
2. Variation of cost ranges between minimum - maximum feasible plant size for energy source
3. 7-10 year payback with "Utilities" financing model
4. Baseload and auxiliary peaker operations at typical capacity factor for source of energy
5. Electricity immediately connectable to grid. Utility may still add 10-15 cts/kwh
6. Waste receipt fee of $60-120/dry ton (PHREG & MWC only)
7. **No subsidies beyond normal tax accounting practices in a "for profit" business. Germany recently announced it was stopping FIT subsidies and would start charging 59cts/kwh for the privilege of using solar power on their grid.**

# SO OUR POINT IS.....

The intent of this presentation is to propose a venue where such a technology can be viably implemented in a mutually beneficial way

PHREG is not the only MSW conversion process but is pretty representative of what should work for a surrounding community of 250K to 1 MM population

7-10 year payback economics does not include the mutual benefit that a refinery could gain with the local municipality, which has to pay a lot of money to rid themselves of trash, maintain a local dump, and/or keep voters and one of their biggest tax bases happy.

The refinery could both make some utilities at a reasonable value AND benefit the surrounding community by allowing them to solve their trash problem and save money at the same time.

# BORON REMOVAL FROM TREATED INDUSTRIAL WASTEWATER

*William Celenza*
*KBR, Inc.*
*Houston, TX*

## Summary

Treated industrial wastewater is increasingly being used for process and cooling water makeup, landscaping and agricultural irrigation, and aquifer recharge. Facilities that source their process water from seawater or brackish / saline reservoirs often employ Reverse Osmosis (RO) membrane desalination equipment that historically have not been very efficient at removing neutral species like boric acid at typical seawater pH. Usually present in seawater from 4 to 6 mg/L, boron can be toxic to both plants and humans at low concentrations. Even though a boron concentration less than 0.5 mg/L in nutrient solutions is essential for plant growth (3), slightly higher levels can prove toxic to sensitive plants. Thus, for an end use application such as "unrestricted irrigation quality," some jurisdictions require maximum boron levels near 0.5 mg/L. For comparison, the World Health Organization (WHO) Guidelines for Drinking Water Quality (6) have a guideline for boron at 2.4 mg/l.

As competition for water resources increase, industrial uses may not receive a high priority. Some industrial facilities, however, can consider seawater sourcing for process water needs, in addition to treated wastewater reuse applications. In this paper, design considerations for boron removal in desalination water treatment and industrial wastewater treatment plants will be reviewed with respect to reuse applications for the treated effluent. KBR has been assisting clients to maximize the reuse potential of treated wastewater effluent to offset the increasing scarcity of water resources and to comply with the more stringent standards for reuse. The first part of this paper examines current regulation of boron in the water environment. In the second part, applications of boron removal by RO and other technologies are presented in relation to overall water treatment and wastewater treatment system design. Opportunities for optimizing boron removal processes are highlighted.

## Introduction

Industrial facilities are devoting considerable efforts to water management as demands on water resources increase. Limited freshwater supply, declining freshwater quality, increasing raw water costs, environmental discharge standards and water reuse programs all impact water resources management. Technological advancements in water and wastewater treatment processes are providing more efficient and economical water management options from non-traditional sources. Developing new water resources via salt water desalination and treated wastewater reuse can be viable options that may result in more dependable raw water supplies and provide for a more self-sufficient facility.

Seawater desalination using thermal processes or membrane technologies has become an important source of drinking water and irrigation water production. Wastewater reuse has been used to provide irrigation water, industrial plant cooling and process water, and groundwater recharge.

In particular, desalinated seawater and treated wastewater reuse, with the wastewater sourced from the desalinated seawater, are potential water resources for some industrial facilities. As a consequence, seawater components such as salts or Total Dissolved Solids (TDS), chloride, sodium, magnesium, sulfate and boron are included along with the wastewater parameters of Biological Oxygen Demand (BOD), Chemical Oxygen Demand (COD), Total Suspended Solids (TSS), oil and grease, and nitrogen compounds as important treatment specifications. Of these parameters, boron may represent the most unique treatment challenge as the relatively newly regulated compound is one of the most difficult seawater components to remove.

Design considerations for boron removal in desalination water treatment and industrial wastewater treatment plants will be reviewed with respect to reuse applications for the treated effluent.

## Boron Concentration

Boron is a naturally-occurring trace element that is widespread in nature. Boron in seawater can range between 0.5 mg/L in the Baltic Sea and 9.5 mg/L in the Mediterranean Sea (1). Boron concentrations in seawater are typically 4 to 6 mg/L (1,2), usually in proportion to seawater salinity. Most removal applications appear to have been designed for a 4.5 to 5 mg/L feed, with consideration for seasonal variability. Boron can be toxic to both plants and humans at very low concentrations, even though a boron concentration up to 0.5 mg/L in nutrient solutions (3) is essential for plant growth. Since just slightly higher levels above the nutrient range can prove toxic to sensitive plants, some jurisdictions require maximum boron levels from 0.3 mg/L to 0.75 mg/L (18,3,4,5) for an end use application such as "unrestricted irrigation quality."

WHO Guidelines for Drinking Water Quality have a guideline for boron at 2.4 mg/l (6). The US EPA (7) determined in June 2008 not to regulate boron with a National Primary Drinking Water Regulation (NPDWR). The EPA concluded that boron was not likely to occur at levels of concern when considering both surface and ground water systems. An NPDWR did not present a meaningful opportunity for health risk reduction. Note that only freshwater, surface, and ground water systems were included in the surveys referenced with this ruling. The US EPA has calculated a Health Reference Level (HRL) of 1.4 mg/L, based on an updated boron oral Reference Dose in August 2004. HRLs refer to exposure levels that will not cause significant risks of non-cancer health effects. Long-term exposure to levels below these levels are assumed to produce no ill effects. Thus, based on the concentrations of boron in the potable water where it occurs relative to the HRL, the US EPA has determined that boron does not present a meaningful opportunity for health risk reduction for persons served by public water systems. Furthermore, boron is not considered mutagenic, and the EPA has determined that there are inadequate data to assess the human carcinogenic potential for boron.

Conversely, states have taken different approaches. The Minnesota Department of Health (MDH), in their Guidance for Groundwater, states that "low doses of boron may be beneficial, but questions about whether it is essential remain. Within an individual, the range between beneficial and toxic quantities may be very narrow. Between individuals, beneficial and toxic quantities may overlap. Because of the difficulty in establishing a concentration value that is low enough that it will not cause toxic effects in any individual, but not so low as to deprive others of the beneficial effects, the MDH does not plan to establish an HRL for boron," (8). Instead the MDH has issued a maximum recommended "chronic" value in drinking water of 1 mg/L.

Table 1 lists the boron levels discussed above as well as those of various state water quality guidelines.

**Table 1 Boron Guideline Levels**

| Guideline | Boron (mg/L), (a) |
|---|---|
| WHO Guidelines for Drinking Water Quality (6) | 2.4 |
| US EPA Health Reference Level (HRL) (7)<br>Florida Groundwater Cleanup Target Level (9) | 1.4 |
| California Department of Public Health Drinking Water Notification Level (NL) (10)<br>Wisconsin Enforcement Standard for Groundwater (11)<br>Maine Maximum Exposure Guideline for Drinking Water (12)<br>Minnesota maximum recommended chronic value in drinking water (8) | 1.0 |
| Long Term Use limit for irrigation water (3,4,5) | 0.5 to 0.75 |
| Nutrient solutions for optimum plant growth (3) | Up to 0.5 |
| Wisconsin Preventative Action Limit for Groundwater (11) | 0.2 |

Notes:
(a) Typical Seawater concentrations are 4 to 6 mg/L.

For practical purposes, a boron concentration of 0.5 mg/L, that represents a conservative value for the unrestricted irrigation use category (3,4,5), will be used along with the WHO Drinking Water Guideline (6) as target treatment specifications for boron removal schemes discussed below.

## Boron Removal

Boron in seawater is predominately in the form of boric acid ($H_3BO_3$). The neutral form of boric acid, at the typical upper range of seawater pH of 8.4, is approximately 85% of the boron species present. The percentage increases at lower pH. Conversely, the hydrated or ionic form predominates at pH over 9.5. A consequence of boric acid's smaller size and absence of an electric charge is increased volatilization into product streams in evaporators (13) and reduced removal by RO membranes (14). The increased size and charge of the ionic form provide for better removal efficiencies. Table 2 presents removal percentages typically achieved by listed technologies used in seawater desalination as well as treating wastewater for reuse. For comparison, Table 3 presents the required seawater boron removal rates to meet irrigation and drinking water use category limits.

**Table 2 Boron Removal Percentages of Water and Wastewater Treatment Technologies when processing seawater**

| Technology | % Removal |
|---|---|
| 1 Pass SWRO – standard membrane (14, 15, 16, 17) | 65 to 80 |
| Evaporator (13) | 70 to 75 |
| 1 Pass SWRO – high-removal membrane (15, 16, 17) | 80 to 85+ |
| 2 Pass – SWRO + BWRO (16) | 75 to 95 |
| 1 Pass SWRO – standard membrane at pH=8.5 (15, 16) | $\geq$90 |
| 2 Pass SWRO + pH$\geq$9.5 + BWRO (2, 16) | 80 to 99 |
| 1 Pass SWRO + IX (2, 16) | >99 |

| | |
|---|---|
| MSF Desalination (21) | >99 |
| Notes:<br>SWRO = Seawater Reverse Osmosis<br>BWRO = Brackish Water Osmosis<br>IX = Boron Selective Ion Exchange<br>MSF = Multistage Flash | |

**Table 3 Boron Removal from Seawater vs. Guidelines**

| Influent Concentration | Unrestricted Irrigation | CA Drinking Water NL | WHO Drinking Water Guidelines |
|---|---|---|---|
| mg/L | | % Removal | |
| 4 | 88 | 75 | 40 |
| 5 | 90 | 80 | 52 |
| 6 | 92 | 83 | 60 |

Removal percentages listed in Table 2 assume adequate pre-treatment, with the evaporators and Seawater RO (SWRO) membranes receiving a standard seawater composition feed at 25°C. As temperature increases, both the water and the salt permeability increase. With RO plants operated at constant pressure flux, the permeate salinity will vary proportionality with temperature. With RO plants operated at constant pressure, the permeate salinity will also increase with temperature due to a greater increase in salt permeability than water permeability (2). An increase of 1°C can increase the salt permeability by 3 to 5% (19). Boron permeability will also increase with temperature.

Evident from comparison of Tables 2 and 3 is that all the listed technologies are potentially capable of meeting the WHO Drinking Water Guidelines while only half may be capable of meeting a nominal 90% removal for Unrestricted Irrigation quality. Irrigation quality requirements have not been satisfied by one-pass SWRO membranes or high-pressure one-pass SWRO membranes (14). In SWRO technology, boron permeate level has been observed to decrease as operating pressure is increased. Notably, the higher pressure differential across the membrane produces more desalinated water, and thus offers more dilution of the boron in the permeate. Also, no dependence of boron removal upon concentration has been observed when the boron level is lower than 35 mg/L (14). Thus, to meet irrigation levels, boron removal has usually required multiple RO passes along with pH adjustment.

One-pass SWRO operation at a pH of 8.5 has been shown cost-effective to produce a permeate of less than 1 mg/L boron with minimal scaling (17). The pH required for 50% ionization of boric acid in seawater is 8.5, while in low salinity water a pH of 9.5 is required (15). Because seawater is supersaturated with respect to calcium carbonate, a small increase in pH may induce scale precipitation. A pH of 8.5 is near the practicable limit for feed into SWRO membranes. A two-pass membrane system offers the option of raising the pH of the SWRO permeate to enhance boron removal across a Brackish Water RO (BWRO). Due to removal of salts in the first pass SWRO, most of the scale precipitation concern should be eliminated for the second pass BWRO. However, since lower salinity or brackish water requires a higher pH for ionization of boric acid vs. seawater, a pH of over 10 in the feed to a BWRO may be required for greater than 90% overall boron removal across the BWRO (16). Also, since removal of both salts and boron depends on salt concentrations, observed BWRO removal is lower than for membranes in SWRO service (16, 2) although higher permeate flux occurs and lower operating pressures are required. Ion exchange with boron-selective resin, producing product below 0.05 mg/L boron, may be used to treat the permeate from an RO membrane to meet boron limits (2). Thermal technologies such as Multistage Flash (MSF) desalination can produce distillate with less than 0.05 mg/L boron that correlates to a removal capacity of more than 99% from the feed seawater (21).

Since the choice of boron removal technology depends on the required product level, the use category with the lowest required boron level controls the options. In seawater desalination plants built to serve municipal customers, drinking water and irrigation uses are most important. The irrigation requirements would control the technology selection due to boron concerns. However, in a seawater desalination plant built to serve only industrial users, utility water requirements would be primary with TDS and chloride levels the main concern.

Sixty percent of the worldwide desalination capacity is devoted to human consumption (20). Agricultural use is included within the 60% both directly and indirectly, since increasing amounts of treated sanitary wastewater are re-used in irrigation. Industrial use represents one third of the worldwide desalination capacity with the majority used for process requirements and pure water for boiler systems (20). Consideration of treated industrial wastewater sourced from desalinated seawater and re-used in process systems is not included in the industrial worldwide desalination capacity number.

A typical industrial seawater desalination water system is shown in Figure 1. The similarities and differences from a seawater municipal drinking water desalination plant are discussed below. An additional BWRO is added to this industrial system for advanced treatment of wastewater and integration of the treated wastewater into the desalination supply system. The connection lines are shown in red.

Figure 1 includes a one pass SWRO to produce permeate of 500 mg/L TDS from pre-treated seawater feed. Permeate of this quality is usable as industrial utility water and is expected to meet the current WHO boron drinking water guidelines. The BWRO serves as a second pass for the SWRO permeate that has not been diverted to utility or potable water use. The BWRO would be designed to produce permeate less than 10 mg/L TDS for use in feed to a demineralizer package. A municipal system may need to process all of the SWRO permeate through the BWRO for production of low boron water to meet alternate drinking water standards or irrigation quality water. A municipal system may also require additional steps than shown for these uses. At this point, the seawater desalination system may have met most of an industrial facility's major water supply needs except maybe that of landscape irrigation.

To offer reclaimed industrial wastewater as a secondary water supply option, a second BWRO is included in Figure 1. Processing secondary treated industrial wastewater through a BWRO should provide utility quality water as permeate. Also, the permeate quality should allow for limited use in higher purity water systems; however, the ratio of mixing the wastewater BWRO permeate to that of the seawater second pass BWRO permeate may need to be limited. Of course, additional treatment of the wastewater BWRO permeate is always possible to increase this ratio. In addition, in Figure 1 the isolation of the wastewater BWRO permeate from the Potable Water production and distribution components is assumed mandatory and integration of these two systems would need to be restricted. The recycling of the treated wastewater using the same RO technology as the water treatment system provides some economy of scale and possible sparing. For instance, if the wastewater BWRO could also be designed for seawater service, the wastewater BWRO could spare the SWRO or second pass BWRO on an as-needed basis. The details on the number of seawater RO trains and relative sizing would need to be considered.

The recycling of the seawater sourced wastewater and treatment by similar BWRO membranes results in the accumulation of seawater constituents that are not readily removed by RO membranes, such as boron. Also, the total volume of SWRO and second pass BWRO permeate would not be collected as wastewater. The wastewater that is collected is likely to be more concentrated in the constituents present in the permeate. For example, permeate that was used in cooling tower systems would be subject to concentrating. The ratio of the wastewater BWRO permeate mixed with the seawater RO permeate will need to be monitored and controlled.

A municipal seawater RO system may entertain low level uses for reject stream water. However, an industrial system with wastewater reuse would need to produce enough seawater RO reject to allow for mixing with and dilution of the wastewater BWRO reject. Since the wastewater BWRO concentrates the constituents in the secondary treated wastewater up to four times in the process of creating the reject, this reject likely would not pass the local marine discharge standards. However, recombining with the seawater RO reject would undo part if not all of the concentrating realized in the wastewater BWRO. The recombining option is another reason why the seawater RO, second pass BWRO and wastewater BWRO may want to be integrated as much as practicable.

**Figure 1 Typical industrial seawater desalination water system**
(From KBR Front-End Engineering and Design (FEED) work for confidential client)

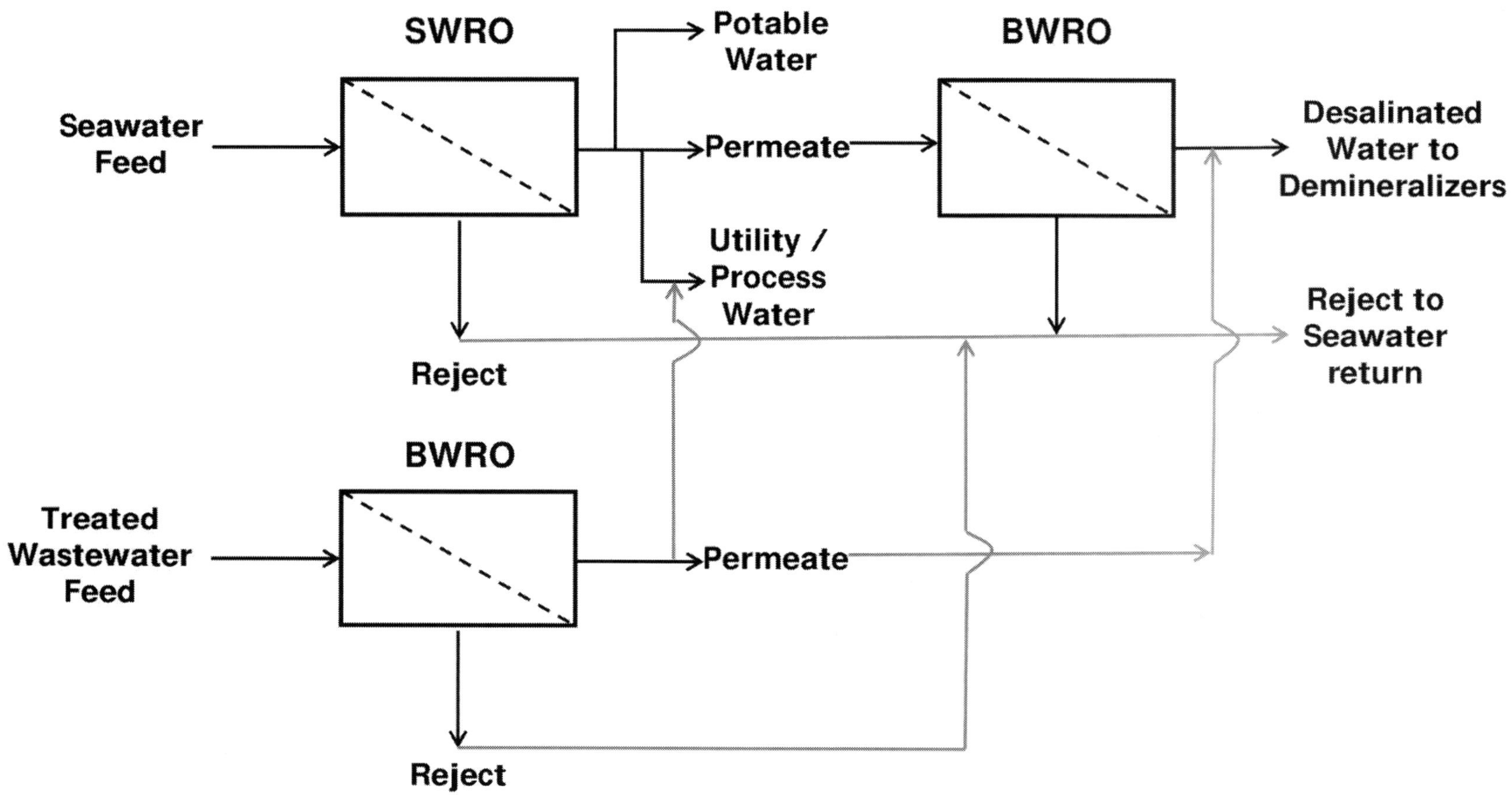

Figure 2 Potential Irrigation Quality Water System
(From KBR Front-End Engineering and Design (FEED) work for confidential client)

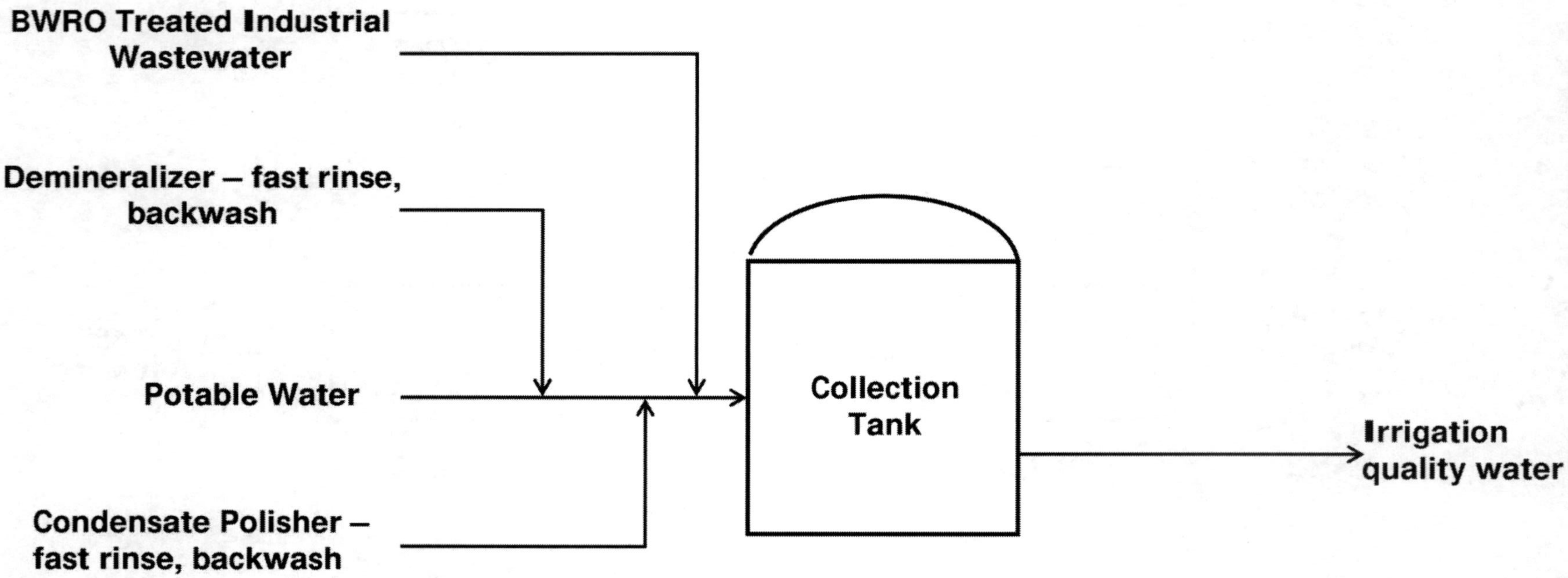

The scheme depicted in Figure 1 will not produce permeate of unrestricted irrigation quality. If landscape irrigation water is needed, Figure 2 shows that isolation and collection of individual waste streams that are properly balanced together may provide a limited volume of irrigation quality water. For instance, if the industrial water demineralizers or condensate polishers are backwashed with product water, and backwash is isolated along with the final fast rinse cycle, then a stream with no boron can be collected. Since these "demineralized" streams themselves would not be suitable for irrigation, conditioning by mixing with the potable water and the wastewater BWRO permeate, if local regulations allow, will produce suitable irrigation water. Additional conditioning is always possible, but the boron content of the potable water and the limited supply of waste demineralized water would limit the total amount. Also an ion exchange unit would be an option to produce low boron level water that could be mixed with other streams for irrigation use.

As more industrial facilities consider seawater sourcing for process water needs, the discussion presented in this paper would offer guidance to expand water supply management options. Design considerations for boron removal in desalination water treatment and industrial wastewater treatment plants can become an integrated solution. Treatment choices should be reviewed with respect to reuse applications for the treated effluent, as well as optimizing overall system design and simplifying operations. Water management supply choices that result in more dependable raw water supplies make a facility more self-sufficient. As one of the greatest challenges facing this century is providing sustainable supplies of clean water, industrial facilities need to examine unnecessary limitations and adjust to benefit from available and developing resources.

## References

1. Parks, JL and M Edwards. (2005), "Boron in the Environment," Critical Reviews in Environmental Science and Technology. 35, pp 81-114.
2. Greenlee, LF; et al (2009), "Reverse osmosis desalination: Water sources, technology, and today's challenges," Water Research 43(9), pp 2317-2348.
3. Fipps, G. (2003), "Irrigation Water Quality Standards and Salinity Management Strategies", Texas A&M Agrilife Extension, B-1667 / 4-03.
4. US EPA. (1986), Quality Criteria for Water, 440/5-86-001.
5. Grattan, SR. (2002), Irrigation Water Salinity and Crop Production, University of California Agriculture and Natural Resources Publication 8066, FWQP Reference Sheet 9.10.
6. WHO. (2011), Guidelines for Drinking Water Quality, 4th ed. World Health Organization, Geneva Switzerland.
7. US EPA. (2008), Health Effects Support Document for Boron, 822-R-08-002.

8. Minnesota Department of Health. (2008), Guidance for Groundwater. Boron: Update to the 2005 Health Based Guidance for Boron in Groundwater
9. Florida DEP. (2005), Final Technical Report: Development of Cleanup Target Levels (CTLs) For Chapter 6-777, F.A.C. Prepared by Center for Environment & Human Toxicology, University of Florida, Gainsville.
10. California Department of Public Health. (2010), Drinking Water Notification Levels Table 1.
11. Wisconsin Administrative Code Chapter NR 140 Groundwater Quality. (Jan 2012), Table 1 Public Health Groundwater Quality Standards.
12. Maine Center for Disease Control and Prevention. (2012), Maximum Exposure Guidelines for Drinking Water.
13. Wong, Joseph M. (1984), "Boron Control in Power Plant Reclaimed Water for Potable Reuse," Environmental Progress. 3(1), pp. 5-11.
14. Magara, Y; et al (1998), "Development of boron reduction system for sea water desalination," Desalination 118, pp. 25-34.
15. Glueckstern, P; Menahem, P. (2003), "Optimization of boron removal in old and new SWRO systems," Desalination 156, pp. 219-228.
16. Kabay, N; et al (2010), "Boron in seawater and methods for its separation – A review," Desalination 261, pp. 212-217.
17. Nir, O; et al (2012), "A novel approach for SWRO desalination plants operation, compromising single pass boron removal and reuse of CO2 in the post treatment step," Chemical Engineering Journal 187, pp. 275-282.
18. Sauvet-Goichon, B. (2007), "Ashkelon desalination plant – A successful challenge," Desalination 203, pp. 75-81.
19. Wilf, M., Klinko, K. (2001). "Optimization of seawater RO systems design," Desalination 138, pp. 299-306.
20. Sidem, Veolia Water Solutions & Technologies. website http://www.sidem-desalination.com/en/Process/FAQ/ (accessed March 6, 2014)
21. Al-Rasheed, R.A., Al-Sulami, S., Hasan, G. (2007). "Survey of Boron Levels in Seawater Desalination Plants in Saudi Arabia," IDA Maspalomas World Congress Proceedings.

# $CO_2$ BACT PREPARTION

*Joseph W. Guida, PE – Fluor, Sugar Land, TX*

*Biography*

*Mr. Guida holds a BSChE and MS Environmental Enigneering from the University of South Florida. He has over 30 years experience in the operation and design of industrial and environmental systems for the refining, petrochemical, heavy metals, and mining industries.*

## Introduction

The US EPA began regulating the emissions of $CO_2$ in 2011 and has since required a BACT analysis (Best Available Control Technology) to be submitted with applications for all new major sources. EPA has provided a GHG Controls Guidance document with procedures for performing a $CO_2$ BACT analysis, however, these procedures are generic in nature and details of each step are left to interpretation. Having recently completed a $CO_2$ BACT analysis that was approved as part of a permit application for a major new grassroots gas to liquids (GTL) complex on the US Gulf Coast, the author shares his experience in performing the analysis along with the details required to prepare a comprehensive $CO_2$ BACT evaluation.

This paper briefly explains the $CO_2$ laws and the current status of the permitting process. In addition, the five step BACT process is discussed in detail with a focus on the hydrocarbon industry. It includes an explanation and strategy for each step along with calculation procedures, and emissions factors. From this paper, the reader will be able to understand how to perform the five step process, what $CO_2$ control technologies should be considered, what is considered good engineering practice with regard to controls, how to screen and rank control technologies based on technical feasibility, and how to assess the energy, economic, and environmental aspects of each control option. Finally, the paper will discuss how energy efficient technologies are considered along with the importance of presenting a comprehensive review of carbon capture and sequestration as an option for BACT.

## Current Laws and Status

The USEPA began regulating emissions of greenhouse gases (GHGs) on January 2, 2011. New sources with a GHG potential to emit (PTE) of greater than 100,000 short tons per year (tpy) are subject to BACT permitting. New sources that fall under the Prevention of Significant Deterioration rules (PSD) must apply for a GHG BACT permit if the GHG PTE is greater than 75,000 tpy (6). GHGs include the compounds shown in Table 1 below.

**Table 1.** Greenhouse Gases and Global Warming Potential (GWP)

| Compound | Global Warming Potential |
|---|---|
| Carbon Dioxide | 1.0 |
| Methane | 25 |
| Nitrous Oxide | 298 |
| Sulfur Hexafluoride | 22,800 |
| HFCs & PFCs | Varies depending on compound |

(From 40 CFR 98 Subpart A, Table A-1, March 7, 2014) (9)

The $CO_2$ equivalent ($CO_2e$) is therefore the sum of the above compounds weighted by their GWP. A facility will typically have numerous sources that will emit $CO_2e$, but only the ones that exceed 100,000 tpy (or 75,000 tpy for PSD) will be required to follow the new BACT rules.

There are currently no New Source Performance Standards (NSPS) for BACT. The BACT is a case-by-case determination, providing discretion to the permitting authority and does not prescribe GHG BACT for any source type. A detailed case and fact specific analysis is therefore necessary for each situation (6).

## Permitting Strategy

In the early stages of design for any new facility (feasibility or Front End Engineering and Design, FEED), thought should be given to the philosophy of dealing with green house gas (GHG) emissions. Identify sources and decide how to best reduce, recycle, capture, or eliminate the $CO_2e$ emissions. There are essentially three strategies for reducing caron emissions:

- Reduce energy intensity
- Reduce carbon intensity
- Capture and sequester the produced carbon

For all fired equipment, this will generally mean utilizing the most fuel efficient design and/or considering fuels with a lower level of carbon emissions per unit of heating value (i.e. natural gas instead of fuel oil).

As part of the $CO_2$ BACT permit, EPA requires that $CO_2$ capture and sequestration (CCS) technology be considered and evaluated for all major $CO_2e$ emission sources. Process vents that may have higher $CO_2$ concentrations and lower temperatures than combustion stacks will generally be more efficient for capture than stack emissions and may present opportunities to be sold or recycled. Consideration should also be given to the unit locations. It is not practical to have multiple CCS units scattered across the facility, so locating the major sources close enough to share a common CCS system makes sense. Currently, CCS technology cannot be justified based on established economics except for a few special cases, however, the analysis must be done for each case where the PTE exceeds the permitting threshold.

## $CO_2e$ Emission Baseline

During the project feasibility phase, a $CO_2e$ emission baseline should be prepared to identify the potential sources and establish the permitting strategy for each. This should be established on the base case facility design without consideration of $CO_2e$ reduction options above what is considered good engineering practice. (Note that the baseline should include all $CO_2e$ sources, however, only the major sources will fall under BACT.) For instance, current engineering practice would typically specify high efficiency combustion burners on a boiler application in addition to economizers, boiler feed water treatment, and condensate collection and reuse. This equipment should be included as part of the base case.

For the $CO_2$ BACT permit, include only direct emissions from process sources. Each $CO_2e$ point source should be presented with emissions stated in units of $CO_2e$ per hour and per annum. For each point source emission reduction that may be identified, a reduction over baseline can be then be quoted in real units. Emissions are recorded on a Potential to Emit basis of 24/7 operation, 365 days a year.

The baseline will be used as a comparison against the options that may be considered in the BACT permit. The BACT analysis must include consideration of the most stringent available technologies, i.e., those that provide the maximum degree of emissions reduction.

### *$CO_2e$ Emissions Baseline – What to Include*

Direct emission sources such as fired heaters, process vents, flares, biological wastewater treatment, etc. including emissions during routine periods of startup and shutdown.

### *$CO_2$ Emissions Baseline – What not to include*

- Emissions that will occur offsite due to the use of fuels or products that are produced in the new facility
- Mobile sources such as rail, truck, and ship traffic that are used to transport raw materials, products, and equipment to and from the site.
- Emissions that may occur at other sites or facilities in which $CO_2e$ emissions are associated with production of offsite raw materials such as oxygen, nitrogen, catalysts, reagents, etc.
- The baseline should not include emissions from utility suppliers. If the facility plans to generate power or steam, these emissions should be included in the baseline. Switching steam turbines to motors and/or purchasing power from a local power source is an option that may be considered to lower $CO_2e$ emissions and impacts should be quantified in the BACT analysis.
- Carbon Capture and Sequestration (CCS) resulting in $CO_2$ transported offsite for use or disposal. The baseline should not include CCS as this will be an option.
- Construction and initial commissioning activities.

## Calculation Methodologies

EPA has provided emission factors for each of the $CO_2e$ compounds for all major types of fuels that can be burned. For $CO_2$ and combustion by-products such as $N_2O$ and $CH_4$, the EPA emission factors should be used, however, be aware that EPA publishes these factors on the basis of higher heating value (HHV). Since the lower heating value (LHV) is typically used to determine the quantity of fuel consumed, it will be necessary to adjust the calculations accordingly to arrive at the heat duty based on the HHV. Refer to 40 CFR 98 Subpart C (General Stationary Fuel Combustion Sources) for calculation procedures and emission factors. Industry specific methodologies are found in the subparts D through UU of 40 CFR 98 (5).

### *Fuel Mixtures*

In most refineries and some petrochemical facilities, fuel gas is produced from various processes and then used in fired heaters. EPA emission factors will not be available for these different mixtures. Use combustion calculations to determine the $CO_2$ emissions. For other $CO_2e$ compounds, use EPA factors by determining the heating value of the mixture. Next, use the 40 CFR values for emission factors and select the fuel that most closely resembles the fuel mixture (i.e. if the primary component of the fuel gas is methane, use the factors for natural gas). The emission factors for the EPA reference fuel are then adjusted based on heating value to arrive at the factors for the fuel gas mixture.

Once all GHG compounds are quantified, the $CO_2e$ rates can be determined by multiplying each compound by the GWP factor discussed above.

## 5 Step BACT Process

The EPA protocol for performing a BACT analysis involves the top down five step process illustrated in Figure 1 and discussed below:

***Step 1 – Evaluate all potential control technologies relevant to $CO_2e$***

For each industry category (utilities, refineries, cement, large commercial /industrial /institutional boilers, pulp and paper, iron and steel, nitric acid), EPA has published a white paper with guidelines for minimizing GHG emissions. For petroleum refining, this document is titled "Available and Emerging Technologies for Reducing Greenhouse Gas Emissions from the Petroleum Refining Industry" (7). This document includes a list of all design considerations that should be evaluated to reduce the GHG emissions. Each applicable item in this list should be addressed. A large number of these items will normally be done as part of good engineering practice and, if your project plans to implement these as part of the base design, it should be stated as such. If they are not to be implemented, a reason should be given. Be creative, don't limit your evaluation to only the items in the EPA list as there will usually be additional things that can be included.

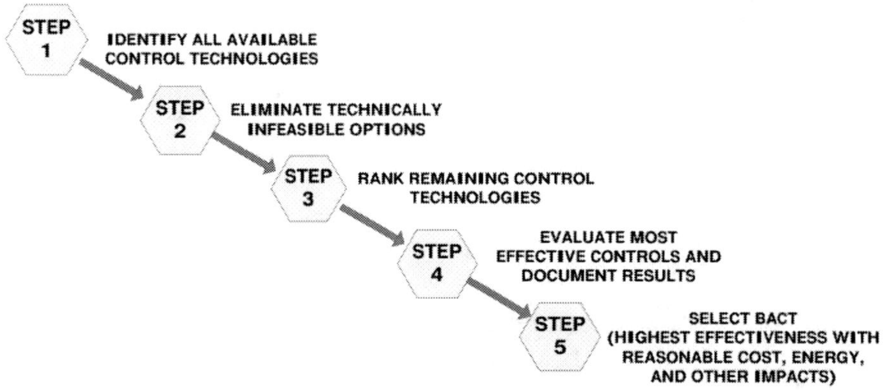


**FIGURE 1.** 5 Step BACT Process

Headers for elevated flares will usually require significant purge gas flow rates (typically natural gas) that can contribute substantially to the $CO_2e$ emissions. Consideration should be given to utilizing a ground flare system for the higher pressure flare headers. These systems operate on a pressure control arrangement and require as little as 10% of the purge gas needed for elevated flares. Ground flares are considered a special type of pressure assisted or non-assisted flare by EPA and are generally accepted by environmental agencies since they don't have the visual stigma that the elevated flares have.

Beware of false $CO_2e$ reduction by switching from steam driven equipment to motor driven equipment. For instance, large refineries and petrochemical plants utilize steam driven cooling tower pumps and other equipment for safety and reliability during power failures. The fired boilers used for steam generation produce $CO_2$ in large quantities. Switching to electric motors reduces or eliminates the need for the boilers, however the burden is shifted to the local power company who will have to generate the electricity. If the power company is burning coal as compared to oil or natural gas which would have been used to generate steam in your facility, the net $CO_2e$ emissions will likely be greater (along with other pollutants such as $SO_2$, $NO_x$, CO, and particulate). On the other hand, if the local power company is a hydroelectric or nuclear facility, then there will be a net $CO_2e$ reduction by using electric motors instead of steam turbines.

For each major source, a table listing all potential control options should be produced with detail surrounding how the control option results in emissions reductions. Available control options, by EPA definition, are those air pollution control technologies or techniques that have the potential for practical application to the emissions unit and the regulated pollutant under evaluation. Specific considerations need to include:

- Technologies that have been demonstrated to achieve the highest levels of control for the pollutant in question,
- Alternative production processes, methods, systems, and techniques, including clean fuels or innovative combustion techniques,
- Technologies in application outside the US should be considered to the extent that the technologies have been successfully demonstrated in practice,
- Energy efficient measures,
- Efficiency improvements to maximize energy efficiency,
- Combined cycle combustion turbines, which generally have higher efficiencies than simple cycle turbines, should be listed as options when an applicant proposes to construct a natural gas-fired facility,
- Carbon Capture and Sequestration. For emission streams with significant $CO_2$ concentrations, CCS must be considered as one of the control technologies.

The table below provides an example of a few responses to the $CO_2$ control options listed in the EPA White Paper that should be considered in Step 1 for boiler control technology. This is only a small portion of the list. Recommendations are also provided for process heaters, combined heat and power systems, fuel gas systems, flares, distillation units, ethylene production, cooling water, and other systems. Don't stop with these items, review other BACT permit applications that are accessible and consider opportunities that may be specific to your project.

EPA is in the process of establishing a benchmarking standard to apply on an industry specific basis. This will require each industry to achieve a minimum efficiency level for the various processes that contribute to $CO_2e$ emissions (1).

**TABLE 2.** Boiler Control Technology Example (7)

| Opportunity Identified By EPA | Description Of EPA Recommendation | Project Response & Statement Of Compliance |
|---|---|---|
| Good Combustion Practices | • Good air/fuel mixing<br>• Sufficient residence time to complete combustion<br>• Proper fuel gas supply system design and operation in order to minimize fluctuations in fuel gas supply<br>• Good burner maintenance and operation<br>• High temperatures and low oxygen levels in primary combustion zone<br>• Overall excess oxygen levels high enough to complete combustion while maximizing thermal efficiency<br>• Filtration of inlet air to reduce heat exchange tube fouling<br>• Cleaning of tubes during regularly scheduled outages | <u>CAPITAL EQUIPMENT</u><br>Combustion equipment shall be specified for optimum thermal efficiency including:<br>(1) sufficient residence time to guarantee efficient operation,<br>(2) fuel gas supply system designed for stable operation,<br>(3) excess air required to support the optimum combustion conditions based on the specific fuel to be burned ,<br>(4) filtration of inlet air.<br><u>OPERATING & MAINTENANCE</u><br>Owner shall operate and maintain the facility for maximum efficiency including:<br>• good burner maintenance<br>• cleaning of tubes during regularly scheduled outages |
| Install heat recovery equipment (feedwater economizers) and/or combustion air preheaters | Recovers available heat from exhaust gases and transfers it back into the system by preheating feedwater or combustion air | Feedwater economizers and combustion air pre-heaters are considered standard equipment on high efficiency boilers and shall be included in the boiler specification. |

***Step 2 – Eliminate Technically Infeasible Technology***

From the list developed in Step 1, pare down the technically infeasible options and the options that were stated to be included and come up with a list of remaining options that require further analysis, including CCS.

A technology is "technically feasible" if it has been demonstrated in practice or is available and applicable to the source type under review. The term "demonstrated" is focused on the technology being used in the same type of source or similar plant producing the same product. A technology is "available" if it can be obtained through commercial channels. An available technology is "applicable" if it can reasonably be installed and operated on the source type under consideration.

Technologies that have not been successfully proven in full scale operation can be eliminated. It will be necessary to make an assessment for each option to determine technical infeasibility. A description of the reasons the option is technically infeasible should be provided. Technical infeasibility can be based on physical, chemical, or engineering principles that would preclude the successful use of the control option in question. In addition, if the control option produces other environmental impacts, it will be necessary to quantify them. The issue of technical feasibility is subtle but very important and the level of review should be clearly documented. The technical infeasibility of CCS may be discussed here, but the analysis must still be performed in Step 4.

***Step 3 – Rank Remaining Control Technologies by Effectiveness***

Effectiveness in the case of Step 3 relates to the amount of $CO_2e$ that can be avoided. At this stage, cost is not a consideration and will be evaluated in Step 4. Options with the largest $CO_2e$ impacts should be placed first. Rank the control technologies that were not eliminated in Step 2 from highest to lowest.

***Step 4 – Evaluate the Most Effective Control Technologies***

Starting at the top of the list from Step 3, perform a cost analysis for each remaining technology along with the impact of $CO_2e$ reduction. BACT guidelines suggest the use of the Annual Cost Effectiveness, ACE, (in \$ per ton $CO_2e$) to evaluate each option.

$$\textit{ANNUAL COST EFFECTIVENESS} = \frac{\textit{COAC}}{(\textit{BE} - \textit{CE})} \qquad \textit{Equation 1}$$

*Where:*
*COAC = Control Option Annualized Cost*
*BE = Baseline Emission Rate*
*CE = Controlled Emission Rate*

In developing the COAC, a higher interest rate and shorter equipment life will lead to much higher annualized costs and result in a greater probability that the equipment will be excluded from further consideration. As a guideline, take the average US 10 year treasury bond interest rate for the previous 6 months, add 2%, and round up to the next integer. The EPA Air Pollution Control Cost Manual can provide additional information and methodology (8).

Each regulatory agency has a different opinion about the maximum economically feasible cost effectiveness value, and many will not tell you what that value is. Just because you believe that the control device is economically burdensome does not mean that the permit reviewer will agree. As such, it is always safer to make sure that your BACT cost analysis is as accurate, and site specific, as possible. One measure which has been suggested is to compare the ACE to the market cost of carbon credits. Before they were closed, California carbon credits traded in the range of \$10 - \$14/ton of $CO_2$ and the Chicago Climate Exchange traded at \$0.05 to \$7.50. Europe has been more volatile at \$4 – \$40/ton of $CO_2$.

Under the "Tailoring Rule", the USEPA considers 100,000 tons of $CO_2e$ equal to 100 tons of criteria pollutant. Therefore, if a criteria pollutant control has a cost effectiveness threshold of approximately \$10,000 per ton, then the equivalent cost effectiveness for $CO_2e$ control should be \$10 per ton (1000:1). (4)

The cost of CCS per ton of $CO_2$ has been estimated by IPCC (2) to be in the range of \$36 to \$82 per ton $CO_2$ (for NG combined cycle plants) and \$65 - \$245 per ton (for pulverized coal plants). The wide cost range is a result of many factors such as the type of fuel burned, location, the final destination of the captured $CO_2$, and the method of transporting it. The recent GTL experience on the US Gulf Coast for installation of a CCS system which received emissions from a Fischer Tropsch process unit resulted in an ACE of \$57/ton $CO_2$ which was based on the capture process shown in Figure 2 at a capture rate of 1,100,000 tons of $CO_2$ per year with disposal to an existing pipeline. Table 2 provides a cost breakdown for this project.

If the $CO_2$ had been disposed to underground geological formations, the cost would have been much greater, adding \$2.5 - \$27 per ton $CO_2$. (2) For a capture facility alone without transportation and storage, the IPCC has estimated the costs to be in the range of \$37 - \$48 per ton $CO_2$.

**TABLE 3.** Annualized Costs for US Gulf Coast CCS Project (Basis – 2012)

| ITEM | $/TON $CO_2$ REMOVED |
|---|---|
| Total Installed CCS Cost | 23.2<br>(annualized over 20 years at 7%) |
| Power | 14.6<br>(94% charged to pipeline compressors) |
| Cooling Water | 2.9 |
| Steam | 11.0 |
| O&M | 5.3 |
| Total | 57.0 |

In evaluating CCS technology, the following considerations should be made:

- Is there a place to dispose of the $CO_2$ that will be captured? In some parts of the country, a $CO_2$ pipeline may be available to tap into. An existing line in Louisiana, for instance, operates at 2000 psig and the owner will impose quality and flow restrictions depending on the downstream user requirements. Deep well injection is also a possibility if geological formations are adequate within a reasonable distance. Lacking a destination to dispose of $CO_2$ will usually make this option impractical. This argument needs to be followed through with as much technical and commercial data as possible to substantiate the conclusion.
- Disposal to geological formations also presents risks to the owner. Risks include the potential for leaks of an asphyxiant gas into an untested formation beneath real estate not under the control of the company, potential liability for any tremors or earthquakes that may occur, potential liability for drinking water impacts (carbonation and lowering of pH plus damages from improper well construction leading to contamination), and legal issues as to who owns the pore spaces in the formation. (1)
- If there is a disposal option, a cost analysis must be performed. EPA, DOE, and other organizations provide costs that can be referenced, however, due to the wide variances in cost bases, a project specific cost estimate should be performed if possible. Removal of $CO_2$ from stack gases is typically done by cooling, absorption in an amine solution, steam stripping of the $CO_2$, condensing the moisture, drying, and pressurization (refer to Figure 2).

As would be expected, the thermodynamics are against this process due to the energy required to take a hot, dilute stack gas and purify, cool, and compress the $CO_2$ for disposal. CCS results in gross process inefficiencies that cannot be overcome by the marginal efficiency increases of the other options presented in Step 1. It has been reported by numerous sources that the parasitic load for CCS can be 30% to 40% for a boiler used in power generation. Based on the recent US Gulf Coast GTL BACT application, the parasitic load was determined to be in the range of 33% and 56% depending on the fuel source.

Increasing the fuel rate by 30% to 56% to achieve the same energy output also increases emissions of other pollutants such as $NO_x$, $SO_2$, particulate, etc. As a result, real air quality impacts can be expected from CCS implementation. Local degradation of air quality and faster depletion of natural resources are outcomes that must be weighed against the perceived global benefits of $CO_2$ reduction.

**TABLE 4.** CCS Parasitic Load For US Gulf Coast GTL Project (Basis – 2012)

| FUEL BASIS | PARASITIC LOAD |
|---|---|
| Natural Gas | 33% |
| Fuel Oil | 45% |
| Coal | 56% |

Parasitic loads include power for compressors, cooling water pumps, and other process motor drives. Also included is the avoided power from steam used in the amine regeneration column which would have otherwise been used to generate power.

The perception from numerous sources is that CCS and GHG emissions will follow a similar track as other pollutants that were regulated in the past. Over time, the pollutants were able to be reduced at a lower cost due to technology improvements. However, $CO_2$ is different. The second law of thermodynamics assigns a high energy penalty for systems that take a high entropy stream (hot, dilute, stack gases) and convert it to a low entropy stream (high pressure, pure $CO_2$). In the author's opinion, the entropy penalty will limit CCS technology to only marginal improvements in cost and efficiency in the future.

In addition, $CO_2$ is a *product* of combustion, not a by-product as is the case for other pollutants. This means that generating energy from hydrocarbon fuels cannot be done without producing $CO_2$. "De-carbonizing energy production" via CCS requires significantly more fuel to achieve the same energy output resulting in a negative air quality impact to the surrounding locality. The best practical way to reduce $CO_2$ emissions and maintain energy production would seem to be by improving combustion and process efficiency, which appears to be the current focus of EPA. Of course, this may change in the future.

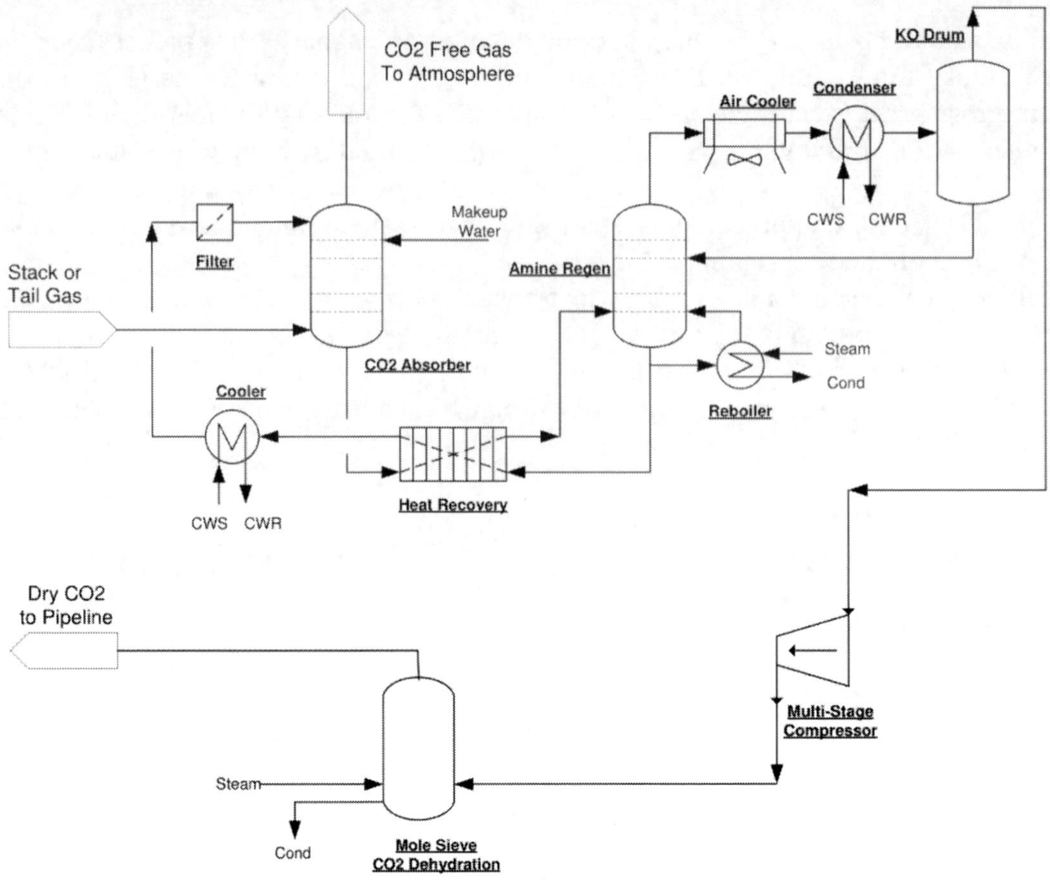


**FIGURE 2.** Typical CCS Project

***Step 5 – Select the Best Option (if any)***

Based on the evaluation from Step 4, select the best option, should one emerge, that meets the technical feasibility criteria and cost effectiveness.

## Summary

New sources with significant $CO_2$e emissions are now required to perform the BACT analysis to substantiate the effectiveness of the $CO_2$e control systems. For the most part, good engineering practice will include the majority of controls now recommended by EPA and these should be documented in the analysis. However, other control options such as fuel switching, optimizing the balance between motor drives and turbine drives, CCS, etc. will require further evaluation and documentation in further detail to complete the analysis.

While most attention regarding the practicality of CCS centers around cost, the dramatic increase in fuel consumption associated with parasitic loads and the increase in other types of air pollution are real impacts that must also be considered and documented in the BACT process. Degradation of air quality and faster depletion of natural resources due to CCS are outcomes that must be weighed against the perceived benefits of $CO_2$ reduction.

The $CO_2$ BACT permit process is still in its infancy and many questions still remain with regard to acceptable costs, process efficiency benchmarking standards, $CO_2$ disposal options, and how EPA will decide to enforce the laws in the future.

## References


1. Crum, Robert and Bruce Broberg (2012), "Update on the Greenhouse Gas Regulation by the US EPA," Hydrocarbon Processing, November, pp. 55-59.
2. Herzog, Howard and Koen Smekens (2005), "IPCC Special Report on Carbon Dioxide Capture and Storage/ Chapter 8 - Cost and Economic Potential" International Panel on Climate Change, pp. 340-362.
3. David, J. and Howard Herzog (2001), "The Cost of Carbon Capture," Proceedings of 5th International Conference on Greenhouse Gas Control Technologies, pp. 985-990.
4. Woodruff, Jeannie (2013), "Preparing a GHG BACT: Lessons Learned," 16th Annual AWMA/NYWEA Joint Seminar, February.
5. United States Code of Federal Regulation, "General Stationary Fuel Combustion Sources", 40 CFR 98 Subpart C.
6. EPA Office of Air and Radiation (2010), "Greenhouse Gas Permitting Guidance,"
7. EPA Office of Air and Radiation (2010), "Available and Emerging Technologies for Reducing Greenhouse Gas Emissions from the Petroleum Refining Industry."
8. US EPA (2002), "EPA Air Pollution Control Cost Manual," EPA/452/B-02-001.
9. US EPA (2014), "Mandatory Greenhouse Gas Reporting", 40 CFR 98 Subpart A

# Optimization of Water Network Design and Dynamic Operation for Agile Batch Manufacturing

Tianxing Cai (Speaker) and Qiang Xu
Dan F. Smith Department of Chemical Engineering
Lamar University

14 AIChE Spring Meeting, New Orleans, LA

LAMAR UNIVERSITY

# Content

- Introduction
- Methodology
- Application
- Conclusion
- Acknowledgement

14 AIChE Spring Meeting, New Orleans, LA

LAMAR UNIVERSITY

# Introduction

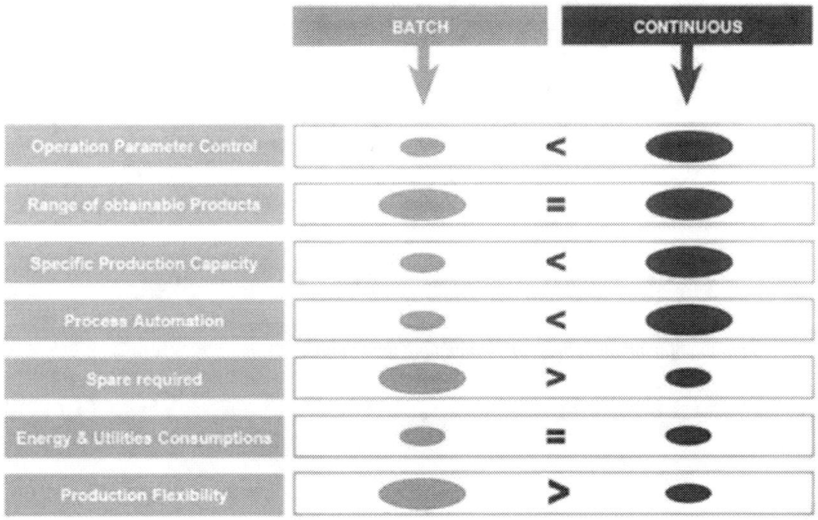


14 AIChE Spring Meeting, New Orleans, LA

LAMAR UNIVERSITY

# Introduction

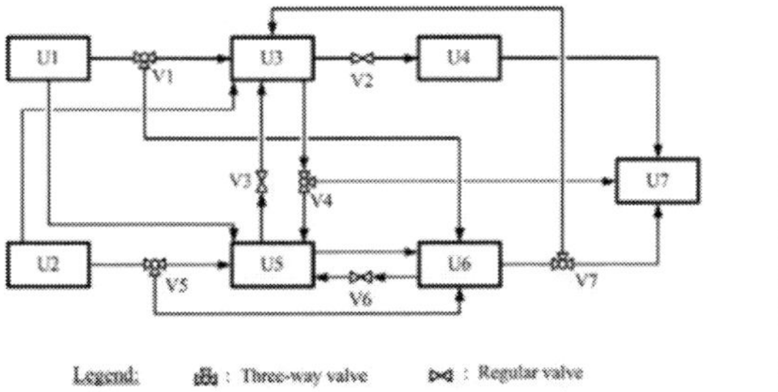

| Networks | Valve Control | | | | | | | | | | |
|---|---|---|---|---|---|---|---|---|---|---|---|
| | V1 | | V2 | V3 | V4 | | V5 | | V6 | V7 | |
| | U1→U3 | U1→U6 | U3→U4 | U5→U3 | U3→U5 | U3→U7 | U2→U5 | U2→U6 | U6→U5 | U6→U3 | U6→U7 |
| 1 | open | close | open | close | close | close | close | close | close | close | open |
| 2 | open | close | open | close | close | close | open | open | close | open | close |
| 3 | close | open | close | open | close | open | open | close | open | close | open |
| 4 | close | close | open | close | open | open | close | close | close | open | close |

**◆a super WRND coupled with all the recipe-based WRNDs;**
**◆the number of regular and three-way valves needed for the super WRND;**
**◆the optimal design of all of the valve placements in terms of the minimum total valve-switching times (*TVST*) to swap one pair of WRNDs at a time;**
**◆the control strategy (valve opening/closing strategy) to switch among WRNDs.**

14 AIChE Spring Meeting, New Orleans, LA

LAMAR UNIVERSITY

# Introduction

14AIChE
Spring Meeting, New Orleans, LA

LAMAR UNIVERSITY

# Introduction

**Assumptions**

- A super agile multiple water networks (AMWN) coupled with all the scenario-based AMWNs;
- The minimal installation number of regular and three-way valves needed for the super AWMN.

**Given information**

- Required processing batches in the specific time period, each of which corresponds to a scenario-based operation of specific AWMN;
- Scheduling requirement for each type of batch productions;
- Interval time for different batch processing.

**Information to be determined**

- Dynamic scheduling of multiple batch processing with the optimal control strategy of the minimal total valve-switching times (TVST) and the minimal total processing delay time.

14AIChE
Spring Meeting, New Orleans, LA

LAMAR UNIVERSITY

# Methodology

**Step 1: Initialize the valve installation location and control strategy for each recipe.**

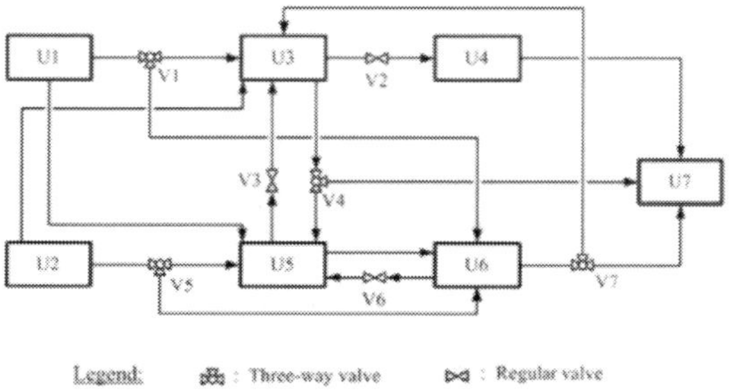


| Networks | Valve Control | | | | | | | | | | |
|---|---|---|---|---|---|---|---|---|---|---|---|
| | V1 | | V2 | V3 | V4 | | V5 | | V6 | V7 | |
| | U1→U3 | U1→U6 | U3→U4 | U5→U3 | U3→U5 | U3→U7 | U2→U5 | U2→U6 | U6→U5 | U6→U3 | U6→U7 |
| 1 | open | close | open | close | close | close | close | close | close | close | open |
| 2 | open | close | open | close | close | close | open | open | close | open | close |
| 3 | close | open | close | open | close | open | open | close | open | close | open |
| 4 | close | close | open | close | open | open | close | close | close | open | close |

14 AIChE® Spring Meeting, New Orleans, LA

LAMAR UNIVERSITY

# Methodology

**Step 2: Generate the valve adjustment matrix for all the production shifts.**

| i | e | V1 | V2 | V3 | V4 | V5 | V6 | V7 | TAN |
|---|---|---|---|---|---|---|---|---|---|
| 1 | 2 | 0 | 0 | 0 | 0 | 1 | 0 | 1 | 2 |
| 1 | 3 | 1 | 1 | 1 | 1 | 1 | 1 | 0 | 6 |
| 1 | 4 | 1 | 0 | 0 | 1 | 0 | 0 | 1 | 3 |
| 2 | 3 | 1 | 1 | 1 | 1 | 1 | 1 | 1 | 7 |
| 2 | 4 | 1 | 0 | 0 | 1 | 1 | 0 | 0 | 3 |
| 3 | 4 | 1 | 1 | 1 | 1 | 1 | 1 | 1 | 7 |

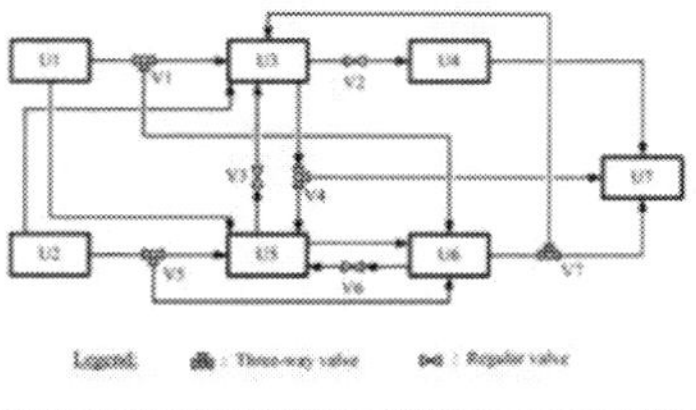


| Networks | Valve Control | | | | | | | | | | |
|---|---|---|---|---|---|---|---|---|---|---|---|
| | V1 | | V2 | V3 | V4 | | V5 | | V6 | V7 | |
| | U1→U3 | U1→U6 | U3→U4 | U5→U3 | U3→U5 | U3→U7 | U2→U5 | U2→U6 | U6→U5 | U6→U3 | U6→U7 |
| 1 | open | close | open | close | close | close | close | close | close | close | open |
| 2 | open | close | open | close | close | close | open | open | close | open | close |
| 3 | close | open | close | open | close | open | open | close | open | close | open |
| 4 | close | close | open | close | open | open | close | close | close | open | close |


| i\e | 1 | 2 | 3 | 4 |
|---|---|---|---|---|
| 1 | 0 | 2 | 6 | 3 |
| 2 | 2 | 0 | 7 | 3 |
| 3 | 6 | 7 | 0 | 7 |
| 4 | 3 | 3 | 7 | 0 |

14 AIChE® Spring Meeting, New Orleans, LA

LAMAR UNIVERSITY

# Methodology

**Step 3: Set up the scheduling model for recipe arrangement.**

**Objectives**

1. Minimize the total waiting time of all the operations
2. Minimize the total number of valve adjustments

**Constraints**

1. Each type of production has its shipment requirement
2. Each unit can only conduct one production at one time
3. The production ratio has been satisfied

14 AIChE Spring Meeting, New Orleans, LA

LAMAR UNIVERSITY

# Case Study 1

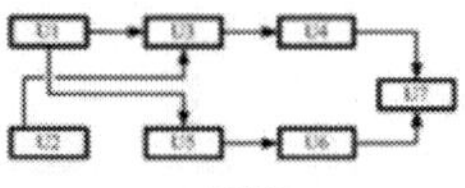


WRND 1

| $J_1$ | U1 | U2 | U3 | U4 | U5 | U6 | U7 |
|---|---|---|---|---|---|---|---|
| U1 | 0 | 0 | 1 | 0 | 1 | 0 | 0 |
| U2 | 0 | 0 | 1 | 0 | 0 | 0 | 0 |
| U3 | 0 | 0 | 0 | 1 | 0 | 0 | 0 |
| U4 | 0 | 0 | 0 | 0 | 0 | 0 | 1 |
| U5 | 0 | 0 | 0 | 0 | 0 | 1 | 0 |
| U6 | 0 | 0 | 0 | 0 | 0 | 0 | 1 |
| U7 | 0 | 0 | 0 | 0 | 0 | 0 | 0 |

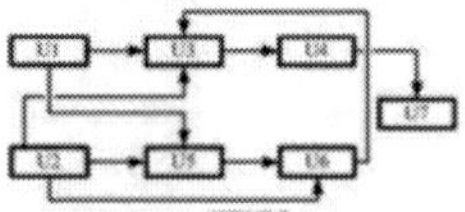


WRND 2

| $J_2$ | U1 | U2 | U3 | U4 | U5 | U6 | U7 |
|---|---|---|---|---|---|---|---|
| U1 | 0 | 0 | 1 | 0 | 1 | 0 | 0 |
| U2 | 0 | 0 | 1 | 0 | 1 | 1 | 0 |
| U3 | 0 | 0 | 0 | 1 | 0 | 0 | 0 |
| U4 | 0 | 0 | 0 | 0 | 0 | 0 | 1 |
| U5 | 0 | 0 | 0 | 0 | 0 | 1 | 0 |
| U6 | 0 | 0 | 1 | 0 | 0 | 0 | 0 |
| U7 | 0 | 0 | 0 | 0 | 0 | 0 | 0 |

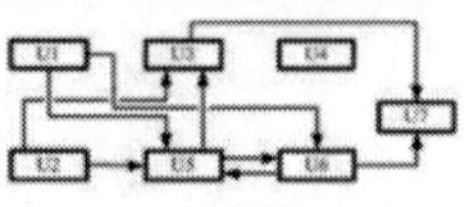


WRND 3

| $J_3$ | U1 | U2 | U3 | U4 | U5 | U6 | U7 |
|---|---|---|---|---|---|---|---|
| U1 | 0 | 0 | 0 | 0 | 1 | 1 | 0 |
| U2 | 0 | 0 | 1 | 0 | 1 | 0 | 0 |
| U3 | 0 | 0 | 0 | 0 | 0 | 0 | 1 |
| U4 | 0 | 0 | 0 | 0 | 0 | 0 | 0 |
| U5 | 0 | 0 | 1 | 0 | 0 | 1 | 0 |
| U6 | 0 | 0 | 0 | 0 | 1 | 0 | 1 |
| U7 | 0 | 0 | 0 | 0 | 0 | 0 | 0 |

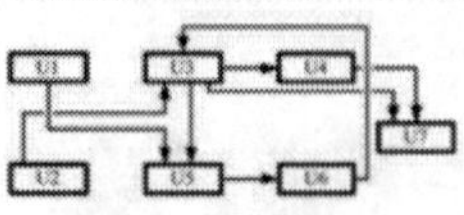


WRND 4

| $J_4$ | U1 | U2 | U3 | U4 | U5 | U6 | U7 |
|---|---|---|---|---|---|---|---|
| U1 | 0 | 0 | 0 | 0 | 1 | 0 | 0 |
| U2 | 0 | 0 | 1 | 0 | 0 | 0 | 0 |
| U3 | 0 | 0 | 0 | 1 | 1 | 0 | 1 |
| U4 | 0 | 0 | 0 | 0 | 0 | 0 | 1 |
| U5 | 0 | 0 | 0 | 0 | 0 | 1 | 0 |
| U6 | 0 | 0 | 1 | 0 | 0 | 0 | 0 |
| U7 | 0 | 0 | 0 | 0 | 0 | 0 | 0 |

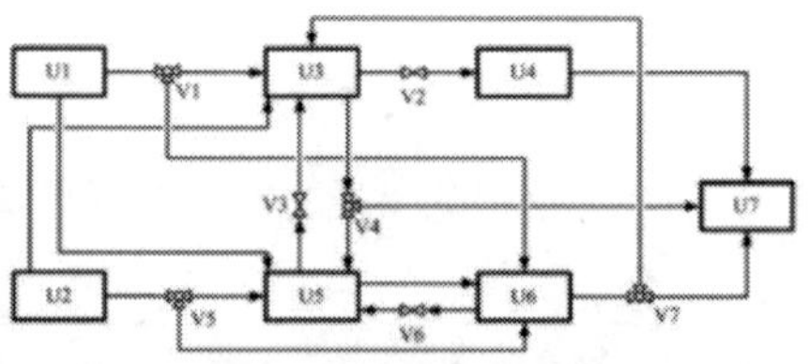


Legend: : Three-way valve ; : Regular valve

| Networks | Valve Control | | | | | | | | | | |
|---|---|---|---|---|---|---|---|---|---|---|---|
| | V1 | | V2 | V3 | V4 | | V5 | | V6 | V7 | |
| | U1→U3 | U1→U6 | U3→U4 | U5→U3 | U3→U5 | U3→U7 | U2→U5 | U2→U6 | U6→U5 | U6→U3 | U6→U7 |
| 1 | open | close | open | close | close | close | close | close | close | close | open |
| 2 | open | close | open | close | close | close | open | open | close | open | close |
| 3 | close | open | close | open | close | open | open | close | open | close | open |
| 4 | close | close | open | close | open | open | close | close | close | open | close |

14 AIChE Spring Meeting, New Orleans, LA

LAMAR UNIVERSITY

# Case Study 1

| i\e | 1 | 2 | 3 | 4 |
|---|---|---|---|---|
| 1 | 0 | 2 | 6 | 3 |
| 2 | 2 | 0 | 7 | 3 |
| 3 | 6 | 7 | 0 | 7 |
| 4 | 3 | 3 | 7 | 0 |

Valve Adjustment Number

| i\e | 1 | 2 | 3 | 4 |
|---|---|---|---|---|
| 1 | 0 | 1 | 2 | 3 |
| 2 | 3 | 0 | 1 | 2 |
| 3 | 1 | 2 | 0 | 3 |
| 4 | 2 | 2 | 2 | 0 |

Transition Time (hr)

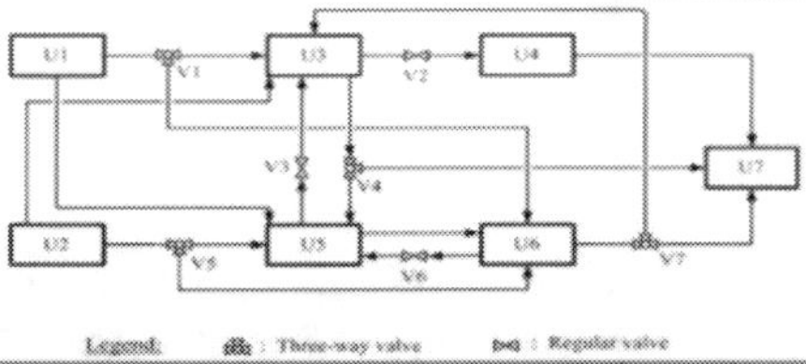


| Product # | 1 | 2 | 3 | 4 |
|---|---|---|---|---|
| Batch Timc(hr) | 4 | 5 | 6 | 7 |

14 AIChE Spring Meeting, New Orleans, LA

LAMAR UNIVERSITY

# Case Study 1

**Scenario 1: Production Ratio=4:3:2:1**

| Production # | P1 | P1 | P1 | P1 | P2 | P2 | P2 | P3 | P3 | P4 |
|---|---|---|---|---|---|---|---|---|---|---|
| Batch # | B1 | B2 | B3 | B4 | B1 | B2 | B3 | B1 | B2 | B1 |
| Penality Coefficient | 1 | 3 | 5 | 7 | 2 | 4 | 6 | 3 | 6 | 4 |

The optimal processing time is 106 hours with the sequence
P1→P1 → P1 → P1 → P2 → P2 → P2 → P4 → P3 → P3

**Scenario 2: Production Ratio=2:2:3:3**

| Production # | P1 | P1 | P2 | P2 | P3 | P3 | P3 | P4 | P4 | P4 |
|---|---|---|---|---|---|---|---|---|---|---|
| Batch # | B1 | B2 | B1 | B2 | B1 | B2 | B3 | B1 | B2 | B3 |
| Penality Coefficient | 1 | 3 | 2 | 4 | 3 | 6 | 9 | 4 | 7 | 10 |

The optimal processing time is 121 hours with the sequence
P4→P4 → P4 → P3 → P3 → P3 → P1 → P1 → P2 → P2

14 AIChE Spring Meeting, New Orleans, LA

LAMAR UNIVERSITY

# Case Study 2

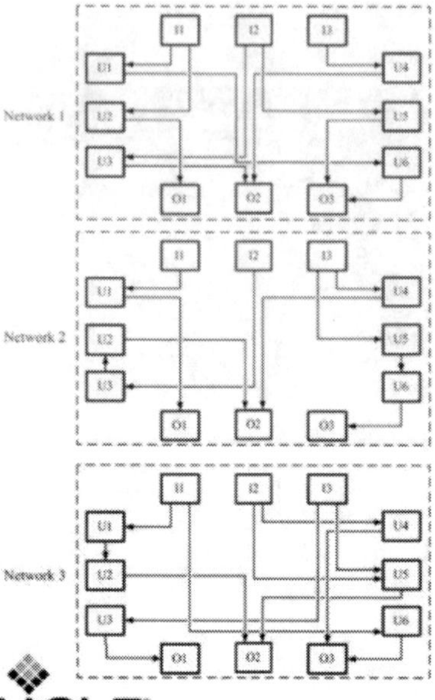


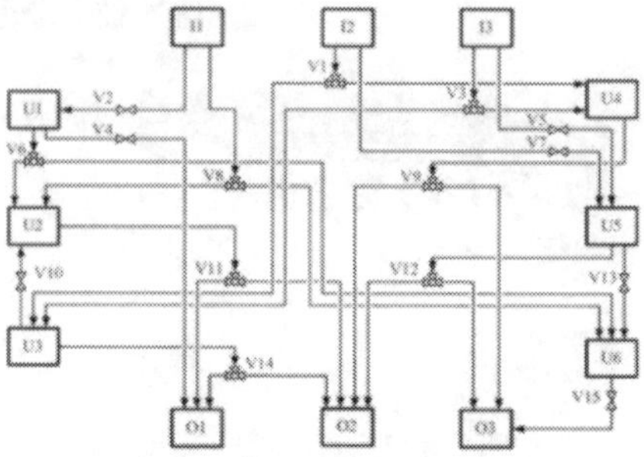


Legend: Three-way valve; Regular valve

Valve Control

| Network | V1 | V1 | V2 | V3 | V3 | V4 | V5 |
|---|---|---|---|---|---|---|---|
| | I2 → U3 | I2 → U4 | I1 → U1 | I3 → U3 | I3 → U4 | U1 → O1 | I3 → U5 |
| 1 | open | close | open | close | open | close | close |
| 2 | open | close | open | close | open | open | open |
| 3 | close | open | open | open | close | close | open |

| Network | V6 | V6 | V7 | V8 | V8 | V9 | V9 | V10 |
|---|---|---|---|---|---|---|---|---|
| | U1 → U2 | U1 → U6 | I2 → U5 | I1 → U2 | I1 → U6 | U4 → O2 | U4 → O3 | U3 → U2 |
| 1 | close | open | open | open | close | open | close | close |
| 2 | close | close | close | close | close | open | close | open |
| 3 | open | close | open | close | open | close | open | close |

| Network | V11 | V11 | V12 | V12 | V13 | V14 | V14 | V15 |
|---|---|---|---|---|---|---|---|---|
| | U2 → O1 | U2 → O2 | U5 → O2 | U5 → O3 | U5 → U6 | U3 → O1 | U3 → O2 | U6 → O3 |
| 1 | open | close | close | open | close | close | open | open |
| 2 | close | open | close | close | open | close | close | open |
| 3 | close | open | open | close | close | open | close | open |

14 AIChE Spring Meeting, New Orleans, LA

LAMAR UNIVERSITY

# Case Study 2

| i\e | 1 | 2 | 3 |
|---|---|---|---|
| 1 | 0 | 10 | 9 |
| 2 | 10 | 0 | 11 |
| 3 | 9 | 11 | 0 |

Valve Adjustment Number

| i\e | 1 | 2 | 3 |
|---|---|---|---|
| 1 | 0 | 5 | 6 |
| 2 | 7 | 0 | 5 |
| 3 | 6 | 7 | 0 |

Transition Time (hr)

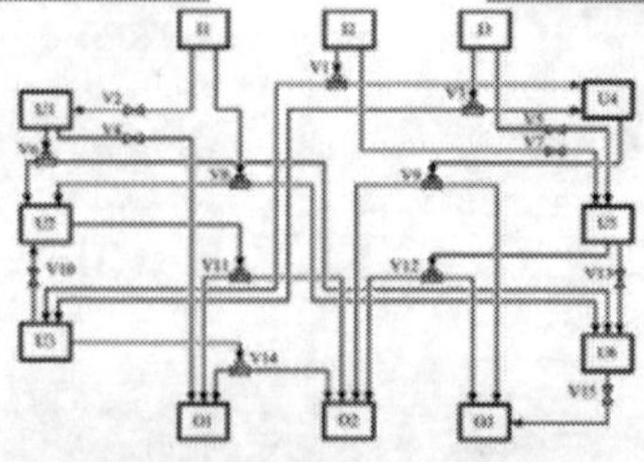

Legend: Three-way valve; Regular valve

| Product # | 1 | 2 | 3 |
|---|---|---|---|
| Batch Time(hr) | 4 | 5 | 6 |

14 AIChE Spring Meeting, New Orleans, LA

LAMAR UNIVERSITY

## Case Study 2

Scenario 1: Production Ratio=3:2:1

| Production # | P1 | P1 | P1 | P2 | P2 | P3 |
|---|---|---|---|---|---|---|
| Batch # | B1 | B2 | B3 | B1 | B2 | B1 |
| Penality Coefficient | 1 | 3 | 5 | 2 | 4 | 3 |

The optimal processing time is 77 hours with the sequence P1→P1 → P1 → P2 → P2 → P3

Scenario 1: Production Ratio=2:2:2

| Production # | P1 | P1 | P2 | P2 | P3 | P3 |
|---|---|---|---|---|---|---|
| Batch # | B1 | B2 | B1 | B2 | B1 | B2 |
| Penality Coefficient | 1 | 3 | 2 | 4 | 2 | 3 |

The optimal processing time is 76 hours with the sequence P2→P2 → P3 → P3 → P1 → P1

Scenario 1: Production Ratio=1:2:3

| Production # | P1 | P2 | P2 | P3 | P3 | P3 |
|---|---|---|---|---|---|---|
| Batch # | B1 | B1 | B2 | B1 | B2 | B3 |
| Penality Coefficient | 3 | 2 | 4 | 1 | 3 | 5 |

The optimal processing time is 81 hours with the sequence P3→P3 → P3 → P1 → P2 → P2

14 AIChE Spring Meeting, New Orleans, LA

LAMAR UNIVERSITY

## Conclusion

- The previous study of superstructure design has been further extended for the agile and comprehensive WND with the integration of multiple WNDs and different purposes
- The maximum operational flexibility of effective network control switch strategy based on production ratio and scheduling requirement
- The potential application to many other network integration problems of agile operation of multiple task-based WNDs with the benefit of minimal valve adjustment number and processing time

14 AIChE Spring Meeting, New Orleans, LA

LAMAR UNIVERSITY

# Acknowledge

- My professor, Dr. Qiang Xu
- Texas Hazardous Waste Research Center
- Graduate Student Scholarship from Lamar University


14 AIChE
Spring Meeting, New Orleans, LA

LAMAR UNIVERSITY

# Q & A

14 AIChE
Spring Meeting, New Orleans, LA

LAMAR UNIVERSITY

Thank you for your attendance!

# ROSS™ Technology
# Field Demonstration Report

An oil well operator wanted to treat produced water containing high concentrations of hardness and free oil for reuse in a water flood application at its facilities in the Western United States. To maintain the formation pressure and help displace the petroleum from the formation it is common to inject produced water down hole back into the formation. Normally the produced water requires some type of treatment to avoid fouling the formation. It is common to filter produced water before injection. However, because of the high hardness and alkalinity in the produced water, the water would be very scaling if injected into the formation without treatment. Occasionally the produced water also contains sulfur reducing bacteria and iron which should not be injected into the formation. Produced water is often warm or even hot. The water temperature is a benefit when the water is injected into the formation but most ultrafilter membranes cannot handle temperatures over 40 degrees C. It would be wasteful to cool the water down only to heat is up again.

In order to address all of the problems with injecting produced water Veolia developed the ROSS™ process. ROSS™ Technology includes Multiflo™ Chemical Softening and Ceramic Membrane (CeraMem®) Ultrafiltration. The ceramic membrane can handle temperatures as high as 250 degrees C. Tests were conducted on a number produced waters as well as blends of produced water containing frac water and fresh make-up. During the pilot testing the ROSS™ Technology effectively achieved the hardness, iron, total suspended solids, and free oil & grease criteria for the treated effluent water.

**The ROSS™ Pilot System**

**The CeraMem® Membranes**

**Typical Produced Water** **Water Samples Taken During Treatment**

In summary we were able to successfully treat each type of produced feed water. Each type of feed water required slightly different types and dosages of chemical additions to achieve the following treatment goals:

| Contaminant | Typical Concentration | Typical Injection Spec |
|---|---|---|
| Suspended Solids, ppm | 10 -100 | <0.5 |
| Free Oil, ppm | 10 - 100 | <0.5 |
| Iron, ppm | 0 -40 | <0.5 |
| Sulfur Reducing Bacteria. CFU | as many as 10,000 | nil |
| pH | 6 - 8.5 | 6.5 - 7.5 |
| Total Dissolved Solids, ppm | 5,000 - 20,000 | |
| Hardness, ppm | 200 - 500 | <20 |
| Alkalinity, ppm | 500 - 3,000 | Non-scaling |

In summary the ROSS™ treatment was able to produce effluent that met the target goals while treating various combinations of surface, produced and frac flowback water.

# 352457 Sustainable Treatment Processes of 1,4-Dioxane in Water

**Abstract:**

For treatment and remediation operations, it is crucial to ensure that contaminants are treated to sufficiently low levels to reduce adverse effects to public health and the environment. Many emerging contaminants may not be completely characterized in terms of their health effects, treatability, and occurrence. 1,4-Dioxane is an emerging contaminant in water which is currently on the candidate contaminant list 3 (CCL 3). This research aims at evaluating various treatment processes for 1,4-dioxane, and analyzing the the treatment processes through a life cycle and green metrics approach.

1,4-Dioxane is a solvent frequently found in ground water. Due to its physical and chemical properties, it is recalcitrant and mobile in the environment. The compound is classified as a likely carcinogen by the EPA. An analytical method has been developed to quantify 1,4-dioxane in water using solid phase micro-extraction with a gas chromatograph-flame ionization detector. Treatability studies are currently underway, and include various advanced oxidation techniques, UV photolysis, ozonation, and adsorption to activated carbon and resins.

This research is exploring the sustainability of these treatment processes used to treat 1,4-dioxane in water by quantifying different green metrics for each process including: water used, hazardous waste generated, energy consumed, global warming potential, and material composition and acquisition. The results will ultimately determine which treatment process is effective as well as sustainable. This general approach can be applied to treatment processes for other contaminants of concern as a means of evaluating the overall sustainabilty.

# 352260 Monetising Gas Hydrates: New Mode for Natural Gas Storage and Transportation Against LNG Storage and Transportation

**Abstract:**

Gas Hydrates are considered in the petroleum industry e.g. natural gas transportation through pipelines, during natural gas production from the crude oil wells. But this paper focuses on converting the nuisance of gas hydrates to an attractive means of transportation. New Technology development has been focusing on using gas hydrates as a way of converting gas to solids.

The Natural Gas Hydrate (NGH) is a clatherate compound in icy state and stabilised around -20 degree Celsius under atmosphere. This paper summarises about the cost effectiveness of the NGH value chain against the LNG (Liquefied Natural Gas) value chain and also in this paper entire processes and operations to create NGH pellets are designed and discussed. The calculations show that this process has a high initial plant cost but on a long term this process of NGH production is cost effective compared to LNG liquefaction process. Moreover, it is easy to handle the NGH at -20 to -30 degree Celsius compared to LNG at around -160 degree Celsius.

Thus, this unique method of transporting the natural gas in bulk solid form leads to reduction of the total investment cost for the entire value chain. Moreover, NGH is expected to be a plausible solution in monetising small to medium gas fields which are considered to be difficult to be developed economically using LNG.

# 352647 Towards the Development of a Control-Relevant Model of the Hydraulic Fracturing Process to Investigate Various Control Strategies

**Abstract:**

Hydraulic fracturing or 'fracking' is an increasingly important technique for developing energy resources across both the U.S. and the world, in which high pressure fluid is injected into the reservoir to crack the rock formation and provide highly conductive channels for oil and gas transport back to the wellbore. During this process, the underground environment is characterized by complex fluid dynamics, geo-mechanics, and multiphase flow phenomena. Surface equipment exerts control over the process from between hundreds of feet to several miles from the moving front of the fracture(s) below ground. The fluid pumped underground is a slurry consisting of water, proppant (often a special type of sand) to keep the created fractures open after the process finishes, and chemical additives such as cross-linkers, acids, and surfactants to maintain fluid viscosity, reduce corrosion, and perform other duties. During the fracturing process, pumping rate and fluid properties (viscosity and the concentrations of proppant and other additives) can be varied at the surface. The main objectives are to keep the fracture growth within a predefined region of the reservoir open by distributing the proppant evenly throughout the fracture so that oil and gas can flow after the fracturing fluids have leaked into the formation. Complications can include excessive growth of the fracture external to the target region that results in the proppant being spread so thinly that it is ineffective to keep the fracture open. Conversely, if a high concentration of proppant is added to the fracturing fluid mix too early in the process the condition called 'screen-out' can occur wherein the proppant enters the advancing fracture front and bridges the tip preventing further fracture growth and causing rapid pressure buildup in the system. A related issue is proppant settling, that is a large sand bank or dune forms near the entrance of the fracture at the wellbore resulting in only a small open region around the wellbore after the fluid leaks off. These three potential proppant-related challenges may have a large negative impact on oil well production.

Applying process control techniques to regulate the hydraulic fracturing process, via setting optimal pumping schedules, is a challenging problem not only due to the remoteness of the control action, but also due to the limited feedback information available because of the sensing system associated with the bottom-hole situation. To better understand these challenges, a mathematical model of the physics of the fracturing system is developed to determine the effect of different pumping schedules on the final fracture conductivity, the most widely used calculated performance metric. The physics-based model incorporates equations governing fracture propagation, fluid dynamics, proppant transport, and bank formation. The resulting system of nonlinear, partial differential equations is solved numerically using a moving grid method to track the fracture growth and particle tracking to predict the proppant movement and settling. Additionally, the 'shut-in' process, during which pumping is stopped and the rock formation walls are allowed to close as the fluid leaks off is a part of this model development. The final geometry of the channel available to oil and gas flow is predicted. Using this model, improved pumping schedules can be developed and tested to achieve near-optimal fracture conductivity.

# 352284 Shale Gas as a Potential Source of Unconventional Reserve for Crude Oil

**Abstract:**

With a decrease in the conventional reserves available for the recovery of oil, the alternative fossil fuel extraction industry is on a growth. As the development of unconventional oil and gas resources has been progressed, the heterogeneity and complexity of shale gas and tight oil reservoirs have become apparent. In the U.S. primarily shale gas and tight oil is expected to continue growing dramatically in coming decades. According to the U.S. Energy Information Administration, an assessment of 137 shale formations are identified in 41 countries other than those identified in U.S. At the local level, though the environmental impacts of shale gas and tight oil development may turn out to be significant.

Shale gas and tight oil are unconventional sources of petroleum that are found within the reservoirs with a very low permeability. The oil and gas contained within these reservoir rocks typically will not flow to the wellbore at the economic rates without assistance from technologically advanced drilling and completion processes. Usually, horizontal drilling coupled with multi stage fracturing is used to access these difficult to produce reservoirs. Unconventional natural gas and oil deposits are difficult to characterize overall, but in general are often lower in resource concentration, more dispersed over large area, and require well stimulation or some other extraction or conversion technologies.

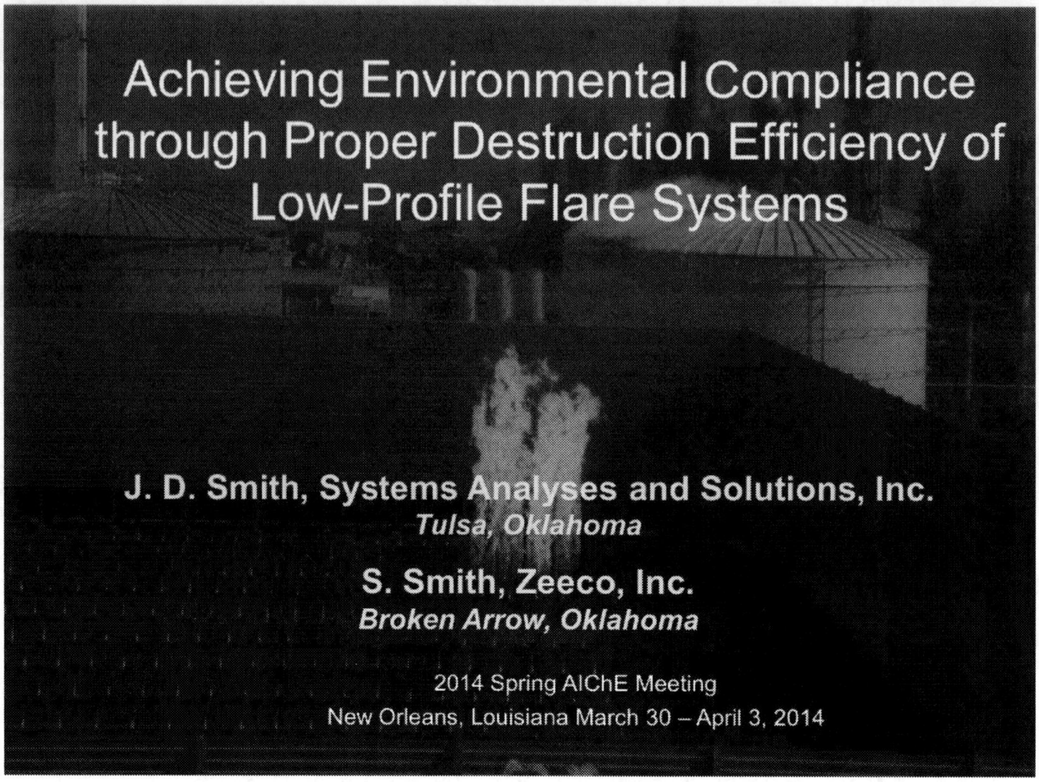
Achieving Environmental Compliance through Proper Destruction Efficiency of Low-Profile Flare Systems
J. D. Smith, Systems Analyses and Solutions, Inc.
Tulsa, Oklahoma
S. Smith, Zeeco, Inc.
Broken Arrow, Oklahoma
2014 Spring AIChE Meeting
New Orleans, Louisiana March 30 – April 3, 2014

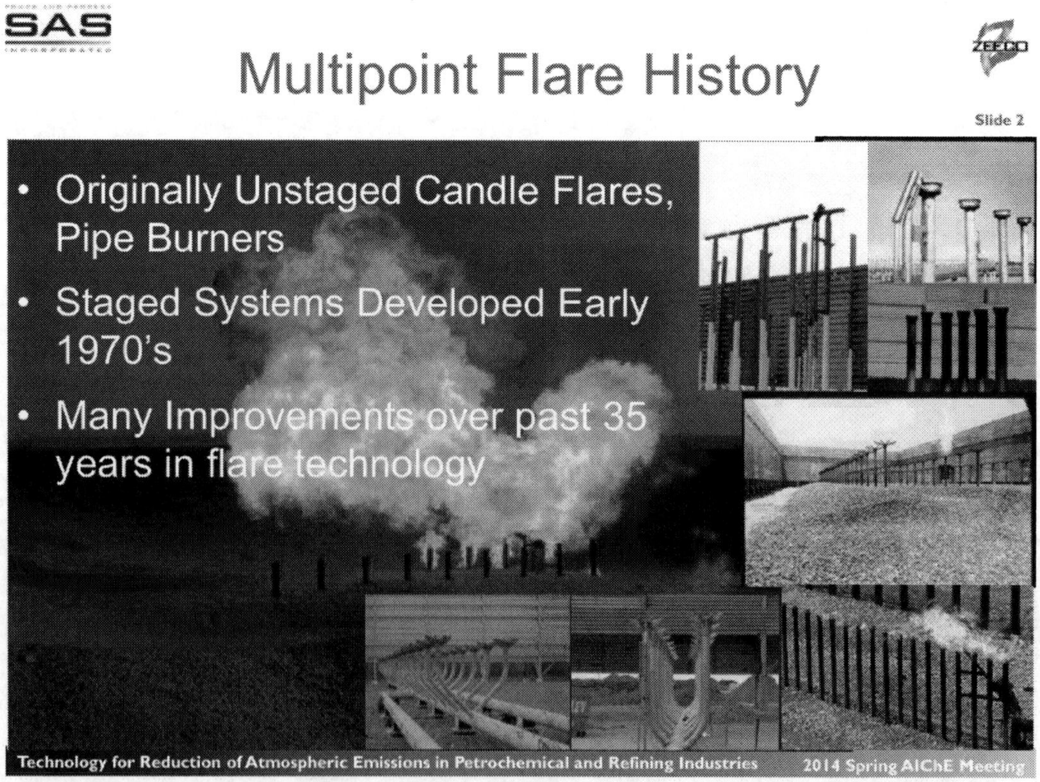
SAS
ZEECO
Multipoint Flare History
Slide 2
• Originally Unstaged Candle Flares, Pipe Burners
• Staged Systems Developed Early 1970's
• Many Improvements over past 35 years in flare technology
Technology for Reduction of Atmospheric Emissions in Petrochemical and Refining Industries
2014 Spring AIChE Meeting

SAS

## Staging Allows Large Variation in Flowrates to Maintain Smokeless Operation

ZEECO

Slide 3

Technology for Reduction of Atmospheric Emissions in Petrochemical and Refining Industries 2014 Spring AIChE Meeting

SAS

## Large Flare System Design Issues

ZEECO

Slide 4

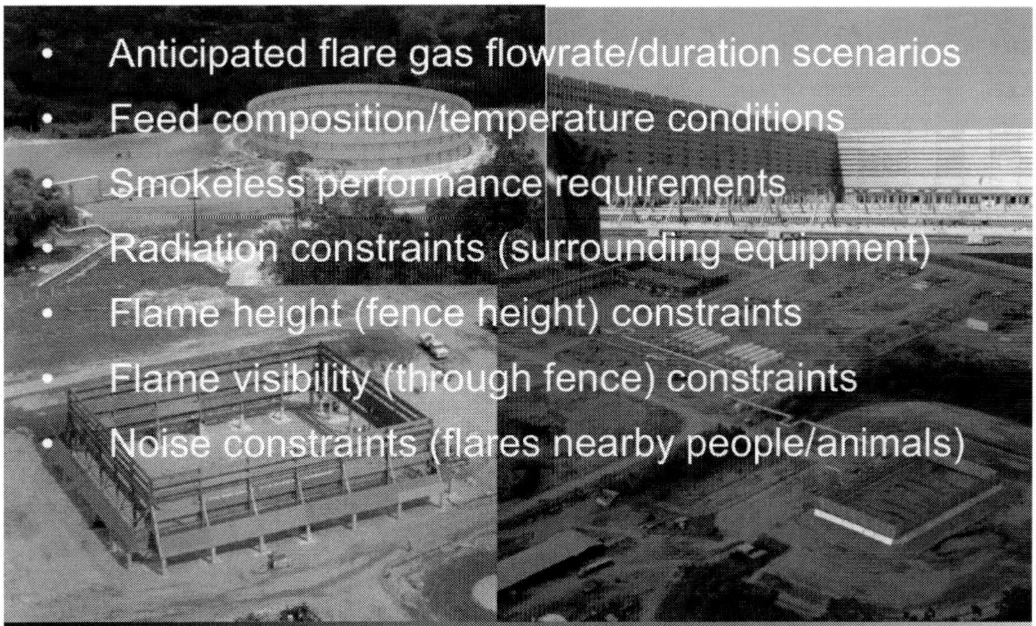


Technology for Reduction of Atmospheric Emissions in Petrochemical and Refining Industries 2014 Spring AIChE Meeting

SAS

# Burner Tip Design based on Multi-arm/ multi-jet configuration

ZEECO

Slide 5

- Variable Arm Area
- Investment Cast
- Pressure Tested at Factory
- 310 SS Cast Material

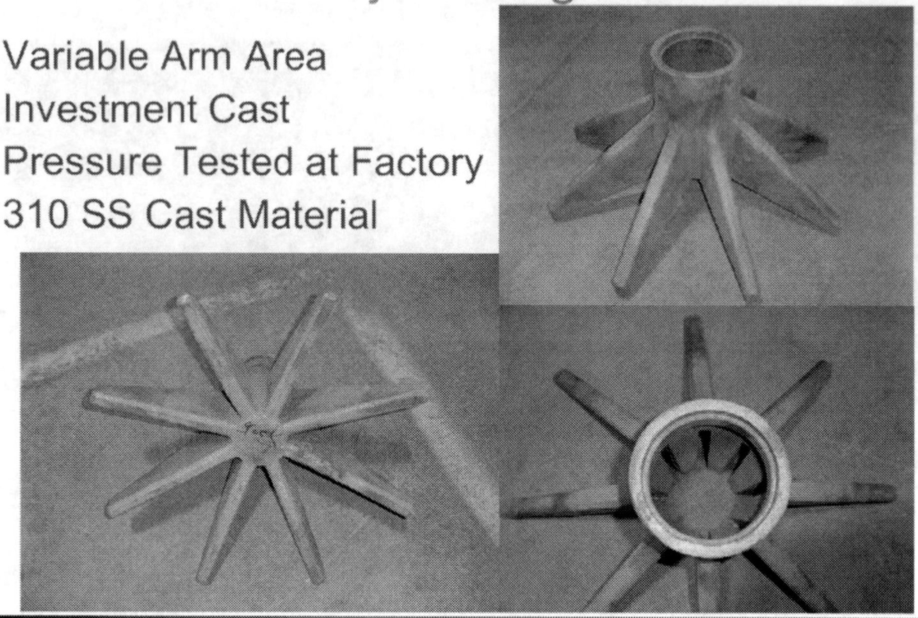

Technology for Reduction of Atmospheric Emissions in Petrochemical and Refining Industries — 2014 Spring AIChE Meeting

SAS

# Flare Regulations and Performance Guidelines

ZEECO

Slide 6

- Flaring regulation (API guidelines) focus on utility flares and assisted flares (little attention to multi-tip ground flares)
- Flare Design/Performance guidelines established for general flares – ground flares have additional performance metrics to consider
- Flare performance difficult to quantify (large flame/high radiation flux, elevated operation)
- Multi-Tip ground flares have higher radiation inside large fenced region (larger plumes/difficult to measure emissions!)

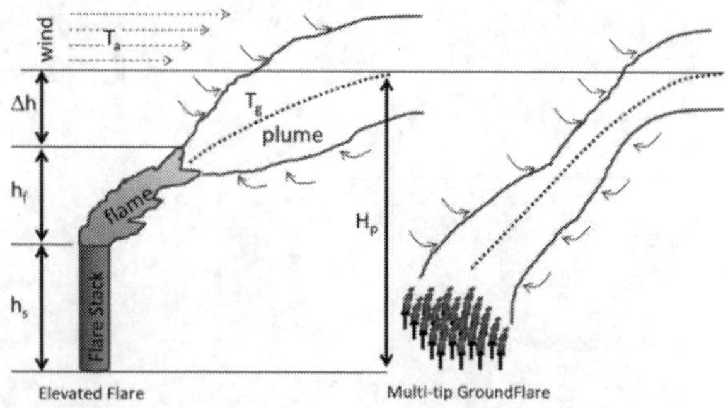


Technology for Reduction of Atmospheric Emissions in Petrochemical and Refining Industries — 2014 Spring AIChE Meeting

# Existing Flare Regulations and Guidelines

Slide 7

- 40 CFR 60.18(b) and 63.11(b)
  - Pilot flame required
  - Sufficiently low exit velocity for stable combustion
  - Heating value high enough for flammable operation
  - Smokeless operation requires adequate mixing energy (air)
- 40 CFR 60.11(d), 60.18(d), 60.482-10(e), 63.6(e), 63.11(b)(1), 63.172(e)
  - Adhere to flare's design documents
  - Use Good Air Pollution Control Practice
- New Source Review

**Federal requirements for flares are found in the New Source Performance Standards (NSPS) in § 60.18 and National Emission Standards for Hazardous Air Pollutants (NESHAP) in § 63.11. At a minimum, these rules require flares to be:**

- Designed and operated with no visible emissions using EPA Method 22 (except for periods not to exceed 5 minutes in 2 hours);
- Operated with a flame present at all times, confirmed by the use of a thermocouple or equivalent device;
- Used only when the net heating value of the gas to be combusted is 300 BTU per standard cubic foot (BTU/scf) or greater (if the flare is steam- or air-assisted), or 200 BTU/scf or greater (if the flare is nonassisted); and
- Designed for and operated with an exit velocity less than 60 feet per second (ft/sec). An exit velocity of greater than 60 ft/sec but less than 400 ft/sec may be used if the net heating value of the gas being combusted is sufficiently high.

Enforcement Alert, "EPA Enforcement Targets Flaring Efficiency Violoations", EPA-325-F-012-002,August (2012).



# Understanding Multi-Tip Ground Flares

Slide 8

- Understanding ground flare performance
  - Where does air come from?
  - How does wind effect flame?
  - How does tip geometry (pressure) affect flame?
- Reacting multi-phase flow in wind
  - Transient flame analysis
  - Near tip mixing
  - 100's of tips in large flare fields
- Model Validation Required
  - Calibrate Smoke and Radiation models

Tip Geometry

SAS

# SAS's C3D Flare Model: Beyond Traditional Flare Analysis

ZEECO

Slide 9

- Extends standard "Steady-State" Flare Analysis:
  - Flame structure interaction with combustion chemistry and radiation flux
  - Fluid interactions with surrounding obstacles
  - Fuel specific soot formation and impact on radiation transfer
- Captures flame dynamics: Momentum vs. Buoyancy, Wind Effect on Flame Structure, Chemical Reactions, Radiation to/from flame
- Assess operating scenarios for specified conditions: wind speed, fuel flow/composition/temperature, impact on/by surroundings
- Key Assumptions:
  - Flame emissivity = F(flare gas composition, soot volume fraction, flame size/shape)
  - Optically thick flames (diffuse radiation) with time dependent view factors
  - Surroundings participate in radiation heat transfer (Re-radiation from ground)
- Overnight analysis on "standard" workstation, animations used to investigate transient flame behavior



SAS

# Multi-Tip Ground Flares: Modeling Issues

ZEECO

Slide 10

- High tip velocity increases air entrainment
  - Tip design critical to air entrainment
  - Local high velocity can translate into high sound levels
- Assist media not available to increase combustion air
  - smoke below certain tip pressure (D-stage pressure)

- Tip spacing critical
  - Flares must cross light
  - Possible Flame merge lengthens flames
  - Adjacent rows compete for air (longer flames, poor performance)

Technology for Reduction of Atmospheric Emissions in Petrochemical and Refining Industries 2014 Spring AIChE Meeting

# Multipoint Ground Flare Modeling Approach

Slide 11

Single Burner Test/Model
- Detailed mesh
- Radiation/soot tests
- Single flame height/size

Multi-Burner Test/Model
- Radiation/soot tests
- Flame interactions (spacing)
- Multi-Flame shape/size

Model Full Flare Field
- Use Calibrated Models
- Predict Flare Performance (Smoke/Air Demand)
- Predict Radiation Flux

Technology for Reduction of Atmospheric Emissions in Petrochemical and Refining Industries 2014 Spring AIChE Meeting

# Transient Flare Analysis: Multi-step Reaction Mechanism

Slide 12

Incomplete Combustion of Flare Gas:

$$1\text{kg } \mathbf{F} + (2.87\text{-}2.6S_1)\text{ kg } \mathbf{O2} \rightarrow S_1 \text{ kg } \mathbf{C} + (3.87\text{-}3.6S_1)\text{ kg } \mathbf{PC} + (50\text{-}32S_1)\text{ MJ}$$

Where: (F); Oxygen (O2); Soot (C); Products of Combustion (PC), $S_1$ soot parameter based on fuel type

Endothermic fuel pyrolysis (cracking reaction – soot producing):

$$1\text{kg } \mathbf{F} + 0.3\text{ MJ} \rightarrow S_2 \text{ kg } \mathbf{C} + (1\text{-}S_2)\text{ kg IS}$$

Where: Intermediate Species (IS), $S_2$ is cracking parameter based on fuel type

Soot combustion:

$$1\text{kg } \mathbf{C} + 2.6\text{ kg O2} \rightarrow 3.6\text{ kg } \mathbf{CO2} + 32\text{ MJ}$$

Where: Carbon Dioxide (CO2) is combustion product

Intermediate Species combustion:

$$1\,\text{kg}\, IS + \frac{2.87-2.6S_2}{1-S_2}\text{kg}\, O_2 \rightarrow \frac{3.87-3.6S_2}{1\text{-}S_2}\text{kg}\, PC + \frac{50-32S_2}{1\text{-}S_2} MJ$$

Coefficients selected so complete combustion of carbon soot (C) and intermediate species (IS) produce same species and thermal energy as direct combustion of fuel; Coefficients in terms of mass weight (not moles)

Technology for Reduction of Atmospheric Emissions in Petrochemical and Refining Industries 2014 Spring AIChE Meeting

SAS

ZEECO

# Flare Burner Tests

Slide 13

Technology for Reduction of Atmospheric Emissions in Petrochemical and Refining Industries 2014 Spring AIChE Meeting

SAS

ZEECO

# Single Tip Burning Propane: wind vs. no-wind

Slide 14

Technology for Reduction of Atmospheric Emissions in Petrochemical and Refining Industries 2014 Spring AIChE Meeting

SAS

## Single Burner Flare Model

Slide 15

- 6 X 6 X 26 m physical domain
- Flare Tip located 2 m above ground level
- Turbulence and Arrhenius kinetics Included for fuel gas
  - Reaction Parameters adjusted to match observed flame characteristics
  - Soot Yield matched flame height (i.e., soot burnout)
- Flare Movies for no wind, 3m/s (7mph) wind conditions
- Predicted results for Air demand and Radiation loss from flame determined

Technology for Reduction of Atmospheric Emissions in Petrochemical and Refining Industries 2014 Spring AIChE Meeting

SAS

## Combustion Models

Slide 16

Propane:

$C_3H_8 + 3.6\ O_2 \rightarrow 3\ CO_2 + 1.6\ H_2O + 0.024$ Soot + 46 MJ/kg propane

Soot + $2.66\ O_2 \rightarrow 3.66\ CO_2$ + 32 MJ/kg Soot

Ethylene:

$C_2H_4 + 0.57\ O_2 \rightarrow 0.93\ C_2H_2 + 0.64\ H_2O$ + 9.4 MJ/kg ethylene

$C_2H_2 + 2.58\ O_2 \rightarrow 2.7\ CO_2 + 0.7\ H_2O + 0.2$ Soot + 34.1 MJ/kg intermediate

Soot + $2.66\ O_2 \rightarrow 3.66\ CO_2$ + 32MJ/kg Soot

Mixed Gases:

$0.572\ C_2H_4 + 0.383\ C_2H_6 + 0.043\ H_2 + 0.982\ O_2 \rightarrow$
$0.53\ C_2H_2 + 0.34\ C_2H_3 + 1.1\ H_2O$ + 14.2 MJ/kg

$0.61\ C_2H_2 + 0.39\ C_2H_3 + 2.66\ O_2 \rightarrow$
$2.66\ CO_2 + 0.813\ H_2O + 0.181$ Soot + 34.4 MJ/kg

Soot + $2.66\ O_2 \rightarrow 3.66\ CO_2$ + 32 MJ/kg

Technology for Reduction of Atmospheric Emissions in Petrochemical and Refining Industries 2014 Spring AIChE Meeting

SAS

## Single-burner Mesh

Slide 17

- Rectangular cells
- Local refinement near burner tip
- 110,000 computational cells

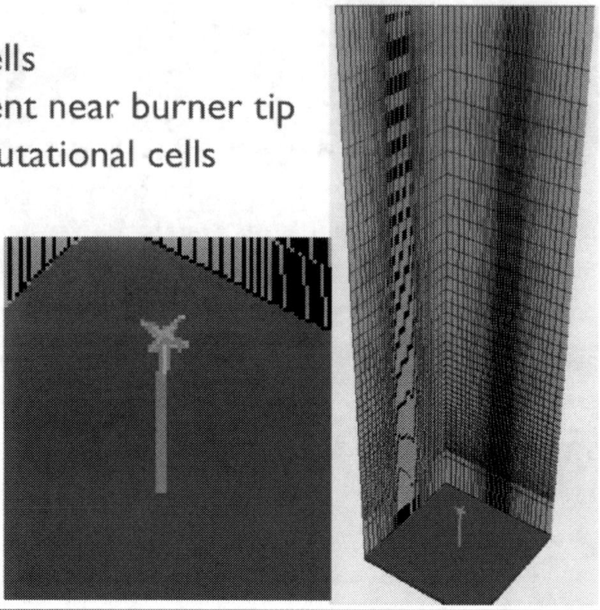



SAS

## Single Tip Burning Propane: 3.0 m/s Wind

Slide 18

Front View

Side View

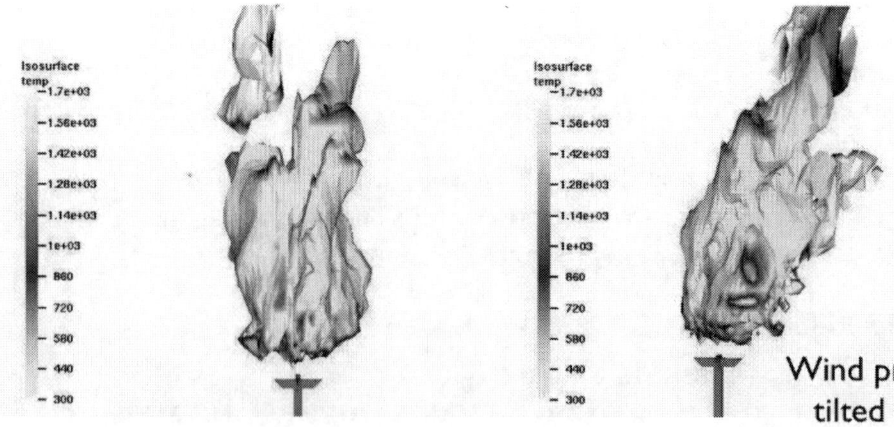


Wind produces tilted bushy, shortened flame

SAS

## Single Tip Burning Propane: wind vs. no-wind

Slide 19

Single Burner

No wind produces tight "pencil-like" flames

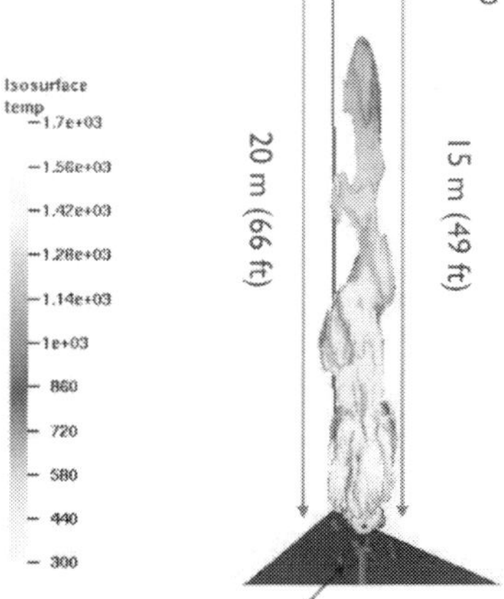


Flare tip located 2 m (~7 ft) above grade

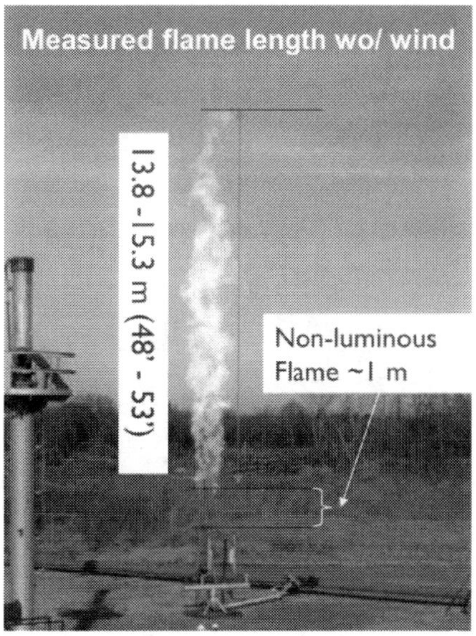




SAS

## Single-Tip Ground Flare Test Results

Slide 20

- 1.4" WC @ 57 °F => 7.3 psig tip pressure
- 12-16 Km/hr (8-10 mph) crosswind
- ~30% flame height reduction
- Minimum flame tilt (~8°)

SAS

ZEECO

# Three-burner Mesh

Slide 21

- Rectangular cells
- Domain size is 30 m X 35 m X 25 m
- Local refinement near burner tips and radiation meters
- 188,000 computational cells

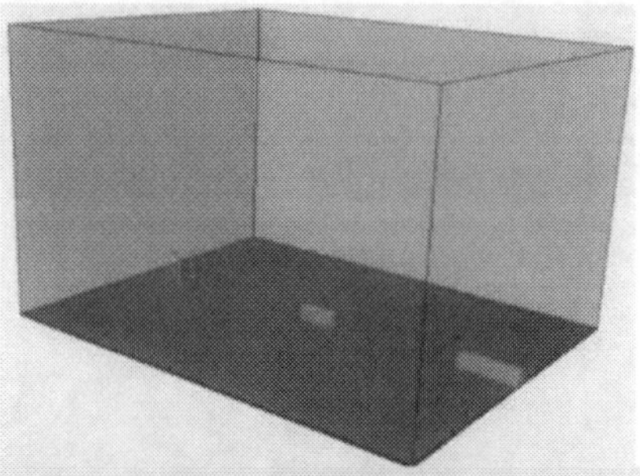

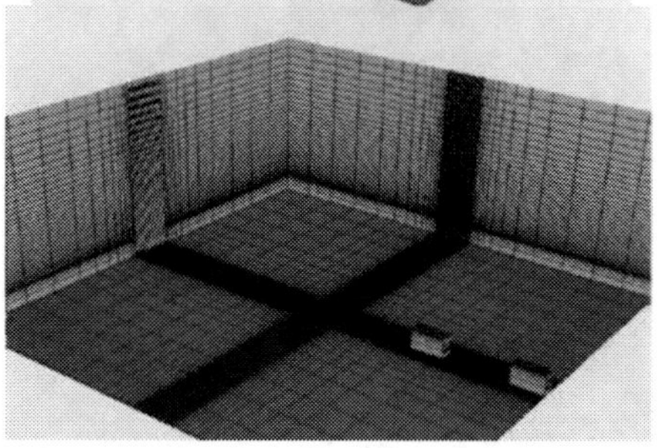

Technology for Reduction of Atmospheric Emissions in Petrochemical and Refining Industries 2014 Spring AIChE Meeting

SAS

ZEECO

# Predicted/Measured Flame for 3-burner Ethylene test

Slide 22

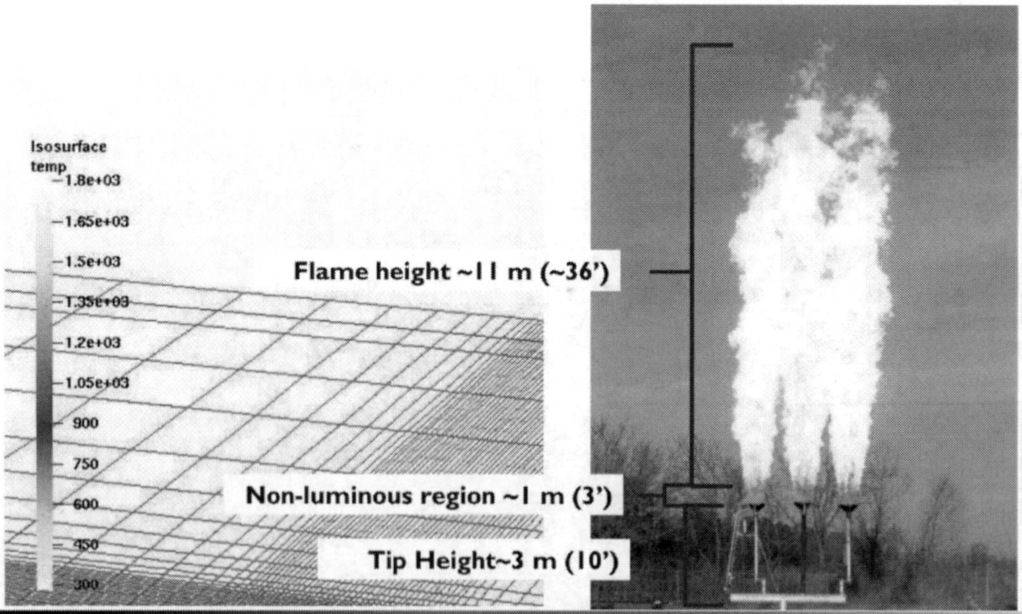


Technology for Reduction of Atmospheric Emissions in Petrochemical and Refining Industries 2014 Spring AIChE Meeting

SAS

## Predicted 3-burner flare with radiation monitors

ZEECO

Slide 23

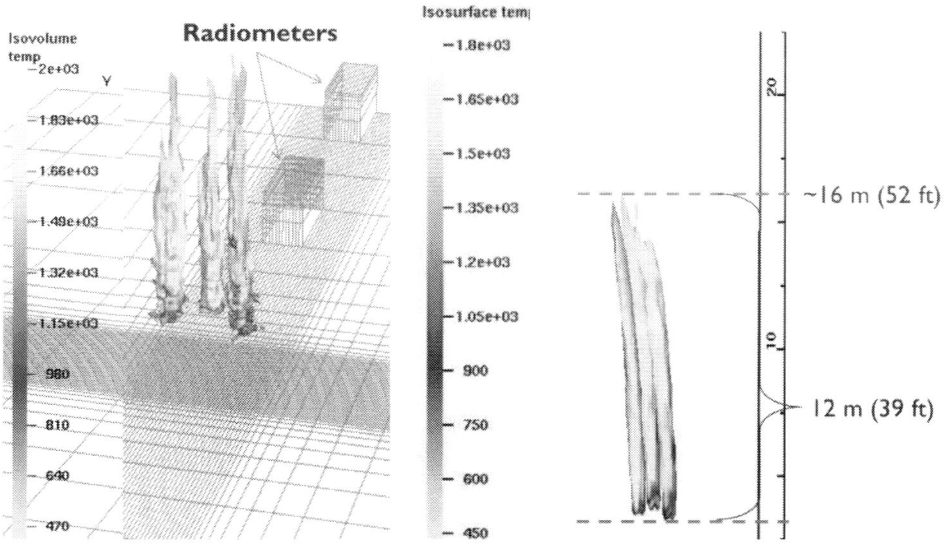


Smith, J.D., Suo-Ahttila, A., Smith, S., and Modi, J., "Estimation of the Air-Demand, Flame Height, and Radiation Load from Low-Profile Flare using ISIS-3D," *American – Japanese Flame Research Committees International Symposium, Advances in Combustion Technology: Improving the Environment and Energy Efficiency*, Marriott Waikoloa, Hawaii - Oct. 22 –24 (2007)



SAS

## Predicted Radiation from 3-Burner Flare with Modifications for Ground and Atmospheric Attenuation

ZEECO

Slide 24

### Radiation Included in Prediction:

1. Ground re-radiating and reflecting incident radiation from flame
   - Assumed ground $\varepsilon = \alpha = 1$; allow ground to heat to steady state temperature
2. Atmospheric attenuation of radiation from flame
   - Model uses ambient/source temperatures with $H_2O$/ $CO_2$ absorption

| WALL (plan view perspective) | Left Wall ($W/m^2$) | Right Wall ($W/m^2$) | Bottom Wall ($W/m^2$) | Flame Optical Thickness |
|---|---|---|---|---|
| **PEAK FLOW** | | | | |
| **Initial Radiation** | 78,000 | 63,000 | 108,000 | 0.275 |
| **Radiation Modification** | 61,000 | 35,000 | NA | |
| **SUSTAINED FLOW** | | | | |
| **Initial Radiation** | 15,000 | 15,000 | 35,000 | 0.28 |
| **Radiation Modification** | 6,600 | 6,600 | NA | |

SAS

## Model Validation: Radiation Predictions Compared to Experimental Data

ZEECO

Slide 25

| Tip Size | Position (m) | Burner Pressure (psi) | Total Predicted Radiation ($W/m^2$) | Measured Radiation ($W/m^2$) | Difference (%) |
|---|---|---|---|---|---|
| 3 | 15 | 2.8 | 2700 | 3344 | -20.0 % |
| 3 | 15 | 7.3 | 4750 | 4803 | -1.0 % |
| 3 | 15 | 11.4 | 6150 | 6192 | -0.7 % |
| 3 | 30 | 2.8 | 650 | 671 | -3.0 % |
| 3 | 30 | 7.3 | 1350 | 1184 | +14.0 % |
| 3 | 30 | 11.4 | 1650 | 1532 | +8.0 % |
| 4 | 15 | 2.8 | 4325 | 6371 | -32.0 % |
| 4 | 15 | 7.3 | 8050 | 8192 | -2.0 % |
| 4 | 15 | 11.4 | 10000 | 9536 | +5.0 % |
| 4 | 30 | 2.8 | 1150 | 1513 | -23.0 % |
| 4 | 30 | 7.3 | 2580 | 2464 | +5.0 % |
| 4 | 30 | 11.4 | 3250 | 2747 | +18.0 % |

Smith, J.D., Suo-Ahttila, A., Smith, S., and Modi, J., "Estimation of the Air-Demand, Flame Height, and Radiation Load from Low-Profile Flare using ISIS-3D," *American – Japanese Flame Research Committees International Symposium, Advances in Combustion Technology: Improving the Environment and Energy Efficiency*, Marriott Waikoloa, Hawaii - Oct. 22 –24 (2007)

Technology for Reduction of Atmospheric Emissions in Petrochemical and Refining Industries 2014 Spring AIChE Meeting

SAS

## Wind Effect on Radiative Flux

ZEECO

Slide 26

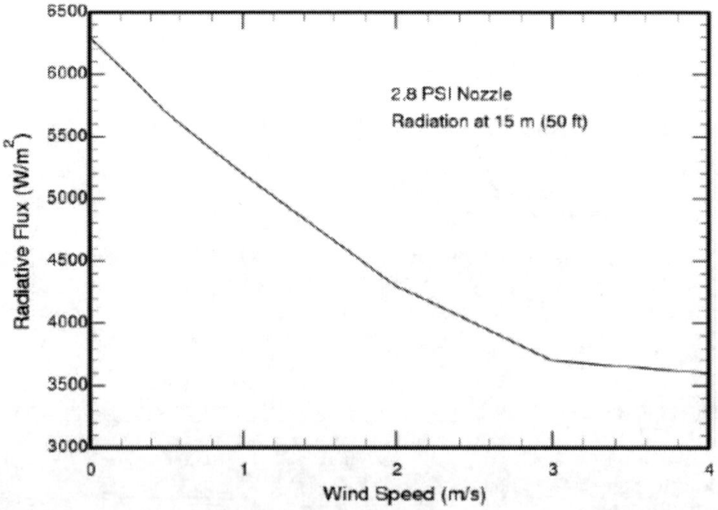


Smith, J.D., Suo-Ahttila, A., Smith, S., and Modi, J., "Estimation of the Air-Demand, Flame Height, and Radiation Load from Low-Profile Flare using ISIS-3D," American – Japanese Flame Research Committees International Symposium, Advances in Combustion Technology: Improving the Environment and Energy Efficiency, Marriott Waikoloa, Hawaii - Oct. 22 –24 (2007)

Technology for Reduction of Atmospheric Emissions in Petrochemical and Refining Industries 2014 Spring AIChE Meeting

SAS

ZEECO

# Multipoint Ground Flare

Slide 27

Technology for Reduction of Atmospheric Emissions in Petrochemical and Refining Industries — 2014 Spring AIChE Meeting

SAS

ZEECO

# Full Field Flare Grid

Slide 28

- Domain size 10 m beyond wind fence; 25 m high
- Local refinement near burner rows/tips
- Sustained Flow Rate: 700,000 kg/hr
- Peak Flow Rate: 1,200,000 kg/hr

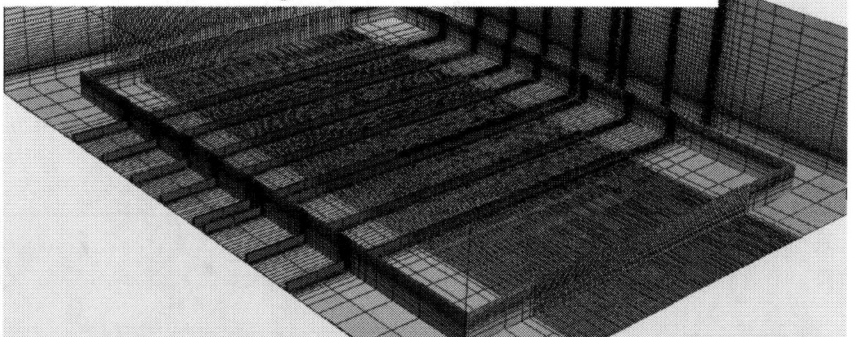

Technology for Reduction of Atmospheric Emissions in Petrochemical and Refining Industries — 2014 Spring AIChE Meeting

SAS

ZEECO

# Simulation of Flare Tests

Slide 29

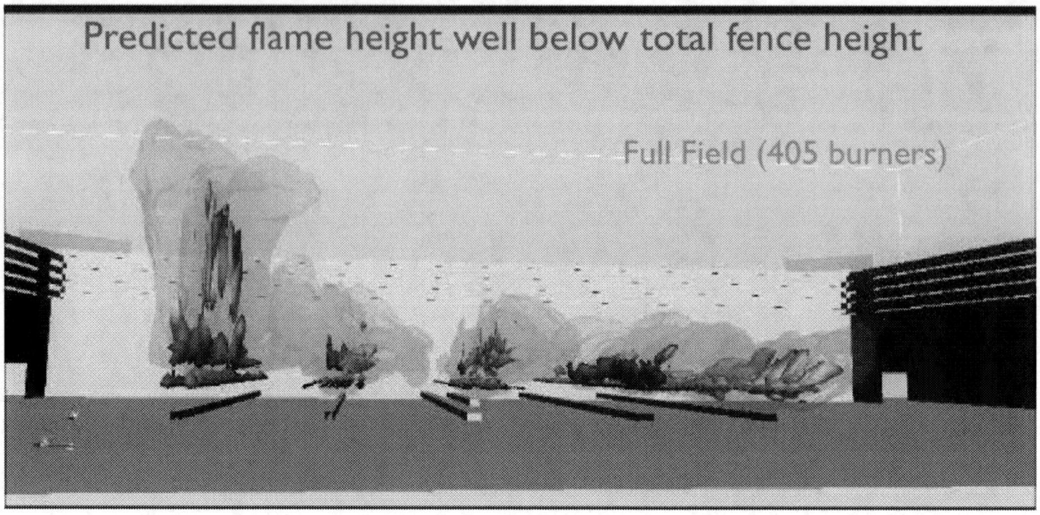




SAS

ZEECO

# Max [$C_2H_4$] along line of sight for peak flow case

Slide 30

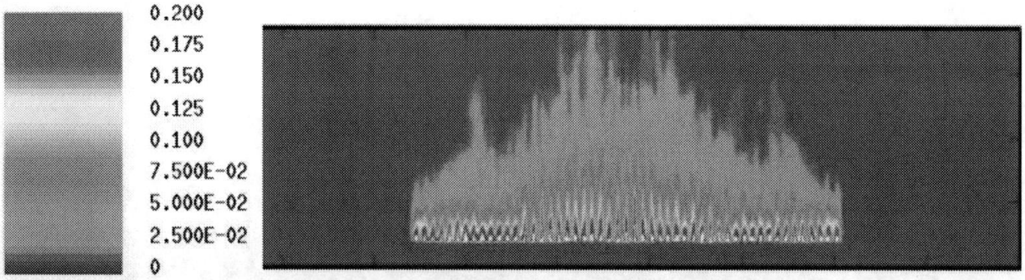


Side view along row showing flame elongation toward center of row

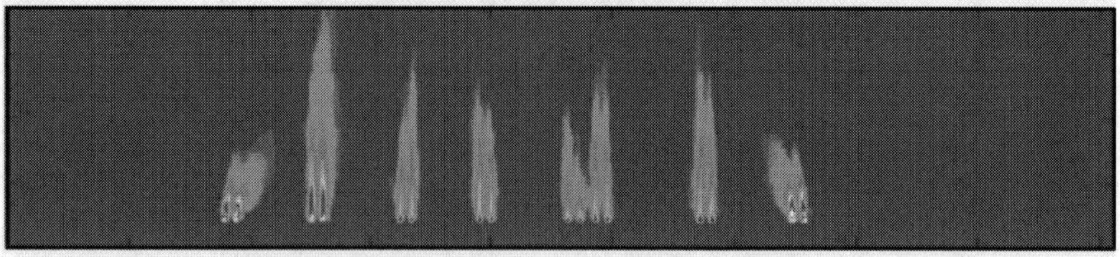

End view of flame height for each row and impact of inflow on outer rows

SAS

# Predicted Air/Fuel Requirements for "no-wind" cases

Slide 31

| Case Description | Evaluation Plane Area (m²) | Total Air Flow (kg/s) | Fuel Flow (kg/s) | Air/Fuel Mass Ratio |
|---|---|---|---|---|
| 1 Burner Propane | 14.63 | 60 | 0.46 | 130 |
| 1 Burner (Tip 3) Ethylene | 13.26 | 52 | 0.94 | 55 |
| 3 Burner (Tip 3 - 7.3PSI) Ethylene | 36 | 134 | 2.88 | 47 |
| 3 Burner (Tip 3 - 11.4PSI) Ethylene | 36 | 144 | 3.84 | 38 |
| 3 Burner (Tip 4 - 7.3PSI) Ethylene | 36 | 150 | 4.26 | 35 |
| 3 Burner (Tip 4 - 11.4PSI) Ethylene | 36 | 160 | 5.79 | 28 |
| Full Field Peak Flow Ethylene | 3226 | 9700 | 262.3 | 37 |
| Full Field Sustained Flow Mixed Gas | 1843 | 4800 | 93.6 | 51.3 |

Technology for Reduction of Atmospheric Emissions in Petrochemical and Refining Industries 2014 Spring AIChE Meeting

SAS

# Conclusions

Slide 32

- CFD Flare Models:
  - Single-burner model: 110,000 cells
  - Three-burner model: 188,000 cells
  - Full-field model: 700,000 cells (Sustained Flow)/1,200,000 cell (Peak Flow) (Combustion chemistry for propane, ethylene, mixed gas considered)
- Analyzed flame shape/size for 3 fuels for single and three burner tests
- Compared predictions to measurements for 12 tests (2 tip sizes, 3 operating pressures, 2 radiation sample locations)
- "*Reasonable*" estimates of radiation heat transfer and air demand for low profile flare
  - Air/fuel ratios range from 28 to 47 for 3-burner test and from 37 (Peak Flow Case) to 51 (Sustained Flow Case)
- Calibrated flare model applied to full-flare field to estimate:
  - Air demand for specified tip/row spacing
  - Radiation load on wind fence for nominal and peak flow cases
  - Expected flame height and smoke production for nominal and peak flow cases

Technology for Reduction of Atmospheric Emissions in Petrochemical and Refining Industries 2014 Spring AIChE Meeting

SAS

# Questions?

Slide 33

# 353103 A Multi-Period Process Integration to the Management of Process Flares with Consideration of Fuel Substitution and Cogeneration Systems

**Abstract:**

A wide variety of process upsets can lead to flaring. Such flaring activities cause environmental emissions and result in economic losses. A possible approach to mitigating these emissions and losses is to utilize the flared gases in combustion systems (boilers and industrial furnaces) that can be used to produce utilities such as steam. Additionally, the generated steam can be further used in a cogeneration system (e.g., through steam turbines) to produce power and heating. This paper introduces a process integration approach to the effective utilization of flared gases into combustion and cogeneration systems. Since the flared gases differ in composition and characteristics from the fuel that is typically used in boilers, there is a need to make sure that the fuel quality is in the acceptable range for the boiler. Various constraints may be used to assess usability of flared gases as part of the fuel blend fed to the boiler. For instance, the Wobbe index (WI) may be used to study fuel properties sent to the boilers. A multi-period optimization formulation is developed to determine the optimal usage of flare gases in the boiler and cogeneration systems. An operational period is discretized into multiple periods corresponding to different operating modes of the process (process episodes). Economic objectives are used while including technical and environmental constraints. Additionally, the cost-benefit analysis for adding a cogeneration system with an optimal size is carried out. A case study using an ethylene plant is used to demonstrate the various aspects of the process.

# PROOF OF CONCEPT TEST
# FOR A REAL-TIME FLARE COMBUSTION EFFICIENCY MONITOR

**Presented at**
**2014 AIChE Spring Meeting &**
**10th Global Congress on Process Safety**

**By**
**Yousheng Zeng, PhD, PE and Jon Morris**
**Providence Photonics, LLC**

**Mark Dombrowski**
**Surface Optics Corporation**

**April 2, 2014**

PROVIDENCE

# A BRIEF REVIEW OF THE CONCEPT PRESENTED AT 2012 FALL CONFERENCE

Ref.: Zeng and Morris, "A New Method to Measure Flare Combustion Efficiency in Real-Time", presented at AWMA Louisiana Section 2012 Fall Conference, Baton Rouge, Louisiana, October 30-31, 2012

2

PROVIDENCE

Flare Combustion Efficiency (CE):

$$CE(\%) = \frac{[C]_{CO2}}{\sum_i n_i[C]_{HCi} + [C]_{CO2} + [C]_{CO}} \quad \text{Eq. (1)}$$

Flare CE – Very difficult to measure

Source: TCEQ/UT



# THE CONCEPT

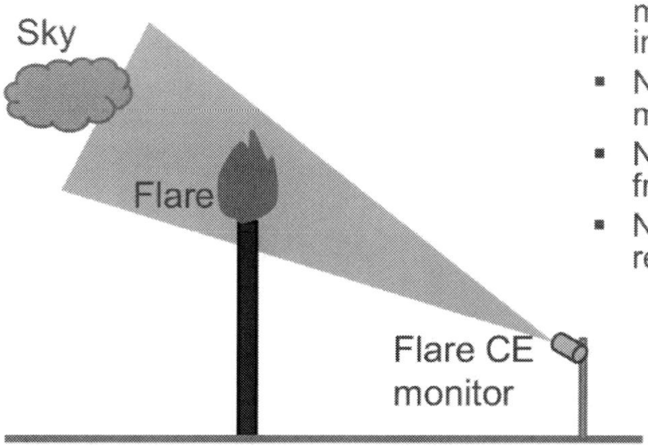


- Monitor and map flare CE in real time through a special multi-spectral IR imaging device
- Not a path measurement
- No scanning; high frame rate
- No operator required

We call it "Flare Efficiency Monitoring System" or "FEMS"

## VISION

- Monitor flare CE in real-time
- Image the full flare flame; measure both overall CE and CE at a pixel level (CE mapping)
- Industrial grade device, suitable for integration with PLC or DCS
- One monitor covers multiple flares on site (step and stare)



## WORKING PRINCIPLE

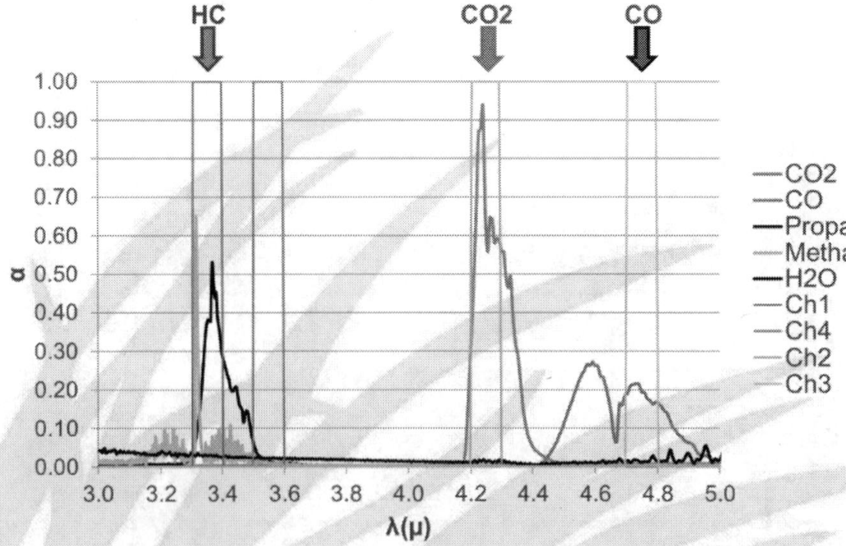

# PROOF OF CONCEPT TEST

The proof of concept test was partially funded by EPA SBIR Phase I grant.

7

PROVIDENCE

# TEST SETUP

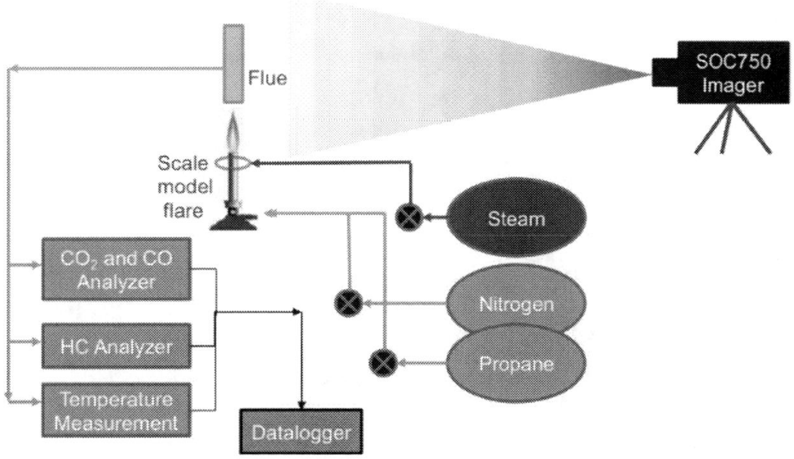


8

PROVIDENCE

Distance from the scale model flare to the Imager: 23 ft.

SOC750 Hyper-Spectral Imager

Scale model flare, Flue w/ Sampling Probe, and Analyzers

9

PROVIDENCE

# SCALE MODEL FLARE

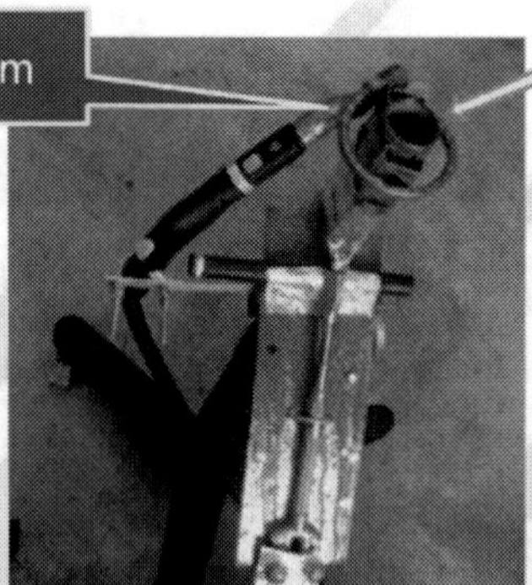

Fuel: approx. 40k Btu/hr.
Nitrogen: varied
Steam: ~ 3 lb./hr.

10

PROVIDENCE

## Flue and Sampling Probe

## Analyzers and Imager

**Testo 350 XL**
(CO2 using NDIR, CO, O2, NOx, H2, Temp, etc.)

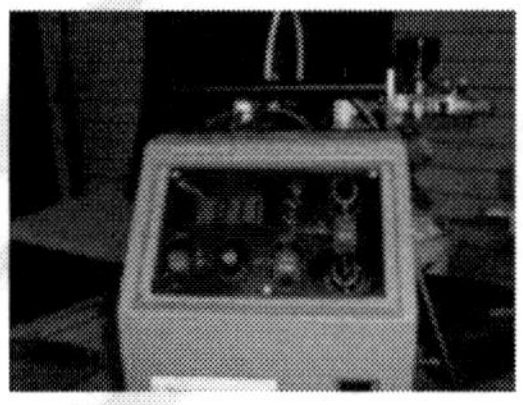

**3010 Mini FID**
Calibrated to Propane

**SOC750 Hyper-spectral imager**
42 spectral channels, operated at 22 cubes per sec.

# TEST RUNS

13

PROVIDENCE

# PRELIMINARY RESULTS - WITHOUT CALIBRATION

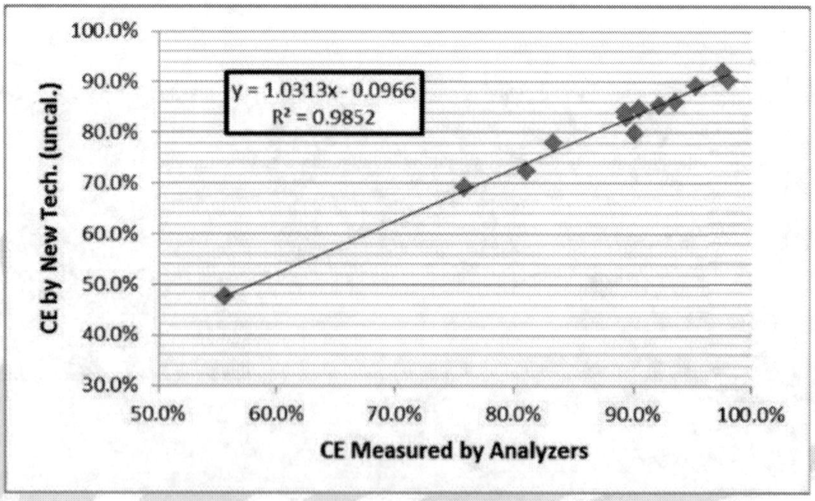


14

PROVIDENCE

## Preliminary Results - with an Initial Calibration

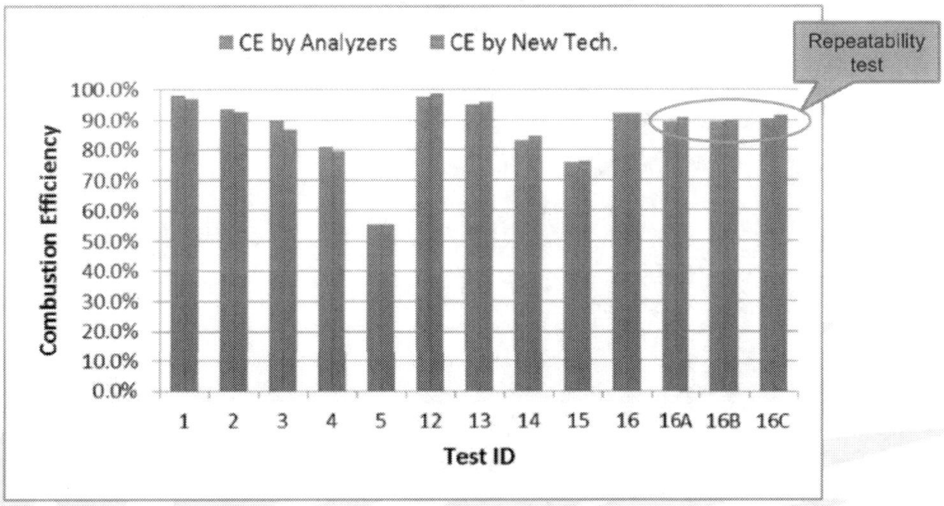


15

PROVIDENCE

## Preliminary Results - Initial Method to Measure Temperature

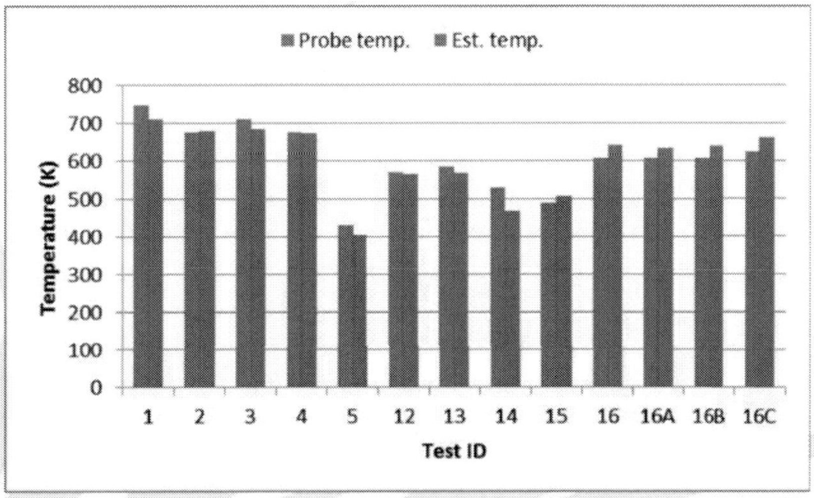


16

PROVIDENCE

## PRELIMINARY RESULTS – CE MAP

30 sec. average (670 frames)

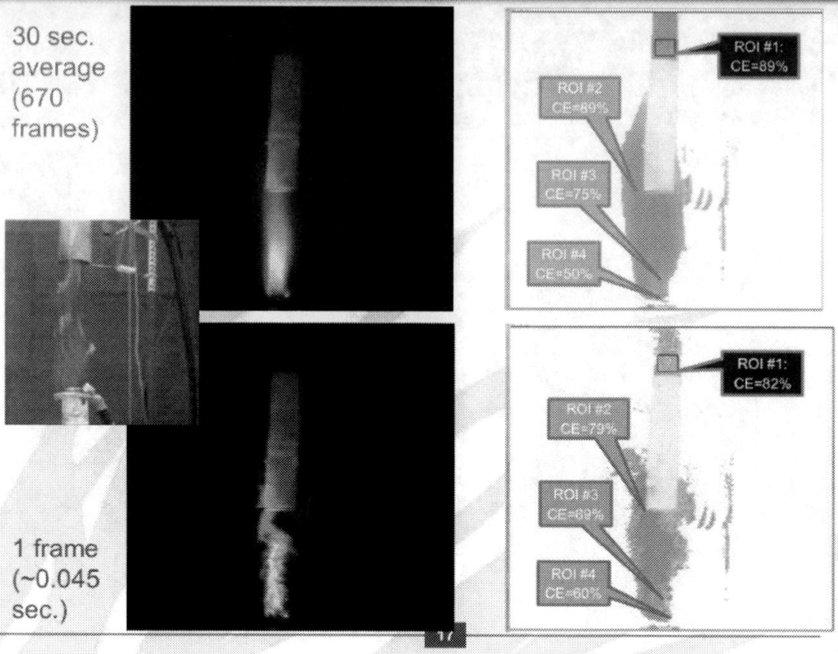


1 frame (~0.045 sec.)



## Three Types of Measurement

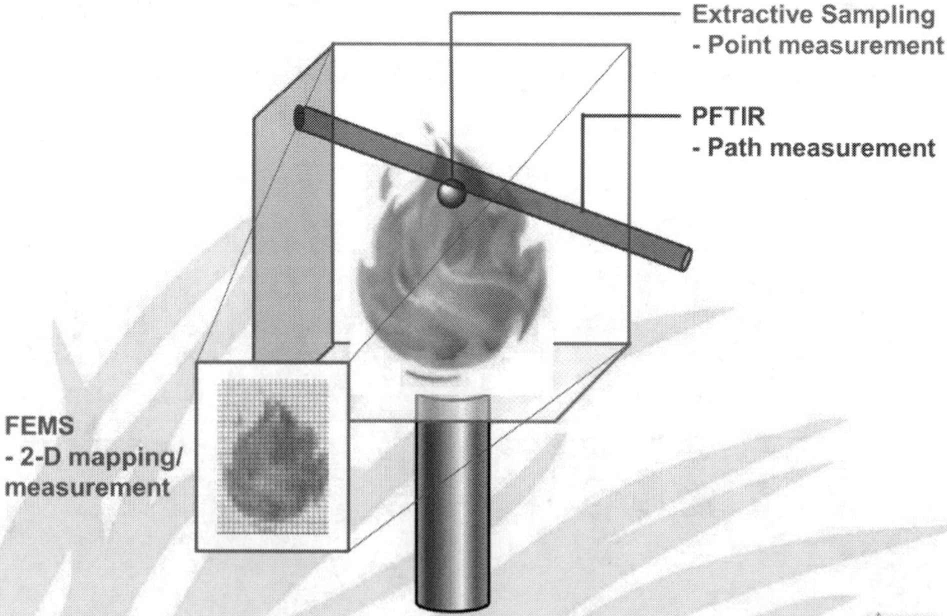

## Comparison with PFTIR

**PFTIR**

- "Scanning" - >1 sec/scan – assuming that flare is static during that time
- Path measurement – aiming required
- Human operation

**FEMS**

- Staring – 20-30 data cubes/sec – match the flare dynamics
- 2-D mapping of CE – no aiming required
- Automatic

19

PROVIDENCE

## Conclusion

- The CE determined by the new technology correlate well with the CE measured by conventional analyzers
- With a further developed calibration method, real-time CE monitoring and feedback for flare optimization is feasible
- The new technology can determine CE at a pixel level, generating a CE map for the entire flare flame. No aiming issue.
- As a side benefit, it can also provide temperature mapping of the flare flame

20

PROVIDENCE

# LOOKING FORWARD…

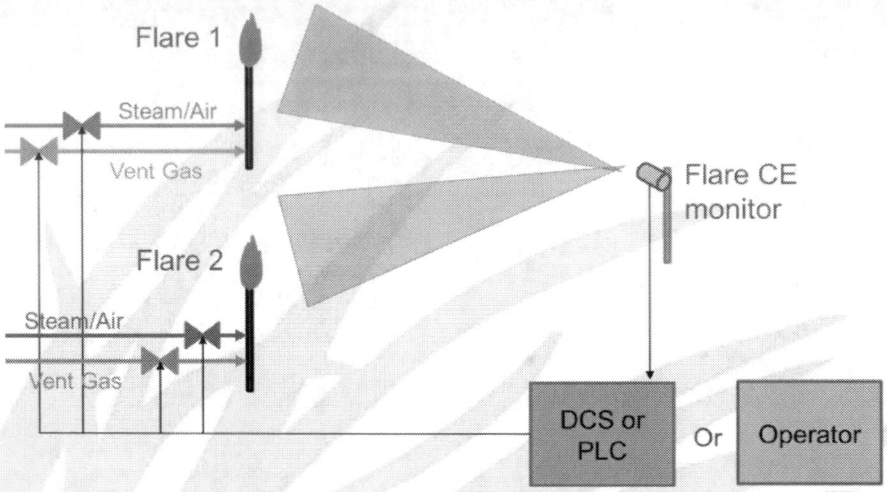


21

PROVIDENCE

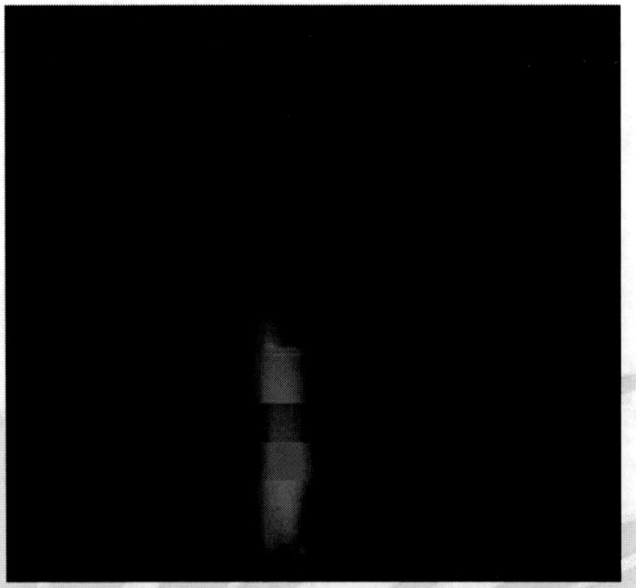

Contact Info. youshengzeng@providenceeng.com

22

PROVIDENCE

# 348171 Flameless Thermal Oxidation

**Abstract:**

Linde's patented Flameless Thermal Oxidizers (FTO) technology oxidizes combustible gases and atomized liquids with air under controlled conditions in a specially designed reactor without a flame. Residence time and reaction temperatures are carefully selected to achieve 99.99 +% destruction efficiency with ultra-low NOx and dioxin emissions at very high thermal energy efficiency. These systems are commercially proven, cost effective and highly effective for the destruction of volatile organic compounds (VOC's) and hazardous air pollutants (HAP's). This paper will present an overview of the FTO technology and market applications with special emphasis on environmental performance and cost considerations.

# 350916 Modeling of the Dynamic Interaction in an LDPE Closed System Purge Bins-Regenerative Thermal Oxidizer (RTO) for Ethylene Emission Mitigation


**Abstract:**

Common designs of LDPE purge bins involve venting the purge and blend air to atmosphere containing small traces of ethylene. To mitigate ethylene emission from these bins, a new process design has been proposed to collect vent air-carrying ethylene via a closed ducting system connected to a regenerative thermal oxidizer (RTO) unit. This necessitates a blower at the downstream end to pull the purge air before feeding into the RTO to incinerate the ethylene. Concerns have been raised regarding the impact of such system on the purge bins, which are not designed to withstand any significant amount of vacuum. While conceptually steady operation of such system is feasible, operation of purge bins is never steady particularly when the purge bins are not designed to withstand any significant level of vacuum pressure. With the minimum safe operating pressure at -0.1 psig (-0.68 kPag), and the design RTO inlet pressure being -1"'WC (-0.25 kPag), the bins are forced to operate under slight vacuum conditions with a very narrow allowable margin for error. The operation is inherently dynamic as LDPE pellets are being conveyed into and out of the bins, purged and mixed in cycle by the interaction of multiple pieces of equipment. In the event of any equipment failure along the train, pressure in the bins can drop below their minimum vacuum level and cause damage to the shell. Dynamic interactions between the various components of the system including the blower with its VFD drive, purge and blend air compressors, capacitances and resistive characteristics of the pellets in the bins, as well as the response of the control system all contribute to interesting dynamic interactions that must be attended to. A model capable of simulating various dynamic phenomena has been built using Aspen HYSYS Dynamics TM v7.3 for the entire system, and used to answer what-if operational scenarios. As an example it was shown that the addition of a pull blower with its variable frequency drive (VFD), cannot maintain the bin pressure above the vacuum pressure limit of -0.1 psig even with the installation of individual pressure vacuum relief valve (PVRV) on each bin. Any interruption to air flows such as the shut-down of blend blower(s), stop/start conveying or swinging the purge air blower poses a non-negligible risk. Interaction between the PVRV's and the blower's VFD both simultaneously trying to control the bin pressure could lead to pressure fluctuations that exceed the design vacuum pressure of the bins. This paper will provide an overview of the system, state the design and operational challenges, basis of the dynamic model, and results of several what-if scenarios.

# 351659 Air Quality Impact of the Startup of a Single Olefin Plant

**Abstract:**

Ground-level ozone is one of six common pollutants defined by US EPA (Environmental Protection Agency). It can cause health effects, including lung diseases. HRVOC (Highly Reactive Volatile Organic Compounds) emissions from industries are highly responsible for ground-level ozone pollution, especially around industrial district. Flare emissions from MSS (maintenance, startup and shutdown) & events are major sources of HRVOC emissions.

Most of air quality modeling treats industrial flares as constant emissions, which means the emissions keep constant for the whole scenario days, and the total amount of emissions equal to the total emissions reported to environmental authorities. However, real flare emissions from MSS and events are time-various. At certain time, flare emissions could be very huge, and at other time, flare emissions could be very small. During emission peaks, the ozone formation could be much higher than a flat emission curve.

In this paper, time-various flare emissions of startup procedures are considered for air quality modeling. Emission data are obtained from dynamic simulation of a typical ethylene plant. The dynamic simulation results were validated with real plant data and emission data of startup events. The time-various flare emission data were added into Houston-Galveston-Brazoria (HGB) 2006 episode base case modeling as a point emission source. Three startup scenarios were simulated, which include a baseline scenario and two flare reduction scenarios. For each scenario, 120 cases were built to simulate different starting time covering the whole episode days.

Simulation results show that for the baseline scenario, the maximum 8-hour average ozone increment could be 15 ppb at certain starting time. It means starting a single plant could temporary increase the regional ozone concentration by

15 ppb, which is 20% of current ozone standard (75 ppb). For the two flare reduction scenarios, the maximum 8-hour average ozone increment could also reach 10 ppb. Simulation results also show that ozone increment is high when starting time is before 8 am. Ozone increment is low when starting time is after 12 pm. For each scenario, air quality simulation gives least impact period, which means ozone impact could be low when starting time is within that period. For the baseline scenario, the least impact period is only 2 hours. For the two flare reduction scenarios, the least impact periods are 4 hours and 8 hours.

In conclusion, dynamic simulation and air quality simulation show that the flare emissions from MSS and events are highly time-various, and may not be treated as flat emissions. Air quality simulations also give least impact periods of startup time, which could help chemical plants to arrange their startup schedule to reduce ozone formation.

# Hydraulic Fracturing

## Water Reuse Challenges & Solutions

Todd Langford
GE Power & Water
Water & Process Technologies
UC Fuels Domain Leader
C: (216) 548-1936
E: todd.langford@ge.com

imagination at work

# Hydraulic fracturing

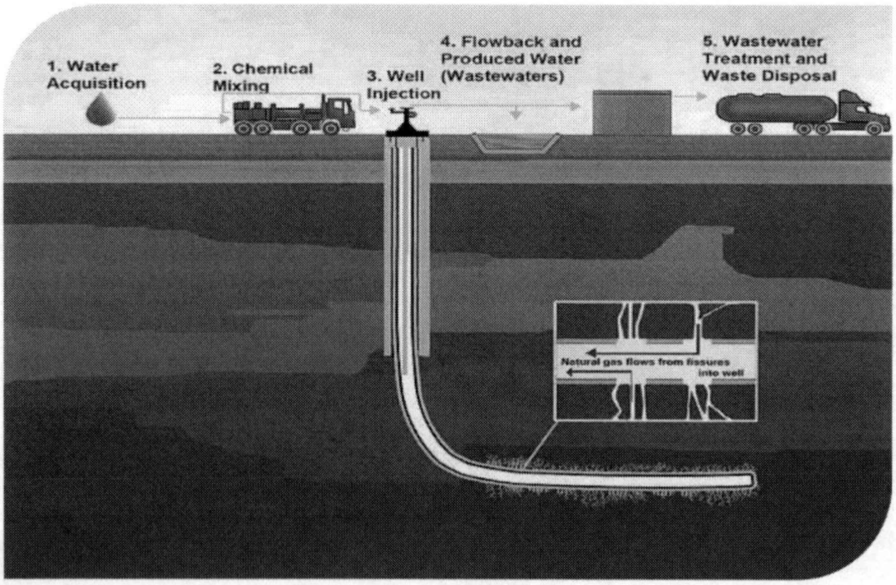


imagination at work

Source: US EPA

2
Langford
March 2014

# Shale gas *Evolution*

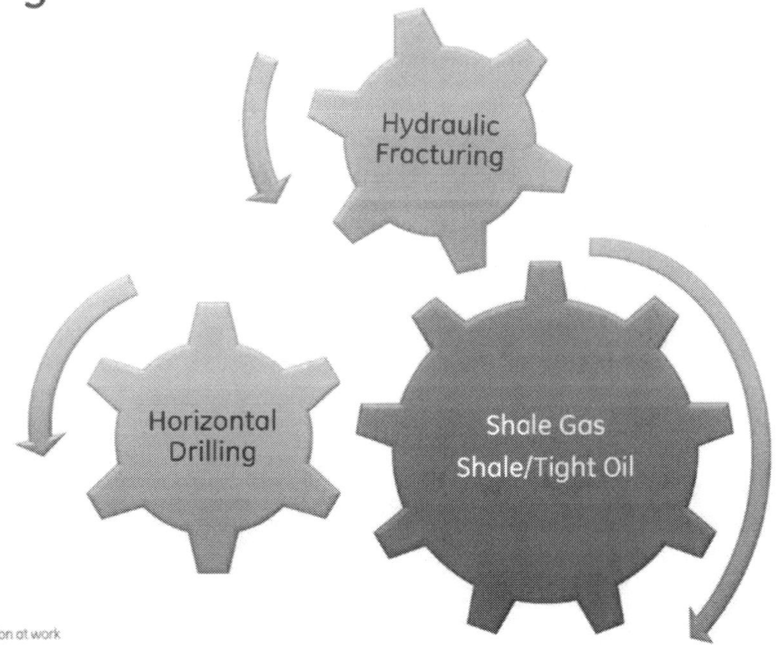


imagination at work

3
Langford
March 2014

# Active shale and tight oil plays

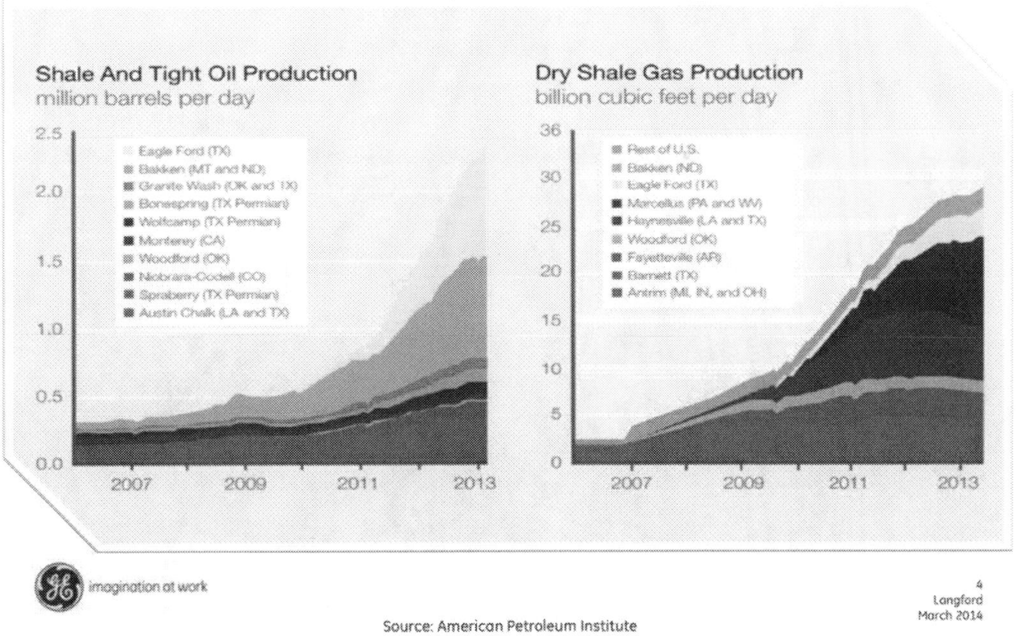


imagination at work

Source: American Petroleum Institute

4
Langford
March 2014

## Energy demands on water resources

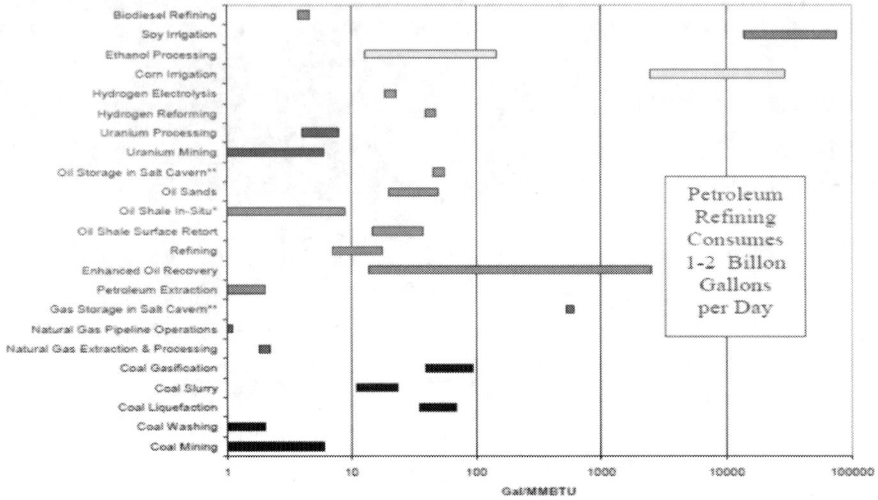


**Figure V-4. Water Consumption Per-Unit-Energy and Current Water Use for Fuel Extraction and Processing**
See Appendix B for Data References

imagination at work

Source: U.S. Department of Energy, 2006
Report to Congress: Interdependency of Energy and Water

5
Langford
March 2014

## What's driving recycle/reuse?

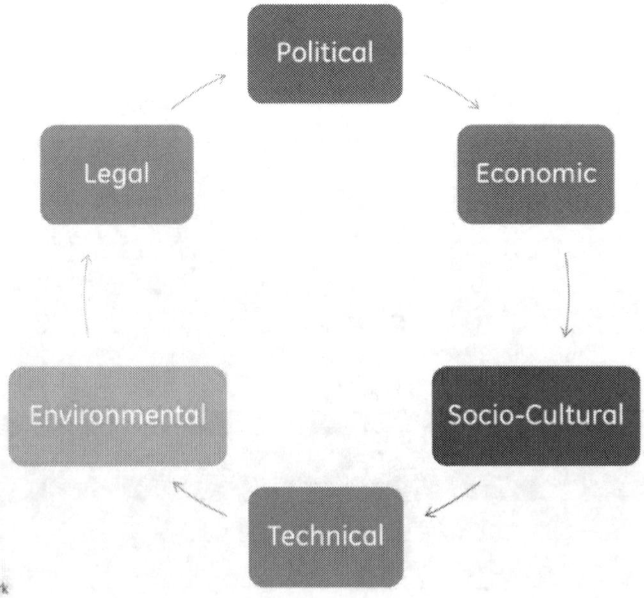


imagination at work

6
Langford
March 2014

# Enviro concerns in unconventional O&G

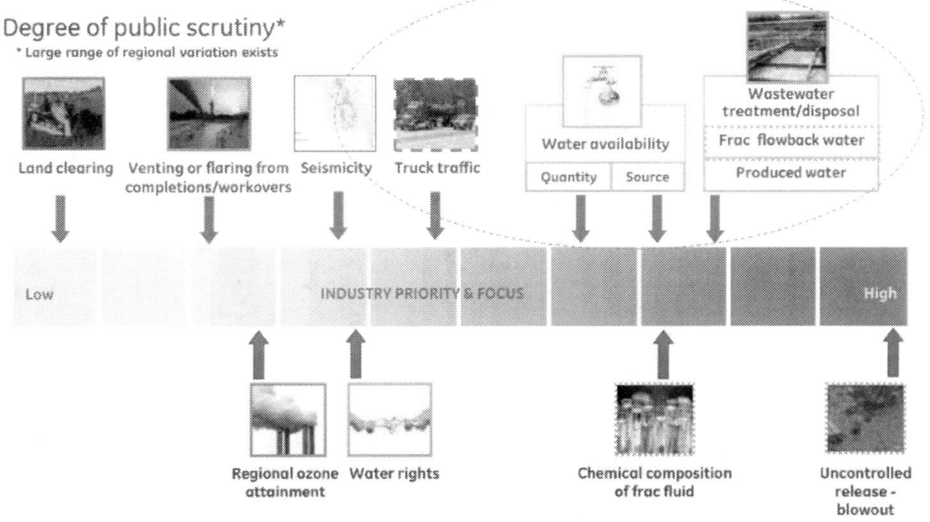


**Water issues elevated in national debate**
**Water issues are LOCAL**

imagination at work

7
Langford
March 2014

Source: GE Energy, Global Strategy & Planning; based on customer discussion not a statistical survey

# Key drivers for projects to move forward

**Regulatory** *...limited disposal + discharge standards*

**Economic** *...expensive trucking to disposal*

**Sustainable** *...multi-generational planning*

imagination at work

8
Langford
March 2014

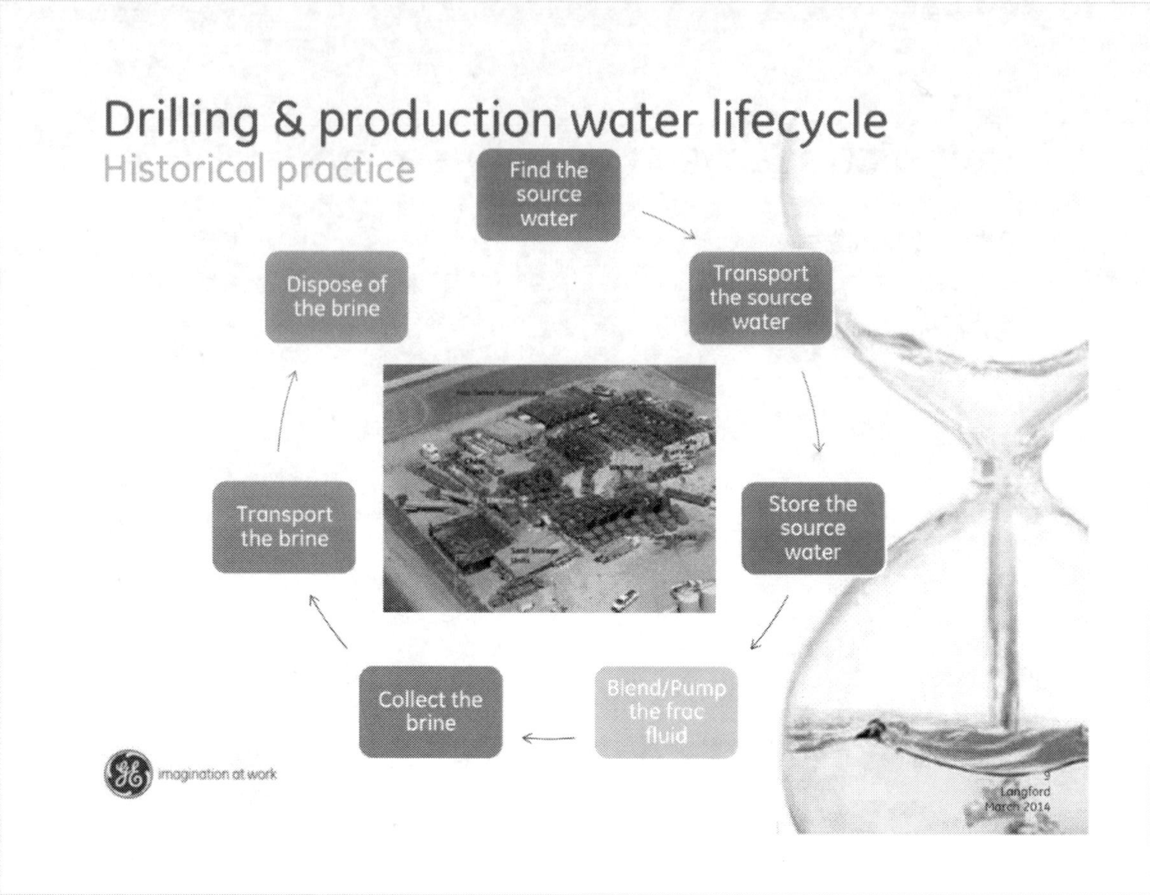
Drilling & production water lifecycle
Historical practice
Find the source water
Transport the source water
Store the source water
Blend/Pump the frac fluid
Collect the brine
Transport the brine
Dispose of the brine
imagination at work
9
Langford
March 2014

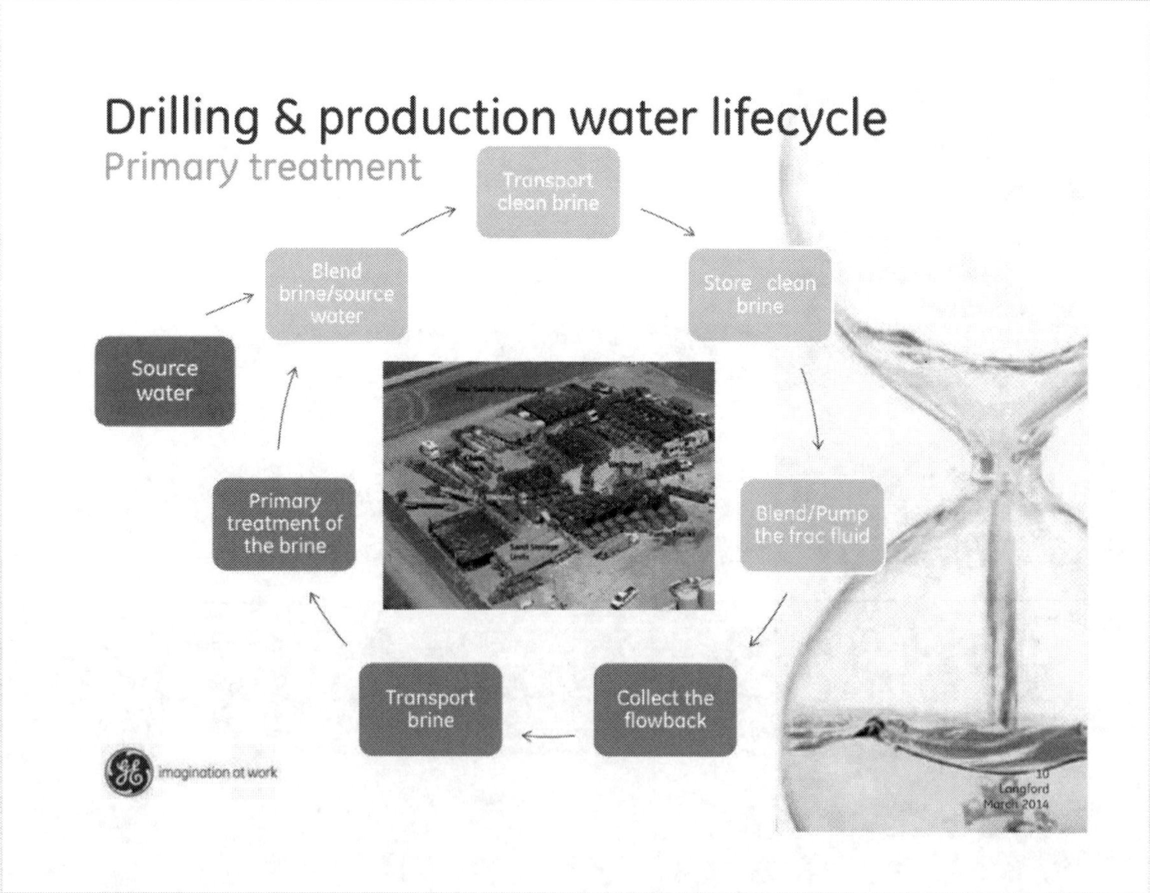
Drilling & production water lifecycle
Primary treatment
Transport clean brine
Store clean brine
Blend/Pump the frac fluid
Collect the flowback
Transport brine
Primary treatment of the brine
Source water
Blend brine/source water
imagination at work
10
Langford
March 2014

# Drilling & production water lifecycle

## Advanced treatment

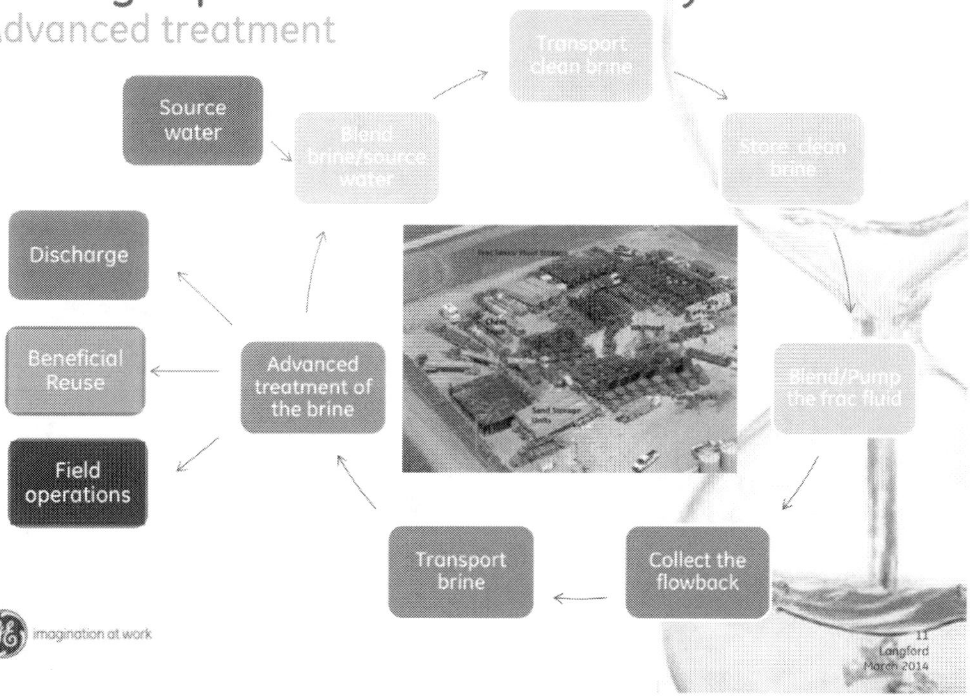


imagination at work

11
Langford
March 2014

imagination at work

12
Langford
March 2014

Source: FracFocus.org

# Flowback/produced water economics

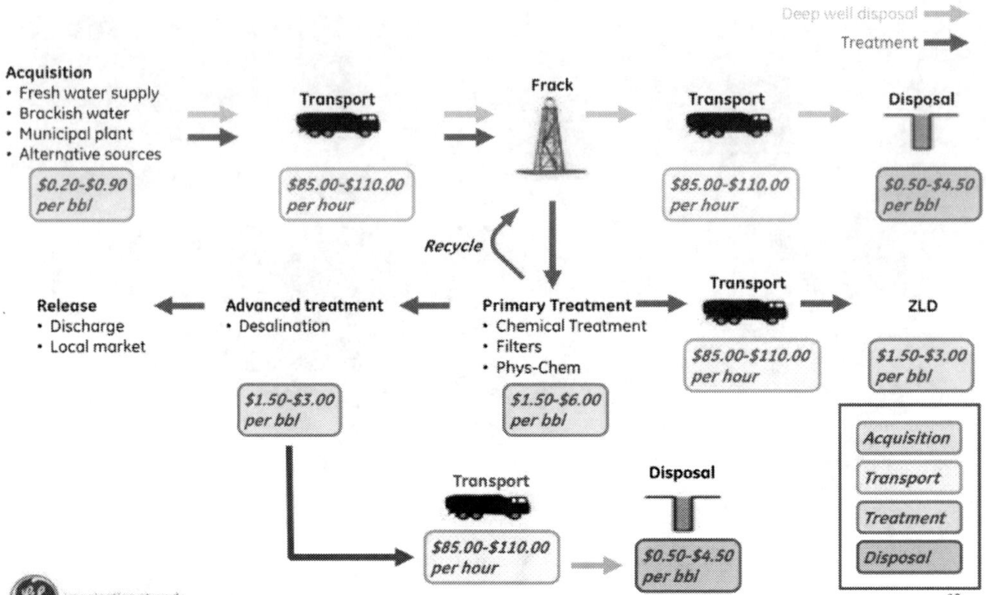


imagination at work

13
Langford
March 2014

# Water spend

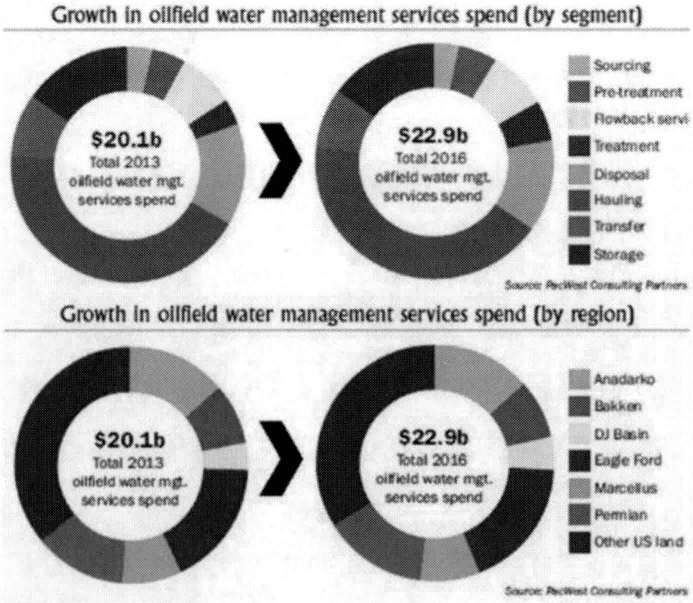


imagination at work

Source: PacWest Consulting Partners/GWI March 2014

14
Langford
March 2014

# Example frac flowback & produced water

**(SINGLE WELL)**

| Day | Percent of Fracture | Barrels |
|---|---|---|
| 1 | 7.56% | 6124.0 |
| 2 | 7.56% | 6119.0 |
| 3 | 7.50% | 6071.4 |
| 4 | 7.44% | 6023.8 |
| 5 | 7.26% | 5881.0 |
| 6 | 4.56% | 3690.5 |
| 7 | 2.88% | 2333.3 |
| 8 | 1.21% | 976.2 |
| 9 | 0.88% | 714.3 |
| 10 | 0.65% | 523.8 |
| 11 | 0.44% | 357.1 |
| 12 | 0.35% | 285.7 |
| 13 | 0.26% | 214.3 |
| 14 | 0.18% | 142.9 |
| 15 | 0.12% | 95.2 |
| 16 | 0.07% | 57.1 |
| 17 | 0.04% | 28.6 |
| 18 | 0.04% | 33.3 |
| 19 | 0.03% | 23.8 |
| 20 | 0.03% | 26.2 |
| 21 | 0.04% | 28.6 |
| 22 | 0.04% | 29.8 |
| 23 | 0.03% | 27.4 |
| 24 | 0.03% | 27.1 |
| 25 | 0.04% | 30.5 |
| 26 | 0.04% | 28.6 |
| 27 | 0.04% | 33.3 |
| 28 | 0.03% | 23.8 |
| 29 | 0.08% | 61.9 |
| 30 | 0.04% | 28.6 |
| 31 | 0.04% | 29.8 |

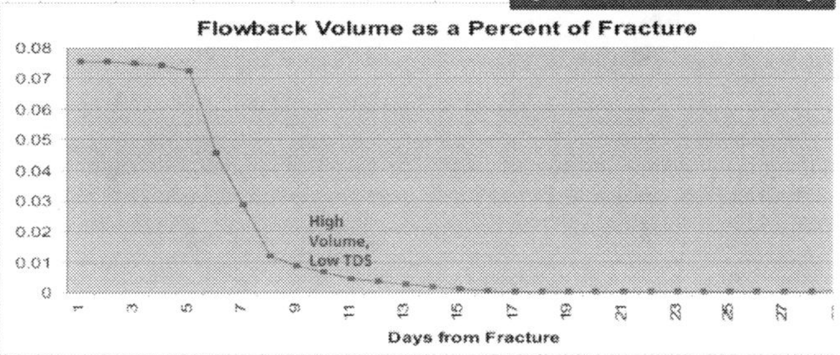


| Day | Specific Gravity | Ionic Strength | TDS | pH | Ca (mg/L) | Mg (mg/L) |
|---|---|---|---|---|---|---|
| 1 | 1.002 | 0.0083 | 559.09 | | 40.00 | 3.72 |
| 1.5 | 1.005 | 0.1025 | 6437.60 | | 379.75 | 31.26 |
| 2 | 1.009 | 0.2067 | 12773.03 | | 935.18 | 77.34 |
| 3 | 1.017 | 0.3727 | 23818.18 | | 1312.51 | 101.69 |
| 4 | 1.028 | 0.6434 | 40803.17 | | 2447.30 | 190.57 |
| 5 | 1.031 | 0.7192 | 45758.81 | | 2679.66 | 209.47 |
| 12 | 1.040 | 0.9371 | 59633.81 | 5.9 | 4395.06 | 381.2 |
| 15 | 1.052 | 1.2300 | 78681.52 | | 4462.10 | 372.03 |
| 21 | 1.102 | 2.4501 | 156570.39 | | 10626.19 | 872.28 |
| 25 | 1.111 | 2.6452 | 170014.09 | | 12084.61 | 1013.11 |
| 31 | 1.121 | 2.9331 | 185728.6 | | 14575.36 | 1141.79 |

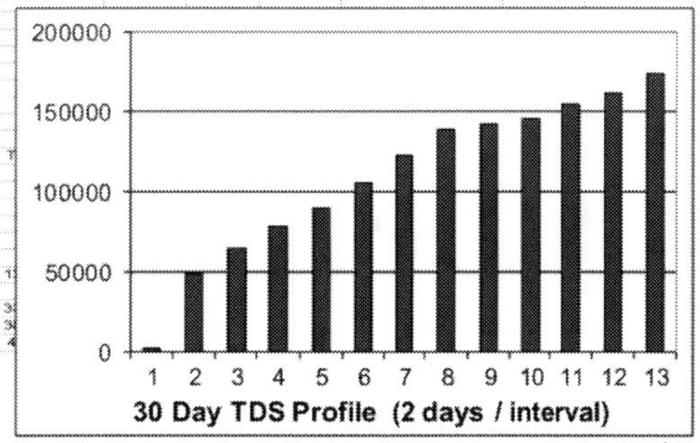


imagination at work

Source: GE Water & Process Technologies



# Multiple wells...

| Assumptions | |
|---|---|
| # wells | 250 |
| Days drilled apart | 15 |
| Volume per frack (gal) | 4,000,000 |
| Volume per frack (bbl) | 95,238 |
| % returned on days 1-5 | 5.00% |
| % returned on day 6-7 | 2.50% |
| % returned on day 8-9 | 1.00% |
| % returned on day 10-30 | 0.25% |
| % returned on day 30-5475 | 0.005% |
| % that returns over time | 64% |

## No reuse + fresh water

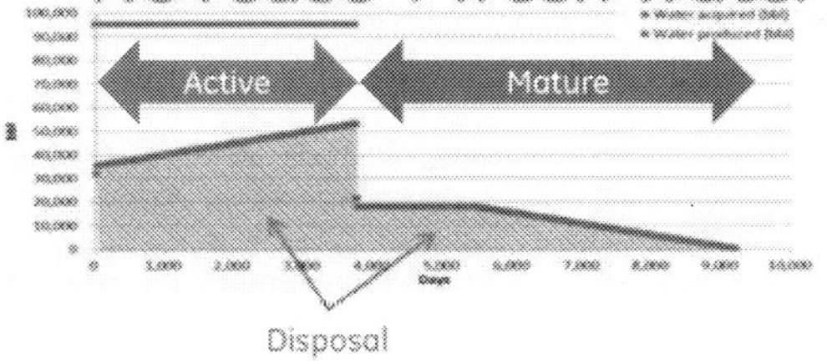


## Reuse + fresh water

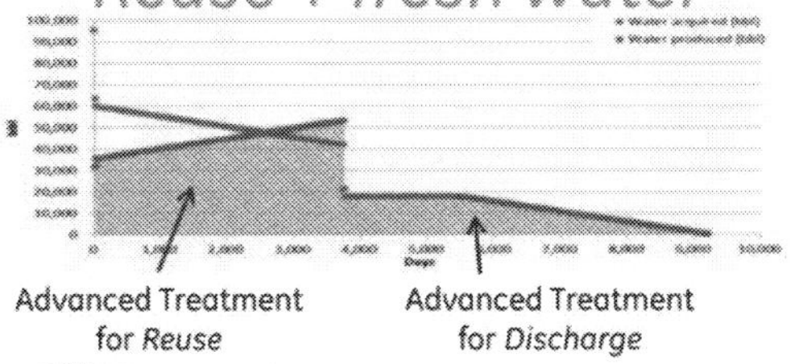


## Reuse + clean brine

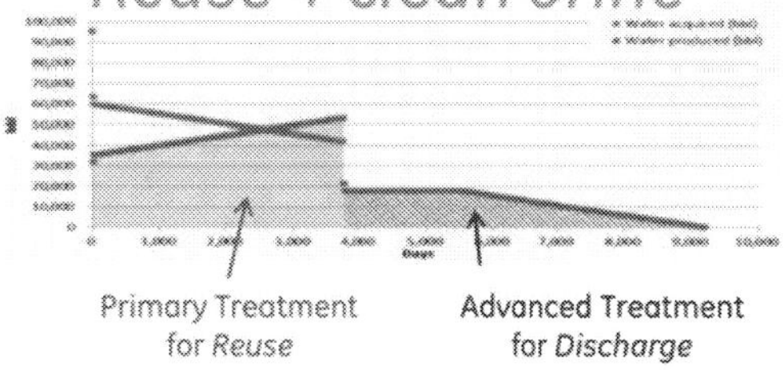


imagination at work

# Recycle/reuse items to consider

**Target quality**
- Fracture fluid
- Discharge

**Source quality**
- Fresh water
- Flowback water
- Produced water
- Blended quality

**Process issues**
- Varying feed quality
- Varying feed quantity
- System mass balance
- System mobility
- System footprint
- Power supply
- Operations
- EHS
- Permitting

**Economics**
- Storage
- Transport
- Processing charge
- Process waste disposal
- Beneficial reuse

imagination at work

17
Langford
March 2014

For illustration only

# Customer economics to determine value

Use Case
Inputs & assumptions in blue

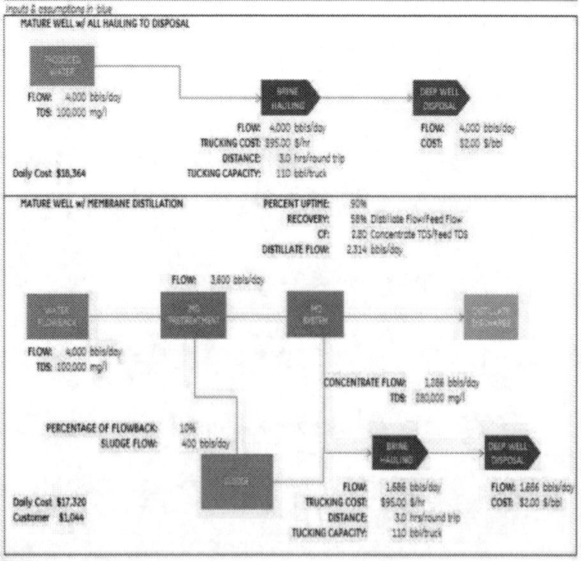


Summary

| | | | |
|---|---|---|---|
| **Annual customer savings** | | | **$381,065** |

Disposal Pathway

| Hauling & disposal | | | |
|---|---|---|---|
| Brine hauling cost | $/bbl | $2.59 | |
| + Deep well disposal cost | $/bbl | $2.00 | |
| = Total hauling & disposal cost | $/bbl | $4.59 | |
| x BBL sent to disposal | bbl/day | 4,000 | |
| **= Daily disposal costs** | | | **$18,364** |

VS.

Treatment Pathway

| Treatment | | | |
|---|---|---|---|
| Volume to treat | bbl/day | 4,000 | |
| x Cost to treat | $/bbl | $2.26 | |
| = Daily treatment cost | | | $9,039 |

| Distillate (revenue)/cost | | | |
|---|---|---|---|
| Volume of distillate | $/bbl | 2,083 | |
| x Distillate (value)/cost | $/bbl | ($0.25) | |
| = Total distillate (revenue)/cost | $/bbl | | ($520.71) |

| Hauling & disposal cost | | | |
|---|---|---|---|
| Pretreatment brine | bbl | 360 | |
| + Concentrate flow | bbl | 1,157 | |
| + Average daily volume caused by downtime | bbl | 400 | |
| = Average daily hauling & disposal volume | bbl | 1,917 | |
| x Total hauling & disposal cost | $/bbl | $4.59 | |
| = Average annual hauling & disposal costs | | | $8,801 |
| **= Average daily cost-to-treat (w/ disposal)** | | | **$17,320** |

imagination at work

18
Langford
March 2014

# Technology

| | Primary Treatment | | | Advanced Treatment |
|---|---|---|---|---|
| | Preliminary | Secondary | Tertiary | Desalination |
| Application | • TSS<br>• Free/suspended oil<br>• Bacteria | • TSS<br>• Emulsified oil<br>• Metals | • Dissolved organics | • Dissolved solids |
| Technology | • Collection/Decant<br>• API Separator<br>• Hydrocyclone<br>• Clarifier/Thickener | • Phys-Chem<br>• MMF<br>• Ultrafiltration<br>• Walnut shell<br>• Electrocoagulation<br>• DAF, EAF<br>• Ceramic membranes | • MBR<br>• GAC<br>• Optipore (DOW)<br>• Biological Treatment<br>• Advanced Oxidation | • EDR<br>• BWRO<br>• SWRO<br>• HERO<br>• Membrane Distillation*<br>• Evaporation<br>• Crystallization<br>• Forward osmosis<br>• Humidification-dehumidification |

* Commercial scale in development

imagination at work

19
Langford
March 2014

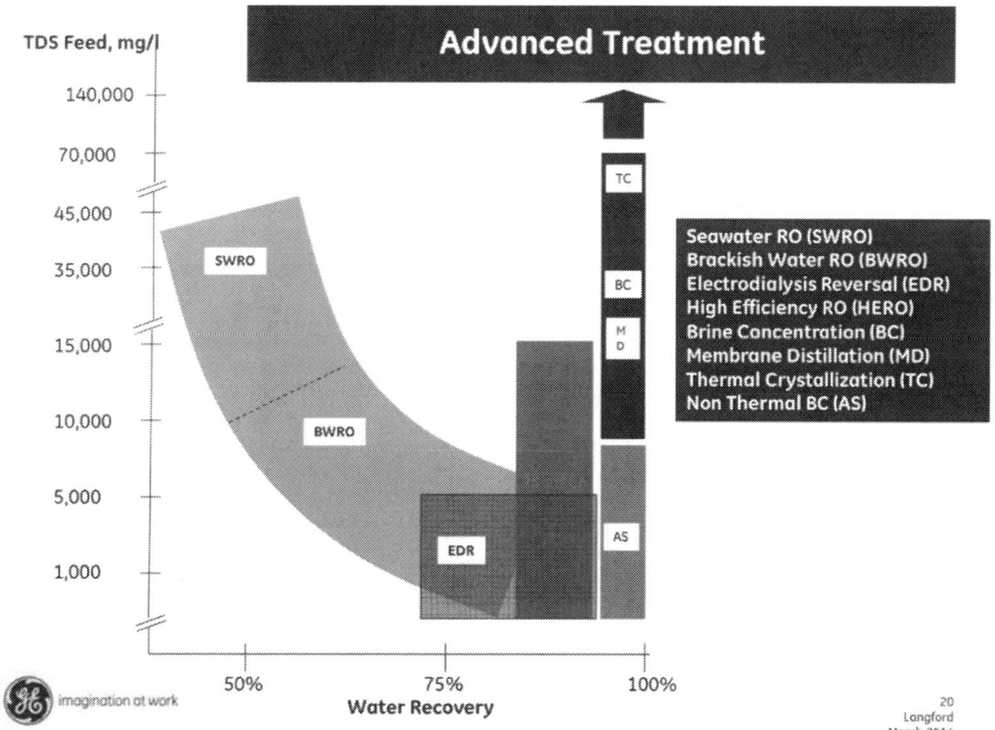


imagination at work

20
Langford
March 2014

## Summary

- Evolving market
- Challenging water
- No well like its neighbor
- No silver bullet
- Evolving solutions

**Need solutions tailored to today and tomorrow**

imagination at work

21
Langford
February, 2014

# 351667 An Integrated Approach to Water Treatment and Recycle in Oil and Gas Production and Processing Via Thermal Membrane Distillation

**Abstract:**

Thermal membrane distillation (TMD) is an emerging technology in the field of water treatment and recycle. It is driven primarily by heat which creates a vapor-pressure difference across a porous hydrophobic membrane. It gained interest because of many advantages; it can work properly with feed temperature below boiling point and can provide high level of separation of the treated water. Because TMD relies on heating the wastewater feed stream, one of the key benefits is that it allows thermal integration with low-quality sources of heat. This thermal coupling can significantly enhance the treatment and recycle costs. The paper presents an integrated approach to the design of TMD networks. First, TMD is modeled based on key design and operating variables. Next, a framework is developed for cost-effective thermal coupling. Several applications in upstream production and downstream processing of oil and gas are presented. The results are compared with conventional separation techniques and with systems lacking heat integration.

# SIMULATION AND SUSTAINABILITY ANALYSIS FOR PRODUCED WATER TREATMENT DRIVEN BY ON-SITE WASTE HEAT

Liwen Chen[1], Parag Saxena[1], Helen H. Lou[1], Lily Xu[2] and Danny Reible[3]
[1.] Dan F. Smith Department of Chemical Engineering, Lamar University
[2.] Environmental and Water Resources Engineering, University of Texas at Austin
[3.] Department of Civil & Environmental Engineering, Texas Tech University



## Topics

- Produced Water Management
- On-site energy source
- Simulation of several thermal evaporation technologies (MSF, MED, MD) in Aspen Plus
- Sustainability analysis in terms of performance, economics and environmental criteria

3

## What's Produced Water?

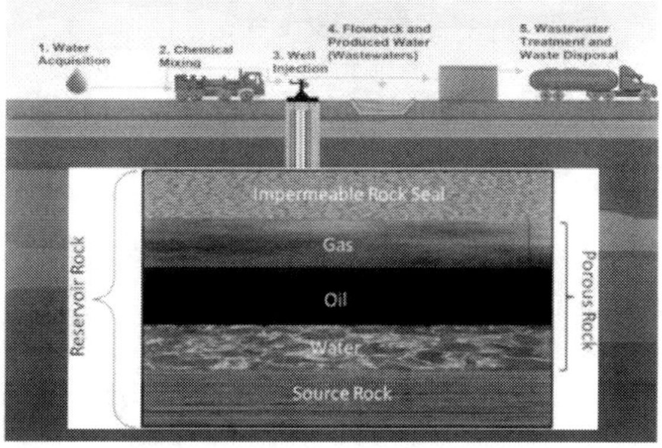


- Water that comes to the surface with oil and gas;
- Rich components (Salt, Oil and grease, Chemical additives, NORMs, HCs, etc);
- Largest volume waste stream from oil and gas production;
- Ratio of water-to-oil in the production
  - World-wide estimate – 2:1 to 3:1
  - U.S. estimate – 7:1
  - Many older U.S. wells have ratios > 50:1!!!

source: http://www2.epa.gov/hfstudy/hydraulic-fracturing-water-cycle; E.T. Igunnu and G.Z. Chen Produced Water Treatment Technologies, International Journal of Low-Carbon Technologies 2012.

4

## PW Composition

√ Salts, Medals,
√ Oil, Grease, HCs
√ Radioactive materials
...

**Total Dissolved Solids (TDS): 8000 ~ 360,000 ppm**

| Parameter | Minimum value | Maximum value | | Heavy metal | Minimum value (mg/l) | Maximum value (mg/l) | |
|---|---|---|---|---|---|---|---|
| Density (kg/m$^3$) | 1014 | 1140 | | Calcium | 13 | 25 800 | Cations |
| Conductivity (μS/cm) | 4200 | 58 600 | | Sodium | 132 | 97 000 | Cations |
| Surface tension (dyn/cm) | 43 | 78 | | Potassium | 24 | 4300 | Cations |
| pH | 4.3 | 10 | | Magnesium | 8 | 6000 | Cations |
| TOC (mg/l) | 0 | 1500 | | Iron | <0.1 | 100 | |
| TSS (mg/l) | 1.2 | 1000 | | Aluminium | 310 | 410 | |
| Total oil (IR; mg/l) | 2 | 565 | | Boron | 5 | 95 | |
| Volatile (BTX; mg/l) | 0.39 | 35 | | Barium | 1.3 | 650 | |
| Base/neutrals (mg/l) | — | <140 | | Cadmium | <0.005 | 0.2 | |
| Chloride (mg/l) | 80 | 200 000 | Anions | Copper | <0.02 | 1.5 | |
| Bicarbonate (mg/l) | 77 | 3990 | Anions | Chromium | 0.02 | 1.1 | |
| Sulphate (mg/l) | <2 | 1650 | Anions | Lithium | 3 | 50 | |
| Ammoniacal nitrogen (mg/l) | 10 | 300 | | Manganese | <0.004 | 175 | |
| Sulphite (mg/l) | — | 10 | | Lead | 0.002 | 8.8 | |
| Total polar (mg/L) | 9.7 | 600 | | Strontium | 0.02 | 1000 | |
| Higher acids (mg/l) | <1 | 63 | | Titanium | <0.01 | 0.7 | |
| Phenol (mg/l) | 0.009 | 23 | | Zinc | 0.01 | 35 | |
| Volatile fatty acids (mg/l) | 2 | 4900 | | Arsenic | <0.005 | 0.3 | |
| | | | | Mercury | <0.005 | 0.3 | |
| | | | | Silver | <0.001 | 0.15 | |
| | | | | Beryllium | <0.001 | 0.004 | |

***Rich components, Wide concentration range, High Salinity!***

1% = 10,000 ppm (mg/L)

Source: Igunnu et al. 2012 Produced water treatment technologies

5

# Federal & State Regulations

*Federal:*

- U.S. Environmental Protection Agency (EPA);
- Department of the Interior's Bureau of Land Management (BLM);
- Bureau of Ocean Energy Management, Regulation and Enforcement (BOEMRE, formerly MMS);

EPA requirement for oil and grease in PW (NPDES):

Offshore [40CFR§435.13]:

- Daily Maximum Discharge (EPA): **42** mg/L
- Monthly average limit: **29** mg/L

Onshore [40CFR§435.32]:

- Zero discharge

*State:*

- Regulations varied widely;

For example, Texas Commission on Environmental Quality (TCEQ)

Rule 3.8 (Water Protection);Rule 3.9 (Disposal Wells)

**Require permits to discharge into surface/subsurface water**

source: http://www.netl.doe.gov/research/coal/crosscutting/pwmis/fed-state-regulations

6

# PW Management

*Three hierarchy:*

**Minimization**

- **Prevent PW entering well** (Mechanical blocking, Water shut-off chemicals)
- **Prevent PW reaching the surface** (Downhole separation, Sea floor separation)

**Recycle/reuse**

- **Injection for enhanced oil recovery** (TX: ~38540 wells (5.3 billion bbl/yr))
- **Injection for disposal** (TX: ~12000 wells (1.2 billion bbl/year), PA: 8 wells!)

**Treatment**

- **Dilution** (blending with freshwater) → ~$1.5-2/bbl
- **Settling** (allow enough residence time for solids to settle) → ~$2-2.5/bbl
- **Filtration** (sand filters, bag filters, disk filters) → ~$2-3/bbl
- **Chemical Precipitation** (pH adjustment, flocculants) → ~$2.5-4/bbl
- **Evaporation** (Highest effluent quality with highest energy demand. Energy cost can be mitigated by waste heat) → ~$5.5-8/bbl

Permeate Standard (WHO): TDS ≤ 500mg/L

source: Veil J.A. 2009 Produced Water Management Options –One Size Does Not Fit All

# Gas Flaring

Why Flare
- Safety concerns:
  - ➢ Equipment malfunction;
  - ➢ Off-spec production;
  - ➢ Depressurizing equipment;
  - ➢ Start ups and emergency shut-downs;
- Lack of infrastructure (pipeline)

1/3 flared in Bakken Shale, $100 million/month!

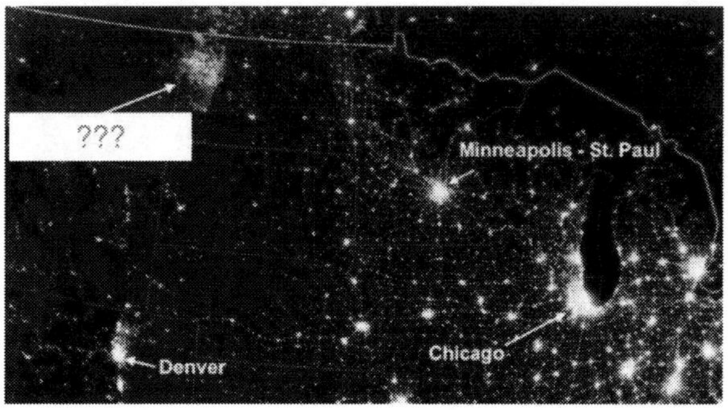


source: http://www.pennenergy.com/articles/pennenergy/2013/07/bakken-shale-natural-gas-flaring-tops-100-million-each-month.html



# Energy Savings

- 5% of global gas production (150 Billion $m^3$/yr)
- 23% US NG use, 30% EU NG use
- 160 MMCFD NG in Bakken Shale flared out of 480 MMCFD production

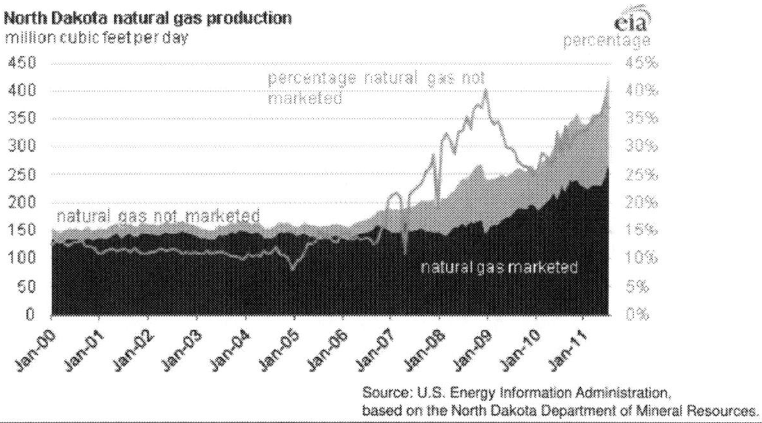


Source: U.S. Energy Information Administration, based on the North Dakota Department of Mineral Resources.

9

## Flare Minimization Strategies

Legislation
- Kyoto Protocol
- Global Gas Flaring Reduction (GGFR);
- Economic Incentives

Reduction
- Reinjection EOR
- Compress to liquid;
- Dynamic Control System

Utilization
- Heat source for co-generation

10

## Energy-Water Nexus Scheme

**Pros:**
Cost becomes profit;
Reduce environmental impacts;
On-site water treatment supply;
…

**Cons:**
Unsteady supply;
Safety issue;
Infrastructure;
Capital cost;
…

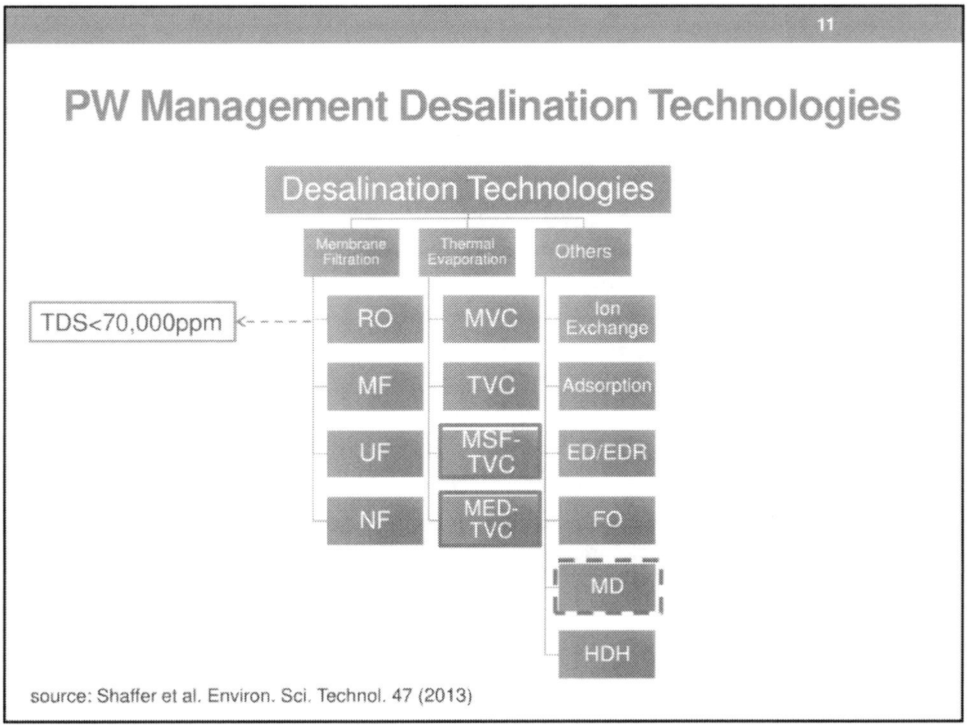
11
PW Management Desalination Technologies
Desalination Technologies
Membrane Filtration
Thermal Evaporation
Others
TDS<70,000ppm
RO
MF
UF
NF
MVC
TVC
MSF-TVC
MED-TVC
Ion Exchange
Adsorption
ED/EDR
FO
MD
HDH
source: Shaffer et al. Environ. Sci. Technol. 47 (2013)

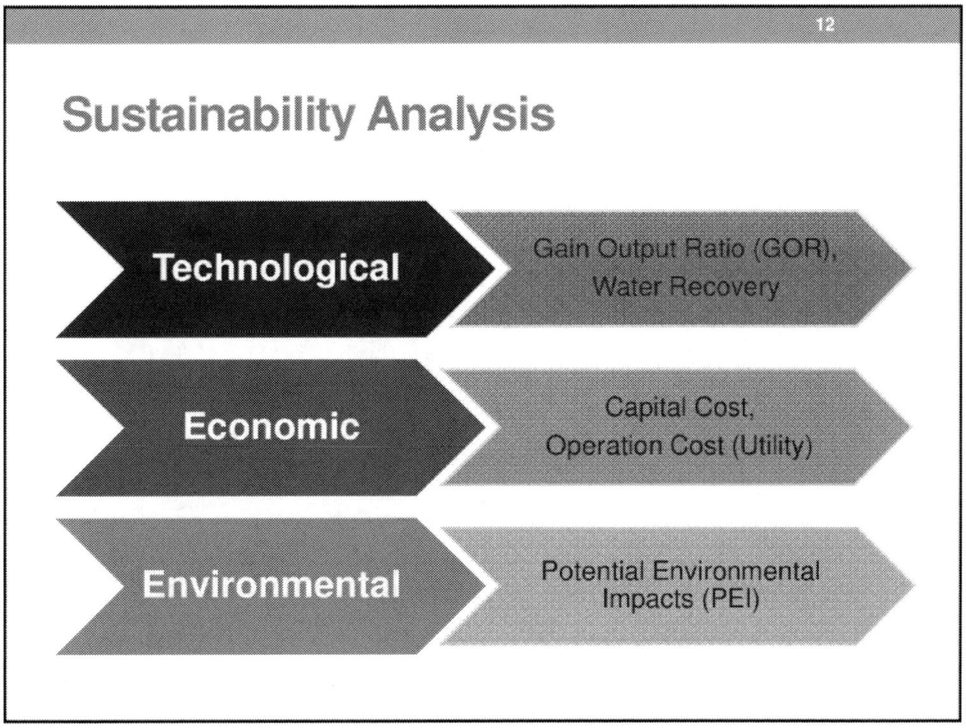
12
Sustainability Analysis
Technological
Gain Output Ratio (GOR), Water Recovery
Economic
Capital Cost, Operation Cost (Utility)
Environmental
Potential Environmental Impacts (PEI)

# Environmental Impact Analysis

- WAR (Waste Reduction) Algorithm
  - Developed by Young & Cabezas (EPA), 1999
  - Database on environmental impacts of chemicals
  - 4 general categories, 8 sub-categories
  - http://www.epa.gov/nrmrl/std/war/sim_war.htm
- PEI (Potential Environmental Impact) index to represent environmental friendliness

  "A low PEI score indicates a lower environmental impact."
- Green house gas emission by $CO_2$ is the major potential environmental impact



# WAR Algorithm: Impact Categories

| General Impact Category | Impact Sub-Category | Measure of Impact Category |
|---|---|---|
| Human toxicity | Ingestion | HTPI |
| | Inhalation/Dermal | HTPE |
| Ecological toxicity | Aquatic toxicity | ATP |
| | Terrestrial toxicity | TTP |
| Global atmospheric impacts | Global warming potential | GWP |
| | Ozone depletion potential | ODP |
| Regional atmospheric impacts | Acidification potential | AP |
| | Photochemical oxidation potential | PCOP |

# Potential Environmental Impact (PEI)

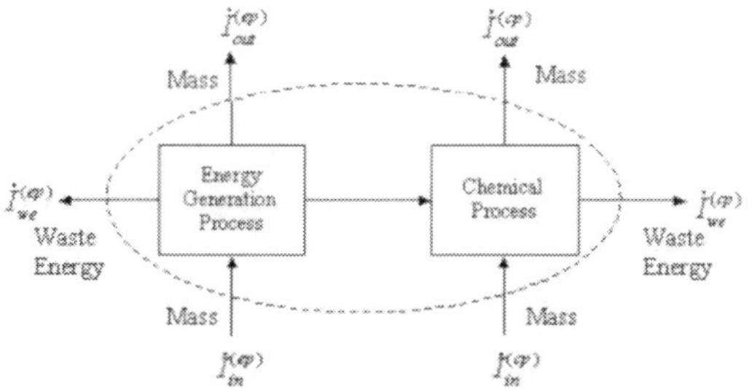


The potential environmental impact balance is as follows:

$$\cancel{\frac{\partial I_s}{\partial t}} = \dot{I}_{in}^{(cp)} + \cancel{\dot{I}_{in}^{(ep)}} - \dot{I}_{out}^{(cp)} - \dot{I}_{out}^{(ep)} - \cancel{\dot{I}_{we}^{(cp)}} - \cancel{\dot{I}_{we}^{(ep)}} + \dot{I}_{gen}^{(t)}$$

$$\boxed{0 = \dot{I}_{in}^{(t)} - \dot{I}_{out}^{(t)} + \dot{I}_{gen}^{(t)}}$$



The total input rate of PEI:

$$\dot{I}_{in}^{(t)} = \sum_{l}^{EnvCat} \alpha_l \dot{I}_{l,in}^{(t)} = \sum_{l}^{EnvCat} \alpha_l \sum_{j}^{Streams} \dot{M}_{j,in} \sum_{k}^{Comps} x_{kj} \psi_{kl}^{s}$$

The total output rate of PEI:

$$\dot{I}_{out}^{(t)} = \sum_{l}^{EnvCat} \alpha_l \dot{I}_{l,out}^{(t)} = \sum_{l}^{EnvCat} \alpha_l \sum_{j}^{Streams} \dot{M}_{j,out} \sum_{k}^{Comps} x_{kj} \psi_{kl}^{s}$$

To evaluate the PEI output index on a production basis,

$$\boxed{\hat{I}_{out}^{(t)} = \frac{\dot{I}_{out}^{(t)}}{\sum_{p}^{ProdStreams} \dot{P}_p}}$$

# Multi-Stage Flash (MSF) Process

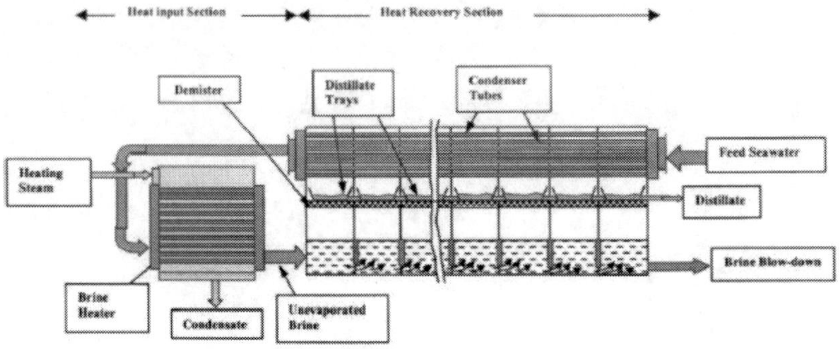


MSF-OT (once-through w/o brine recirculation)

Source: El-Dessouky et al./ Chemical Engineering Journal 1999



# Specification

**Feed Components:** $Na^+$, $K^+$, $Ca^{2+}$, $Mg^{2+}$, $Cl^-$, $HCO_3^-$, $SO_4^{2-}$ ($CaSO_4 \downarrow$)
**Feed Flowrate:** 10,000 ton/day

Three Cases: (Low TDS, Medium TDS, and High TDS)

| Location | Powder River (MT) | Barnett Basin (TX) | Bakken Basin (ND) |
|---|---|---|---|
| Compnents (mg/L) | Low | Medium | High |
| Na+ | 6022 | 20935 | 71466 |
| K+ | 75 | 61 | 512 |
| Ca++ | 327 | 5441 | 8993 |
| Mg++ | 93 | 1734 | 1467 |
| Cl- | 9074 | 45989 | 129461 |
| SO4-- | 1529 | 1394 | 1885 |
| HCO3- | 1329 | 408 | 322 |
| TDS | 18447 | 75962 | 214106 |

Low TDS (Powder River) (<35,000ppm)

Medium TDS (Barnett) (35,000~200,000ppm)

High TDS (Bakken) (200,000~400,000ppm)

Source: US Geological Survey Produced Water Database

# Multi-Stage Flash w/ Thermal Vapor Compression (MSF-TVC) Simulation in Aspen Plus

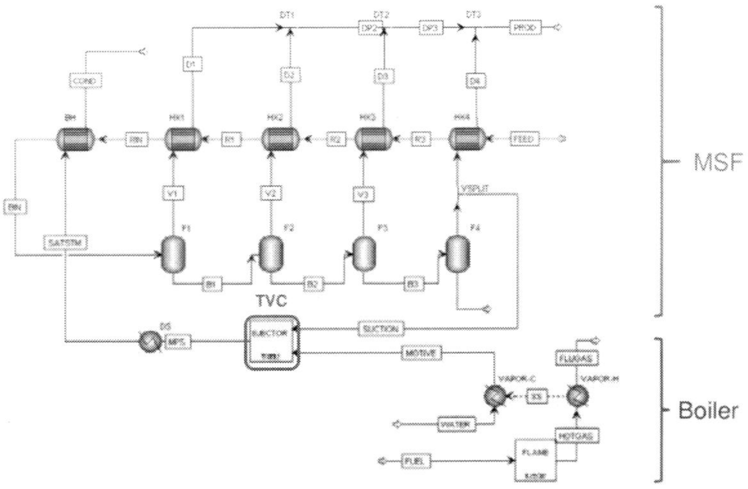




# Simulation Results of MSF-TVC

| MSF-TVC stages | | F1 | F2 | F3 | F4 |
|---|---|---|---|---|---|
| Temperature (°C) | | 90 | 88 | 86 | 84 |
| Pressure (bar) | | 0.671 | 0.618 | 0.568 | 0.521 |
| Vapor (ton/day) | | 970 | 873 | 786 | 707 |
| Prod (ton/day) | | 2840 | | | |
| Steam Supply | (°C) | 133.6 | | | |
| | (bar) | 3 | | | |
| | (ton/day) | 440 | | | |
| GOR (Prod/Steam) | | 6.5 | | | |
| Product Recovery (Prod/Feed) | | 28.4% | | | |
| Energy Consumption (MMBTU/day) | | 1099 | | | |
| Utility Cost ($/day) | | 4130 | | | |
| $CO_2$ Emission (ton/day) | | 62 | | | |

5-year (2009-2013) average Henry Hub NG Price = $3.76/MMBTU (http://www.eia.gov)

# Case Results

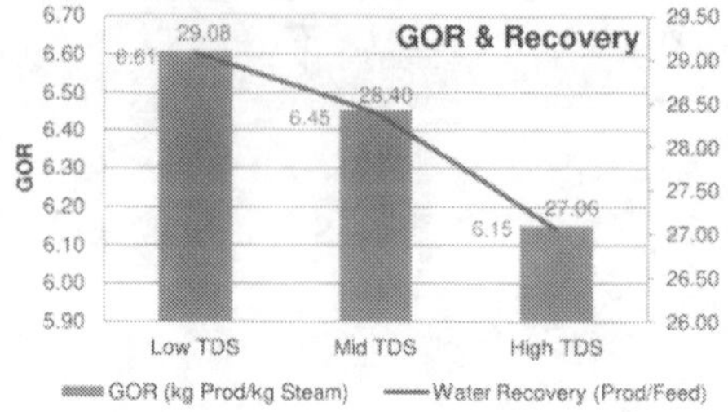


| | Utility Cost |
|---|---|
| $/ton | 1.42-1.53 |
| cents/gal | 0.54-0.58 |
| cents/bbl | 16.94-18.21 |

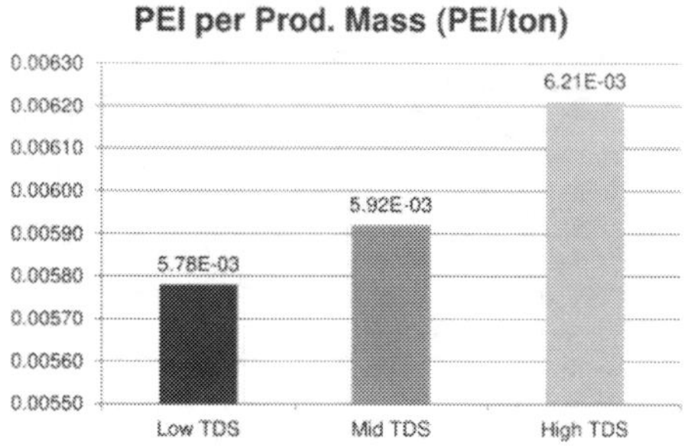


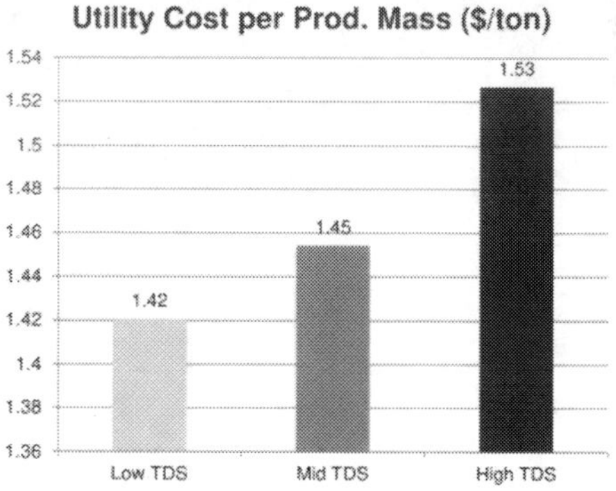




# PEI in each categories (MSF-TVC)

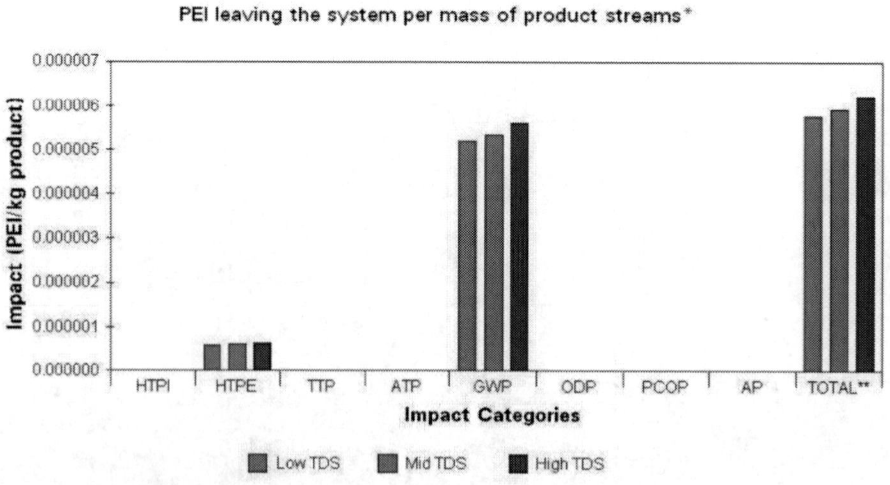


HPTE: Human Toxicity Potential by Inhalation or Dermal Exposure
GWP: Global Warming Potential

# Multi-Effect Distillation (MED)

Features:

- Higher heat transfer coefficient;
- Max Brine Temperature 70°C (avoid scalants);
- Economic and durable material (aluminum alloy vs. copper);
- Possibility of using low grade heat;

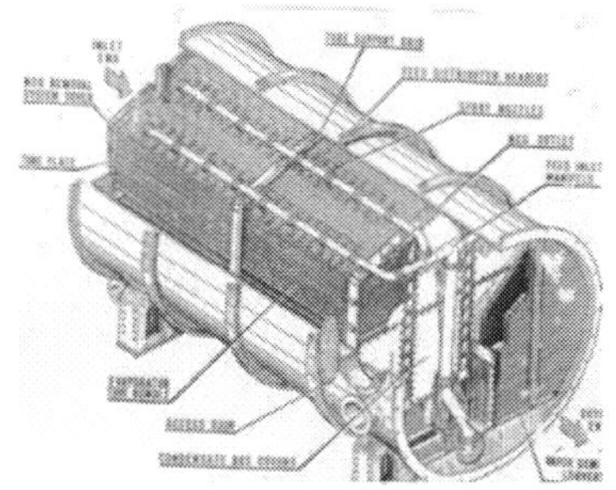

Horizontal tube assembly
A. Ophir, F. Lokiec et.al. Desalination 182 (2005)

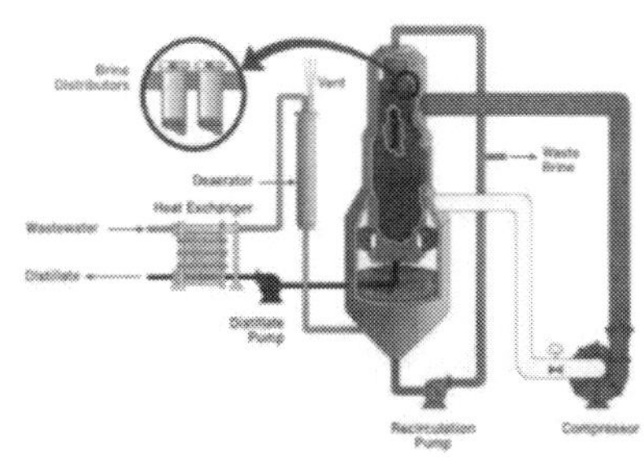

Falling Film Evaporator
(GE Power & Water 2012)



# MED-TVC Schematic Diagram

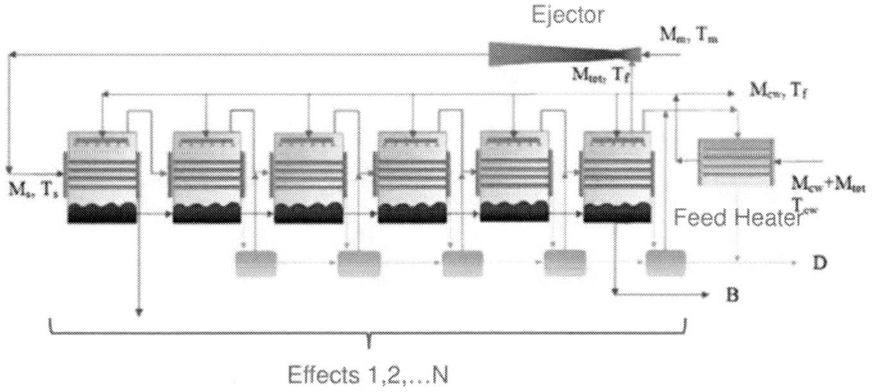


source: Manesh et al. Desalination 316 (2013)

## MED-TVC Simulation in Aspen Plus

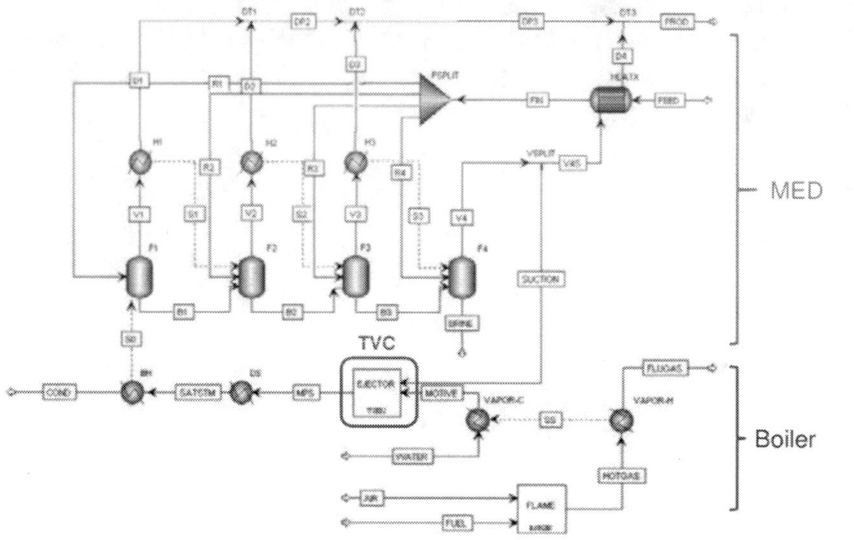




## Results for MED-TVC

| MED-TVC stages | F1 | F2 | F3 | F4 |
|---|---|---|---|---|
| Temperature (°C) | 60 | 56 | 52 | 48 |
| Pressure (bar) | 0.182 | 0.152 | 0.125 | 0.103 |
| Vapor (ton/day) | 1156 | 1099 | 1067 | 1059 |
| Prod (ton/day) | 3640 | | | |
| Steam Supply (°C) | 133.6 | | | |
| Steam Supply (bar) | 3 | | | |
| Steam Supply (ton/day) | 500 | | | |
| GOR (Prod/Steam) | 7.3 | | | |
| Recovery (Prod/Feed) | 36.4% | | | |
| Energy Consumption (MMBTU/day) | 1249 | | | |
| Utility Cost ($/day) | 4693 | | | |
| $CO_2$ Emission (ton/day) | 71 | | | |

5-year (2009-2013) average Henry Hub NG Price = $3.76/MMBTU (http://www.eia.gov)

# MED-TVC Results Summary

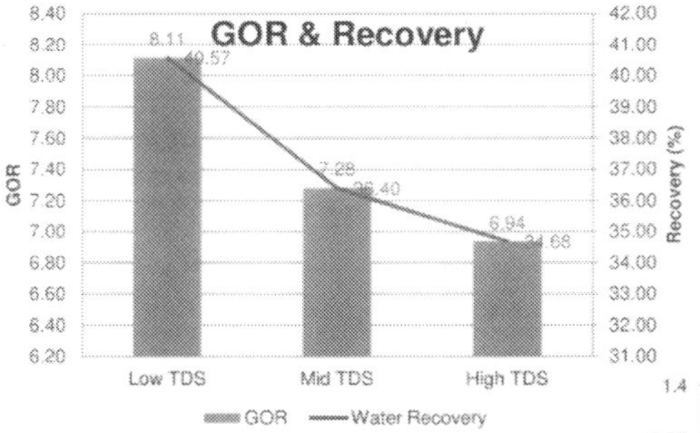


| | Utility Cost |
|---|---|
| $/ton | 1.16-1.35 |
| cents/gal | 0.44-0.51 |
| cents/bbl | 13.80-16.14 |

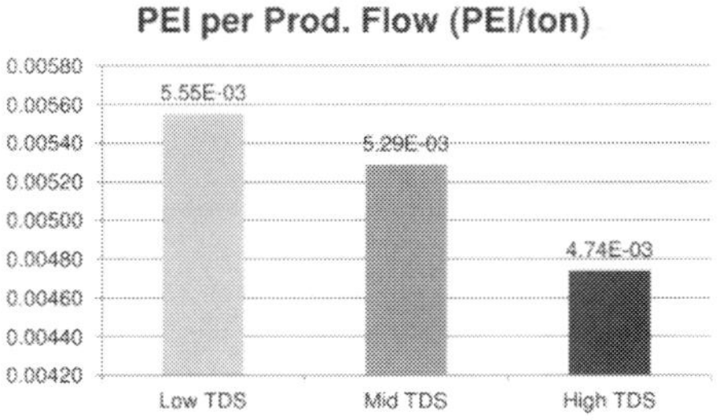


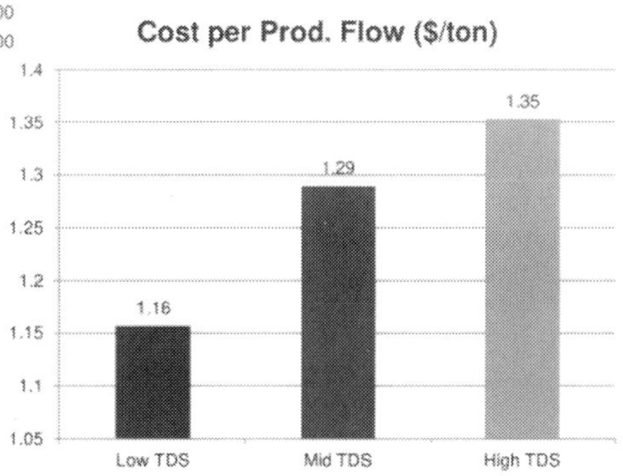




# Membrane Distillation (MD)

Advantages:

- Lower operating temperatures and vapor space than MSF & MED;
- Lower operating pressure than RO;
- 100% rejection of non-volatile solute;
- Performance not limited by high osmotic pressure or concentration polarization;

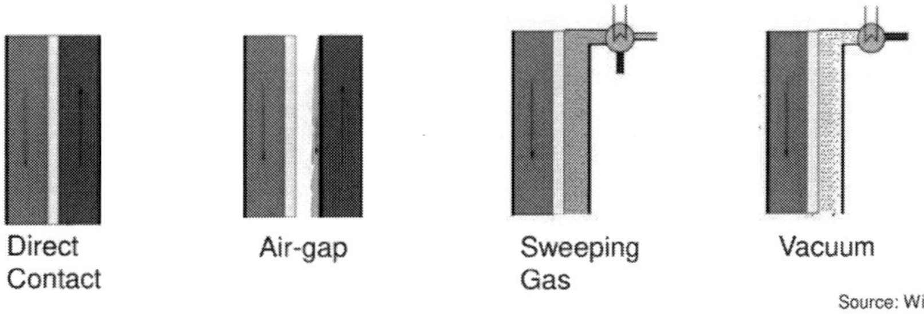


Source: Wikipedia

29

## DCMD Concept

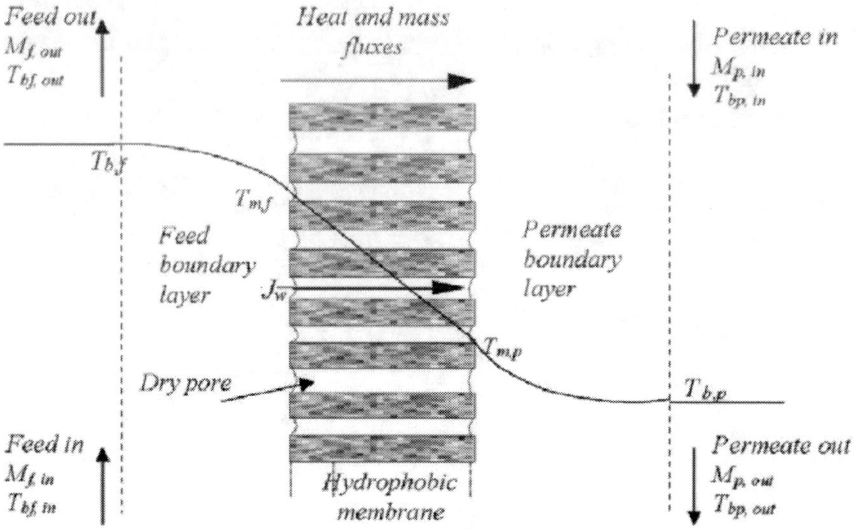


Source: M. Qtaishat et al. / Desalination 219 (2008) 272–292

30

## Simulation of DCMD in Aspen Plus

***15-20 effects for good approximation***

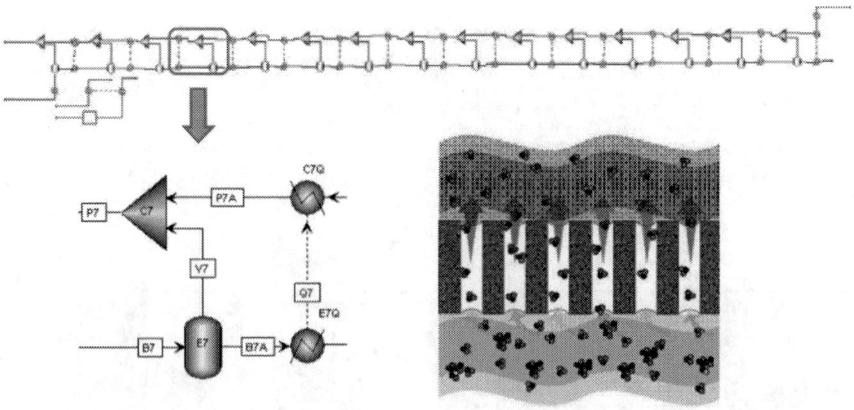


Source: http://www.nap.edu/openbook.php?record_id=12027&page=128

31

## Hot Feed (E1, E2)

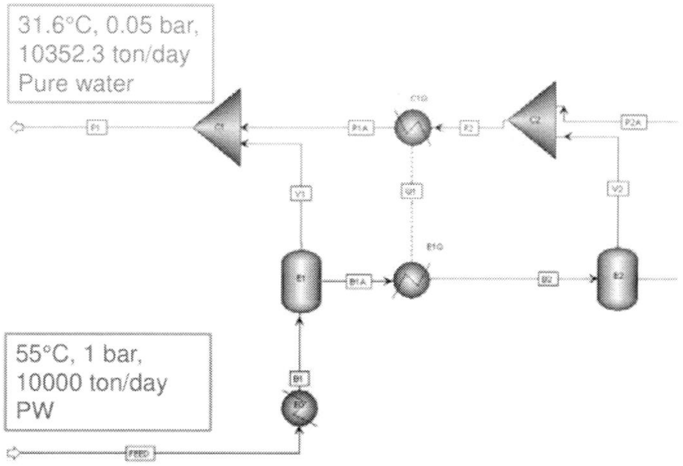


32

## Cold Feed (E14, E15)

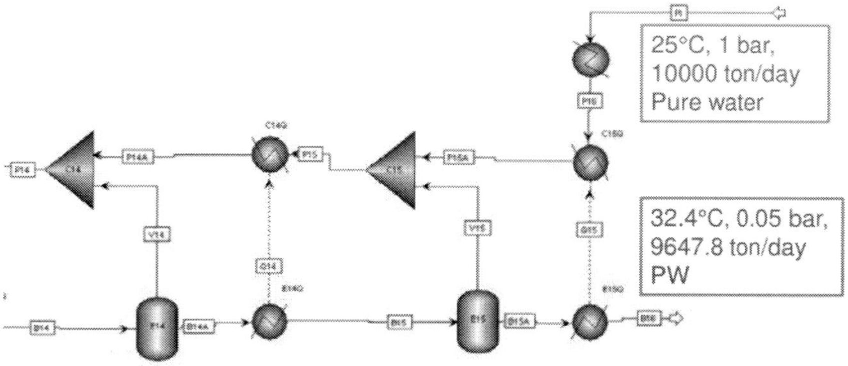

33

## Conclusion Remarks

- All 3 desalination technologies (MSF, MED-TVC and MD) have potentials to treat PW with pros and cons;
- Coupling flared gas as heating source is beneficial but challenging;

34

## Acknowledgements


- Professor Chau-Chyun Chen
  (Department of Chemical Engineering, Texas Tech University)
- Professor John L. Gossage
  (Department of Chemical Engineering, Lamar University)
- Texas Hazardous Waste Research Center
- Lamar University Graduate Scholarship

35
감사합니다
Natick
Grazie
Danke
Ευχαριστίες
Dalu
Obrigado
Thank You
Köszönöm
Tack
Спасибо
Dank
Gracias
谢谢
Merci
Seé
ありがとう

# * DESMAR SOLAR

NATALY AGUDELO
YURANI GUTIÉRREZ
ANDRES OLAYA
FELIPE SASTOQUE

**Energetic and exergetic analisys from a seawater solar desalinator**

DESMAR SOLAR

# *SCHEDULE

- Introduction
- Justification
- Objectives
- Resources available at the university.
- Main processes of water desalination.
- Material and energy balances.
- Preliminary test.
- Environmental considerations.
- Results
- Bibliography.

DESMAR SOLAR

# *INTRODUCTION

*Water is the main element that humans use in their daily work. Of all the water available on the planet only 3% is fresh water and is on the surface. The remaining water is either frozen at the poles or distributed in the oceans.

DESMAR SOLAR

# *JUSTIFICATION

Desalination processes are quite costly and cause pollution to the environment because their main source of energy is at its most fossil fuels.

Therefore, this investigation is intended to use efficient and clean energy sources to develop the process

DESMAR SOLAR

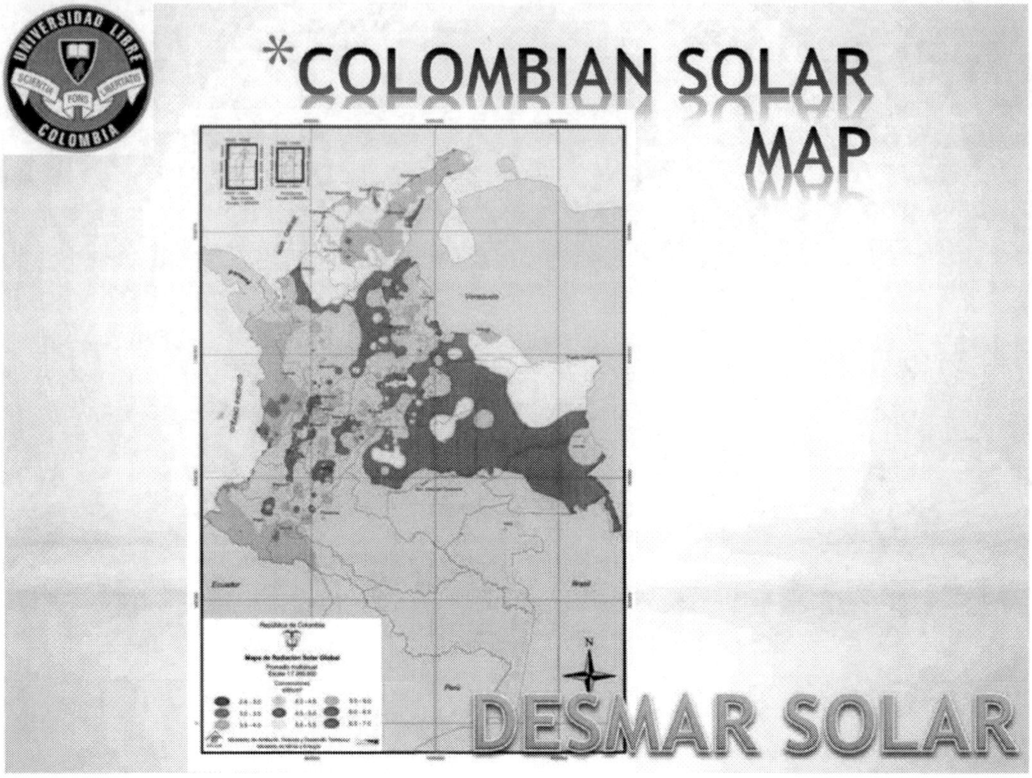
UNIVERSIDAD LIBRE
COLOMBIA
*COLOMBIAN SOLAR MAP
DESMAR SOLAR

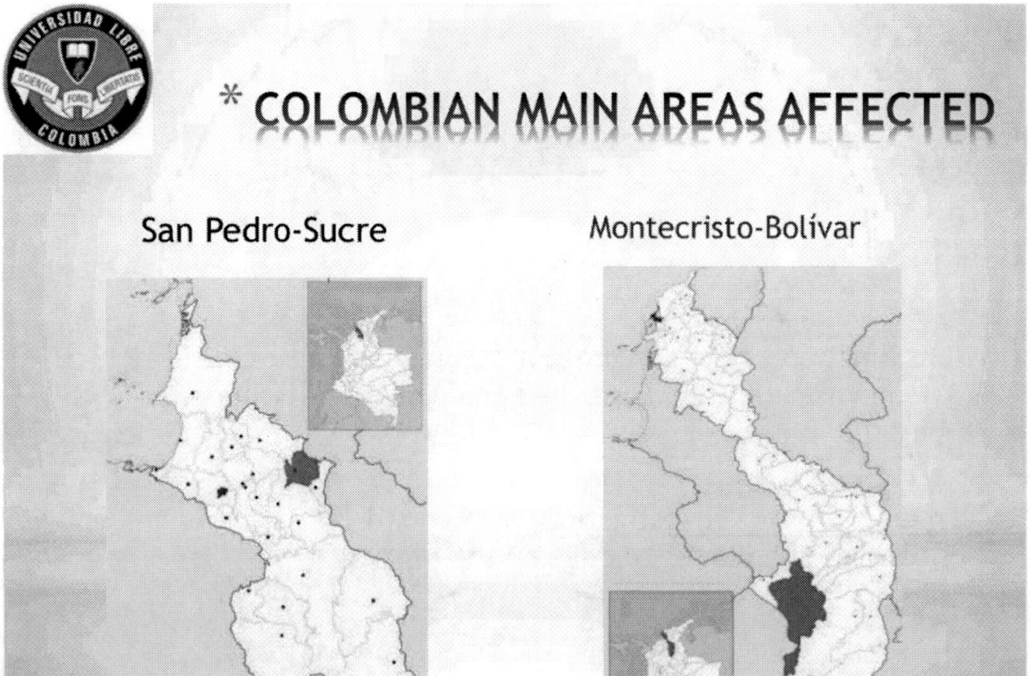
UNIVERSIDAD LIBRE
COLOMBIA
* COLOMBIAN MAIN AREAS AFFECTED
San Pedro-Sucre
Montecristo-Bolívar

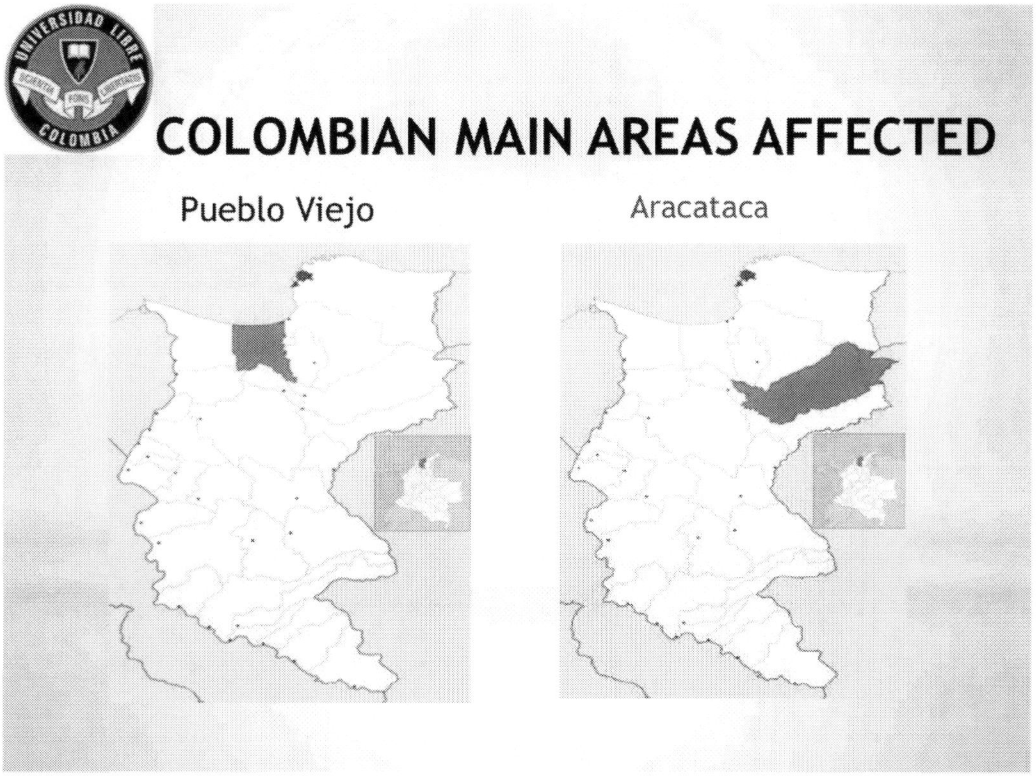
UNIVERSIDAD LIBRE
COLOMBIA
COLOMBIAN MAIN AREAS AFFECTED
Pueblo Viejo
Aracataca

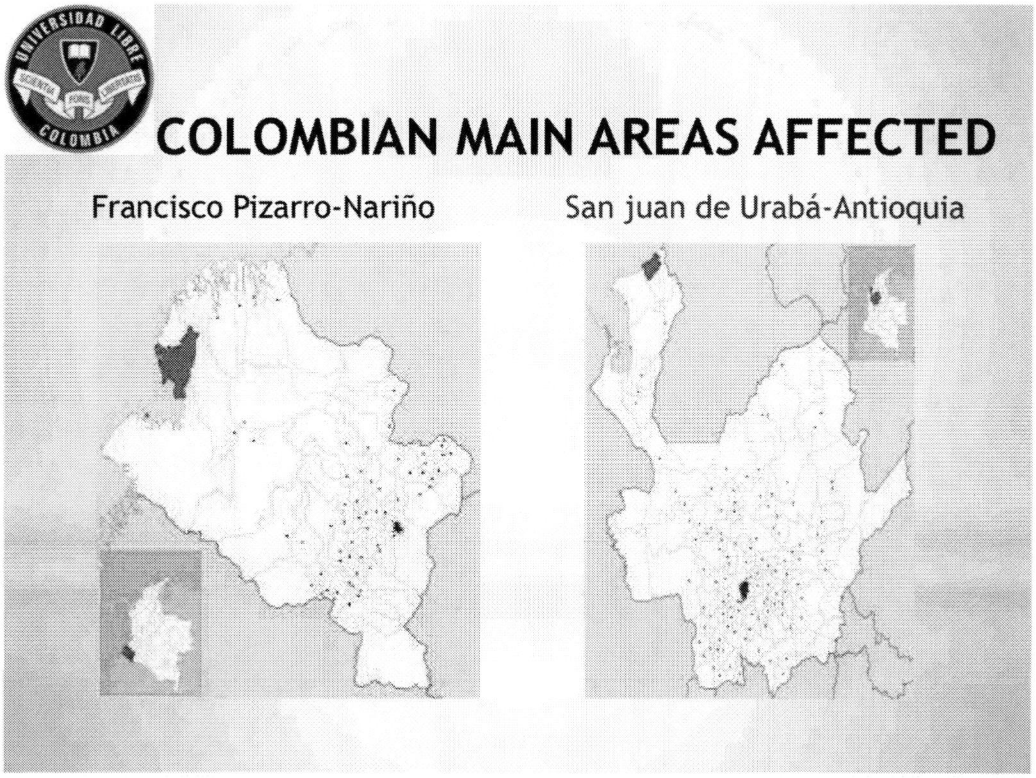
UNIVERSIDAD LIBRE
COLOMBIA
COLOMBIAN MAIN AREAS AFFECTED
Francisco Pizarro-Nariño
San juan de Urabá-Antioquia

## COLOMBIAN MAIN AREAS AFFECTED

Timbiquí

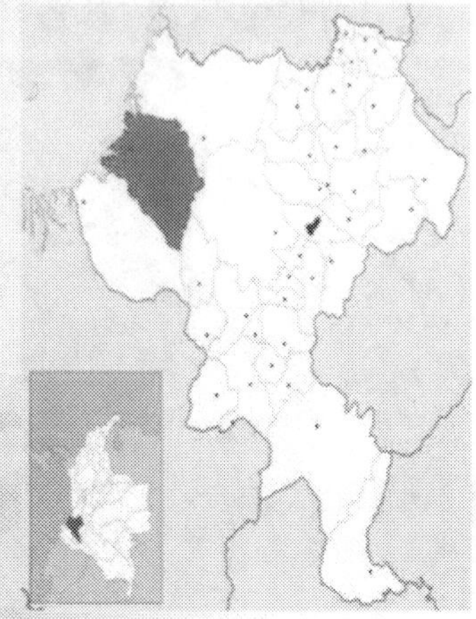

Villa rica

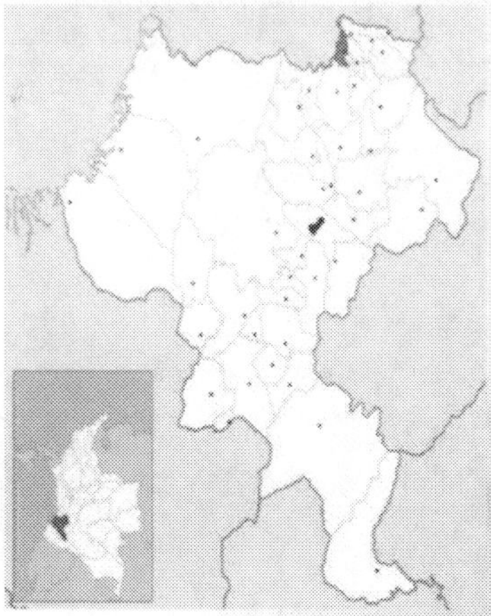

## * GENERAL PURPOSE

Establish the energy and exergy analysis of a desalination system using solar energy.

DESMAR SOLAR

## *SPECIFIC PURPOSE

*Determine the energy requirements of each of the stages of a solar desalination process.

*Establish the efficiency of different models of solar energy collection system to be used in the desalination process.

*Evaluate the exergy of the process and establish improvements in the design selecting the most suitable model system for the process

*Determine physical properties of the materials.

*Quantify the effect of temperature and the properties of the materials for their selection and use in a solar desalination.

*Establish the lifetime of the material chosen from the tests already performed.

# DESALINATION PROCESS

*Desalination by reverse osmosis

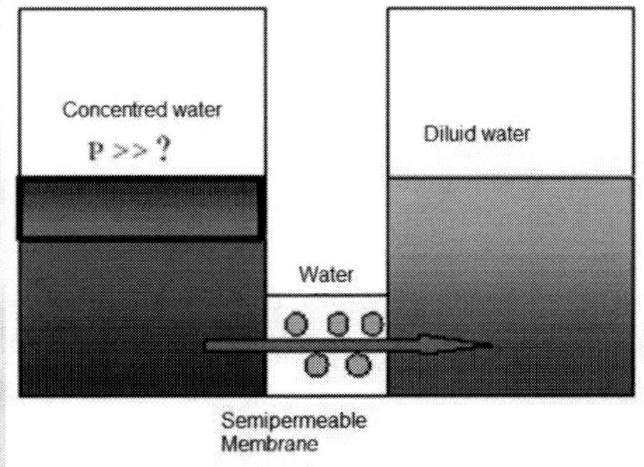


DESMAR SOLAR

*Desalination by Freezing

*Desalination by hydrate formation

*Flash point desalination

DESMAR SOLAR

*Desalination by distillation

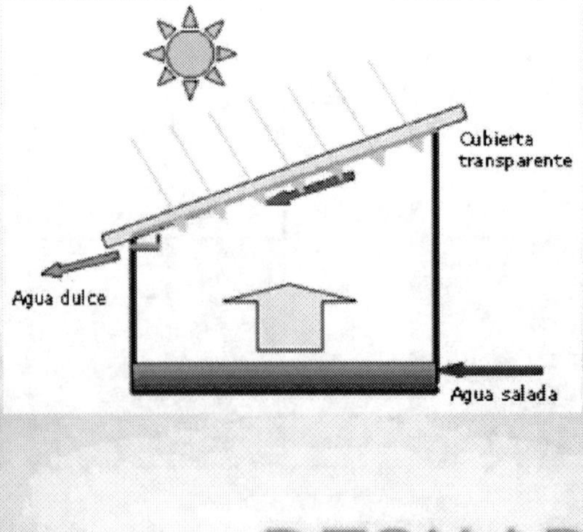


DESMAR SOLAR

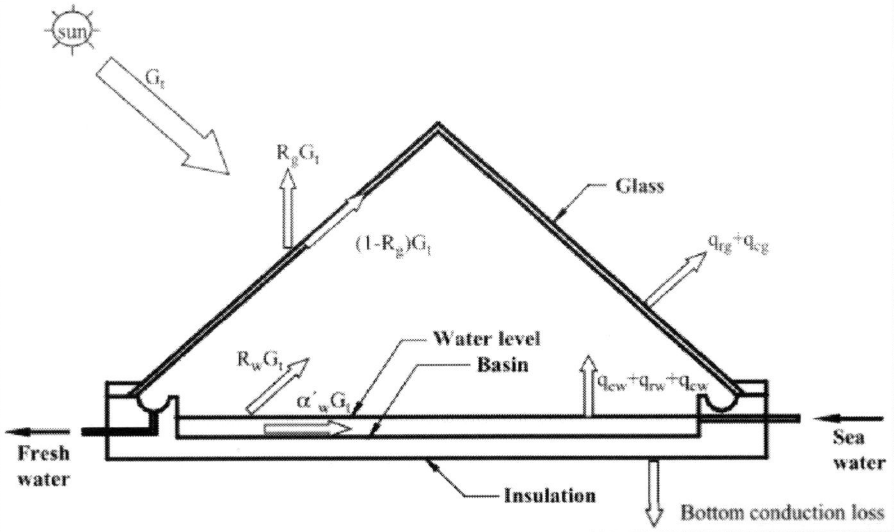


DESMAR SOLAR

## *Energetic and exergetic models

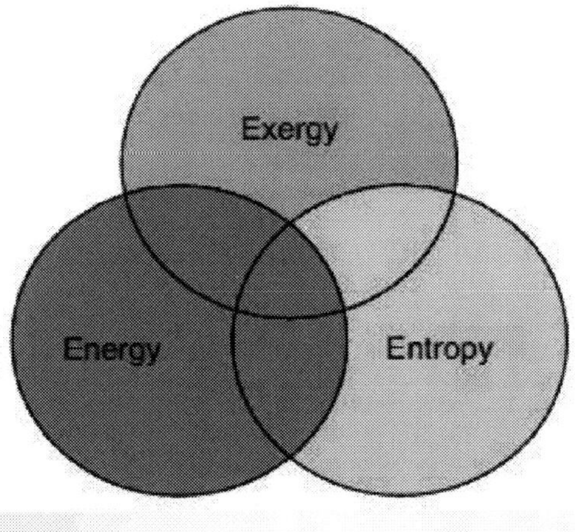


DESMAR SOLAR

# ENERGY AND EXERGY DEFINITION?

* Energy: The ability that have the matter to produce work in the form of motion, light, heat, etc..

* Exergy: A thermodynamic property that determines the useful work potential of a given amount of energy that can be achieved by the spontaneous interaction between a system and its environment. Reports the system's potential usefulness as a source of labor.

DESMAR SOLAR

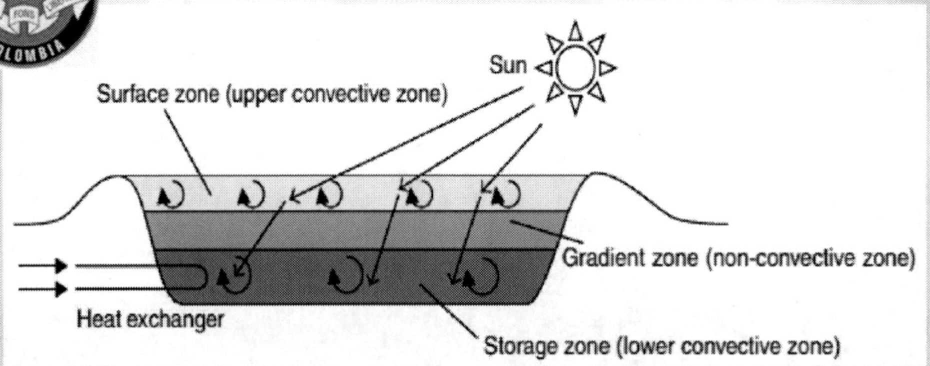


$$n = \frac{m_s h_s + m_w h_w}{G_t}$$

# *EXERGY

$$Q_{net} = Q_{HSZ,solar} - Q_{bottom} - Q_{up} - Q_{side}$$

$$n_{HSZ} = 1 - \frac{(Q_{bottom} + Q_{up} + Q_{side})}{Q_{HSZ,solar}}$$

DESMAR SOLAR

# *ENTROPY

$$\begin{aligned} s &= x_s s_s + x_w s_w \\ &= x_s[s_{s,\text{pure}}(T,P) - R\ln(x_s)] + x_w[s_{w,\text{pure}}(T,P) - R\ln(x_{\mathbf{w}})] \\ &= x_x s_{s,\text{pure}}(T,P) - R[x_s \ln(x_s) + x_w \ln(x_w)] \end{aligned}$$

DESMAR SOLAR

# *EXPERIMENTS

*EXPERIMENT 1.

Distillation through heating plate

DESMAR SOLAR

# * HEATING DATA

| | TEMPERATURE (°C) | TIME (min) |
|---|---|---|
| Evaporated | 67 | 02:46 |
| Finished | 91 | 43:33 |

## FLOW COOLING

| Measure | Volume (mL) | Time(s) | Cooling flow(mL/seg) | Cooling flow(L/min) |
|---|---|---|---|---|
| 1 | 1000 | 25,57 | 39,1 | 2,35 |
| 2 | 1000 | 25,86 | 38,7 | 2,32 |
| 3 | 1000 | 25,26 | 39,6 | 2,38 |
| | PROMEDIO | 25,56 | 39,1 | 2,35 |

DESMAR SOLAR

# RESULTS

| Solution | Temperature (°C) | Volume (mL) | ppm |
|---|---|---|---|
| Initial | 20 | 200 | $90x10^3$ |
| Final | 20 | 128 | $148x10^3$ |
| Desalinated | 20,4 | 68 | 0 |

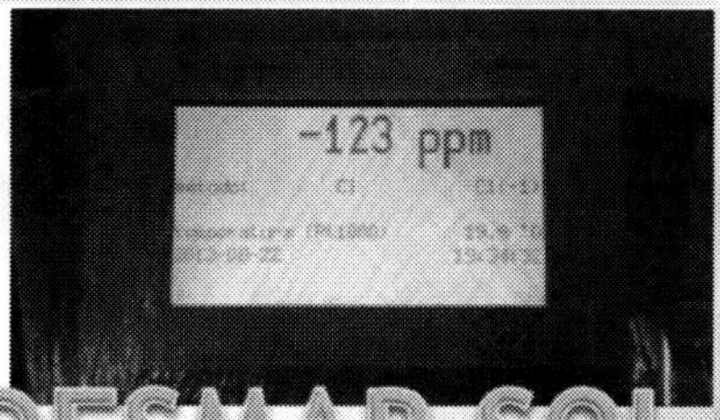


DESMAR SOLAR

# BALANCE RESULTS

1

| Matter balance | | |
|---|---|---|
| Input | Output | |
| 200 ml | 128 mL | 68 mL |
| $90 \times 10^3$ | $148 \times 10^3$ | 0 |

| | |
|---|---|
| me = m1+ m2 | $90x10^{3} * 0{,}200 = 0 * 0{,}068 + 0{,}128$ *ppm output |
| ppm output= | $140 \times 10^{3}$ |

**PERCENTAGE OF**

$$\frac{ppm_{plancha} ppm_{salida}}{ppm_{plancha}} * 100 \qquad \frac{148000 - 140000}{148000} * 100 = 5.40\%$$

DESMAR SOLAR

## 2

| Energy Balance | | |
|---|---|---|
| | w*t | |
| Input | $2{,}8 \times 10^7$ | W * seg |
| | 7,8 | kWh |

$$\rho v C_{H2O}\Delta T = m\Delta H + m C_{H2O}\Delta T$$

| | | |
|---|---|---|
| ρ | 1037,5 | kg/m^3 |
| v | 0,0002 | m^3 |
| C H2O | 4184 | kJ/(seg * K) |
| ΔT | 70,68 | K |
| m | 0,2075 | kg |
| ΔH | 173 | kJ/(kg * K) |
| | | |
| 61362 | = | 61398 |
| | 36 J | |

DESMAR SOLAR

## EXPERIMENT 2

Distillation through incandescent bulbs

DESMAR SOLAR

# HEATING DATA SAMPLE

| | TEMPERATURE (°C) | TIME(min) |
|---|---|---|
| Evaporated | 38 | 46 |
| Finished | 90 | 380 |

DESMAR SOLAR

# BALANCE RESULT

| Matter balance | | |
|---|---|---|
| Input | Output | |
| 200 ml | 148mL | 52 mL |
| $90 \times 10^3$ | $115 \times 10^3$ | 0 |

| | |
|---|---|
| me = m1+ m2 | $90 \times 10^3 * 0,200 = 0 * 0,052 + 0,148$ *ppm output |
| ppm output= | $121,62 \times 10^3$ |

## PERCENTAGE OF ERROR

$$\frac{ppm_{plancha} ppm_{\ salida}}{ppm_{plancha}} * 100 \qquad \left|\frac{115000 - 121620}{115000}\right| * 100 = 5,76\%$$

DESMAR SOLAR

# RESULTS

| Solution | Temperature (°C) | Volume (mL) | ppm |
|---|---|---|---|
| Initial | 20 | 200 | $90x10^3$ |
| Final | 20 | 148 | $121,62x10^3$ |
| Desalinated | 54 | 52 | 0 |

DESMAR SOLAR

# *MATERIALS FOR THE DESALINATOR

DESMAR SOLAR

# *TYPES OF MATERIALS

*Metallic Materials

*Ceramic Materials

*Plastics Materials

DESMAR SOLAR

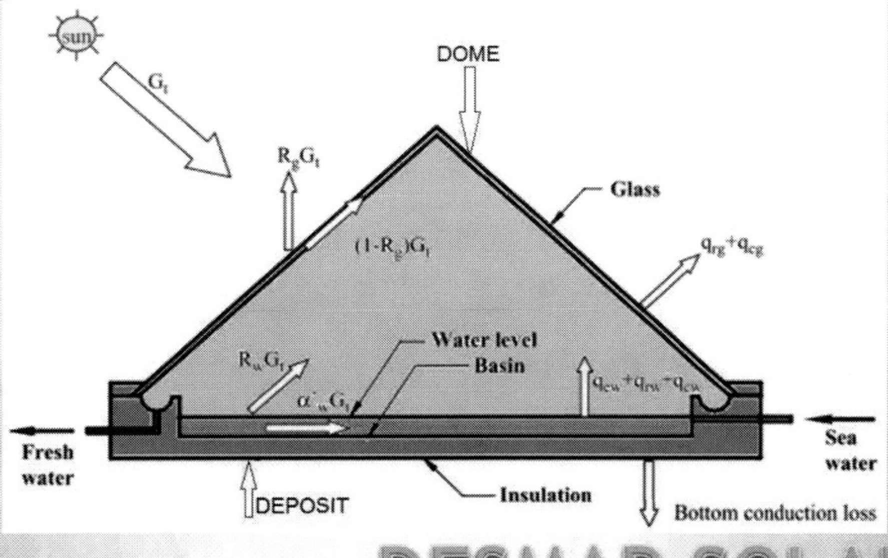


DESMAR SOLAR

UNIVERSIDAD LIBRE
SCIENTIA FONS LIBERTATIS
COLOMBIA
*MATERIALS FOR THE DOME
DESMAR SOLAR

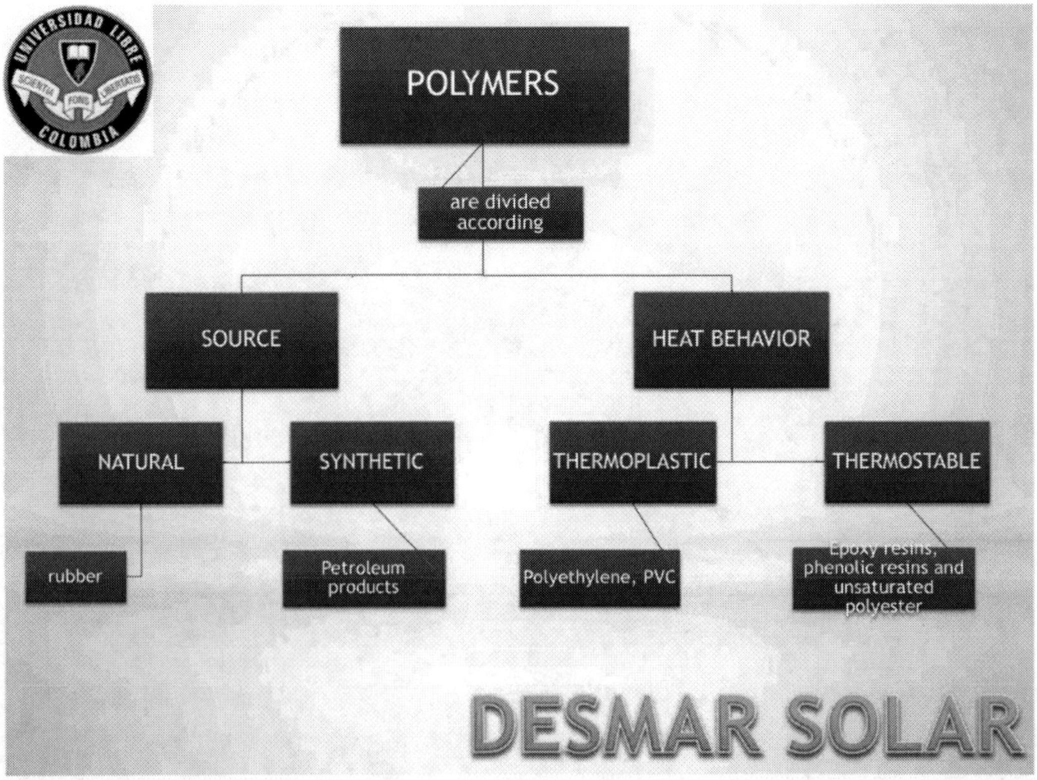
UNIVERSIDAD LIBRE
SCIENTIA FONS LIBERTATIS
COLOMBIA
POLYMERS
are divided according
SOURCE
HEAT BEHAVIOR
NATURAL
SYNTHETIC
THERMOPLASTIC
THERMOSTABLE
rubber
Petroleum products
Polyethylene, PVC
Epoxy resins, phenolic resins and unsaturated polyester
DESMAR SOLAR

# *PROPERTIES

* **Elongation:** Quantity that measures the increase in length of a material
* **Thermal conductivity:** Ability of a material to conduct heat
* **Specific heat:** The heat to be supplied to the unit mass of a substance to raise its temperature by one unit
* **Coefficient of thermal expansion:** Coefficient that express the length change when a body undergoes a temperature change

**DESMAR SOLAR**

| | •Polietileno de baja densidad (LDPE) | •Polietileno de alta densidad (HDPE) | •Polipropileno (PP) | •Cloruro de polivinilo rigido(pvc) | •Cloruro de polivinilo plastico(pvc) |
|---|---|---|---|---|---|
| Elongation [%] | 90-800 | 50-1000 | 200-700 | 002-40 | 200-450 |
| Thermal conductivity [kw/mk] | 3,35 | 4,6-5,2 | 1,17 | 1,3-2,9 | 1,3-1,7 |
| Specific heat [kj/kgk] | 2,3 | 2,3 | 1,9 | 0,84-1,17 | 1,26-2,09 |
| Coefficient of linear thermal expansion [K^-1] | 1e-4 - 2e-4 | 1,1e-4 - 1,3e-4 | 5,8-10,2 | 5-18,5 | 7-25 |
| Continuous use temperature [K] | 355,37-373,15 | 394,26 | 394,26-433,15 | 338,71-352,59 | 338,71-352,59 |
| Deflection temperature [K] | 100-121 | 140-190 | 200-230 | 135-180 | ………… |
| Solvents | N/a | N/a | N/a | Cetonas, ésteres,aromaticos,hidrocarburos | Cetonas, ésteres,aromaticos,hidrocarburos |
| Elasticity in flexure limit [pa] | ………… | 6894,76 | 41368,56-55158,08 | 68947,6-110316,16 | N/a |

## *APPROPRIATE MATERIALS

- *Low density polyethylene (LDPE)
- *High density polyethylene (HDPE)
- *Polypropylene (PP)
- *Polyvinyl chloride (PVC)

DESMAR SOLAR

## DEPOSIT MATERIALS

DESMAR SOLAR

# CERAMIC MATERIALS

* Nonmetallic inorganic solids produced by heat treatment.
* Compared with metals and plastics are hard, non-combustible and non-rusting.
* They can be used in environments with high temperature tribological and corrosive.

DESMAR SOLAR

# CERAMIC´S PROPERTIES

* High melting temperature.
* Low density.
* Strength, hardness, structural strength.
* Corrosion resistance.
* Good thermal and electrical insulators.

DESMAR SOLAR

# ENVIRONMENTAL CONSIDERATIONS

- By using solar energy production CO2 is removed.?
- The use and advancement of renewable energy is implemented. (In our case solar energy is used).
- Drinkable water in coastal areas.

DESMAR SOLAR

# *BIBLIOGRAPHY


- Robert Cook (1986). Computer Simulation of Stress-Strain Behavior in Polymeric Materials. MRS Proceedings, 79, 409 doi:10.1557/PROC-79-409.
- David L. Edwards and Mircea Chipara (2004). Polymeric materials for Solar Sail: The combined effects of polymer thickness, radiation, and temperature. MRS Proceedings, 851, NN1.6 doi:10.1557/PROC-851-NN1.6.
- Alice B. Paterakis (1992). The Consolidation and Desalination of Ceramic Impregnated with Sodium Sulphate. MRS Proceedings, 267, 1071 doi:10.1557/PROC-267-1071.
- T. Y. Wu and D. L. Questad (1992). Mechanical Characterization and Constitutive Description of Thin-Film Polymeric Materials. MRS Proceedings, 264, 143 doi:10.1557/PROC-264-143.(Jr., n.d.)
- Jr., W. D. C. (n.d.). *Materials Science Engineering, An Introduction.* (third edit., p. 785).


DESMAR SOLAR

- Suzana P. Nunes, Ali Reza Behzad and Klaus-Viktor Peinemann (2013). Self-assembled block copolymer membranes: From basic research to large-scale manufacturing. Journal of Materials Research, 28, pp 2661-2665. doi:10.1557/jmr.2013.253.
- Paterakis, A. B. (1992). The Consolidation and Desalination of Ceramic Impregnated with Sodium Sulphate. *MRS Proceedings, 267.*
- Paul C. Painter, M. M. C. (1996). *Fundamentos de ciencia de polimeros: un texto introductorio* (p. 467). Cambridge University Press.
- Zalba, B. (2003). Review on thermal energy storage with phase change: materials, heat transfer analysis and applications. *Applied Thermal Engineering*, *23*(3), 251-283. doi:10.1016/S1359-4311(02)00192-8


DESMAR SOLAR

# SOLID FUEL TO NATURAL GAS CONVERSIONS FOR EXISTING BOILER APPLICATIONS

*William Gurski*
*Director Power Sales*
*Zeeco Inc.*

*John Guarco*
*Technical Director Boiler Burners*
*Zeeco Inc.*

*Nando Nunziante*
*Manager Power Controls*
*Zeeco Inc.*

Prepared for Presentation at the Year 2014 Spring National Meeting
New Orleans, LA, March 30 – April 3, 2014

## Abstract

Recent discoveries of vast natural gas reserves in the United States have resulted in increased production of Natural Gas, resulting in lower prices. Petrochemical facilities are performing solid fuel conversions on their boilers to natural gas as a cost-effective and efficient fuel solution. Natural Gas is not only economically beneficial but also environmentally efficient with cheaper prices and reduced $SO_2$, $NO_x$, and $CO_2$ emissions. As more facilities consider boiler fuel conversions, it is important to understand all facets of the conversion, from the thermal evaluation of the boiler, to the complete design, supply and installation of the new firing system.

We will provide specific details and recommended practices from a recent Circulating Fluidized Bed (CFB) Boiler fuel conversion application designed for 1.4 billion Btu/hr of heat input for the maximum continuous steam rating. We will also review thermal performance results, design and supply of the complete new gas firing system, and installation conversion assistance for the boiler modifications and firing system installation details.

## Introduction

Zeeco was contacted by a large industrial plant owner to assist them with the fuel changeover of a utility designed solid fuel CFB. Due to current contract requirements and the competitive nature of the industry, the plant owner will remain confidential.

The primary goals of the project are as follows:

- Full Conversion from Petcoke firing to Natural Gas
- Maintain boiler steam design capacity (~1200 kpph)
- Meet permitted $NO_x$ requirements – 0.07 lbs/MMBtu
    - 25-100% heat input for $NO_x$ emissions
- Minimal or no impact on the boiler waterside design

The owner provided the original boiler design information for Zeeco's review of potential boiler burner retrofitting options. This was the first of two boilers that would require retrofitting Two sister units were built/commissioned on solid fuels approximately three years previously. The units were originally designed with natural gas start-up burners and commissioned for main fuel support via the petcoke. The petcoke was to be fired and combusted via a circulating fluid bed system. Upon initial advisement from the plant owner, they noted that their petcoke fuel contract was no longer cost effective, as this fuel feedstock was significantly higher in price than originally estimated in the plant system's budget. The boilers were only in operation for less than two years when this decision to switch to natural gas was determined.

Zeeco will review the major facets of this boiler conversion—from the initial proposal phase to the commissioning and startup of the boiler and firing system. We will also discuss the combustion parameters of the technology behind the project's success–Zeeco's DT Ultra-Low $NO_x$ Free-Jet Burner technology.

## Combustion Parameters

In order to understand why the ZEECO® Free-Jet technology was chosen, formation of thermal $NO_x$ emissions must first be examined. For gaseous fuels with no fuel-bound nitrogen ($N_2$), thermal $NO_x$ is the primary contributor to overall $NO_x$ production. Thermal $NO_x$ is produced when flame temperatures reach a high enough level to "break" the covalent $N_2$ bond apart, allowing the "free" nitrogen atoms to bond with oxygen to form $NO_x$.

Stoichiometric Equation describing typical combustion in a natural gas fired burner:

Methane & Air with Excess Air

$$2CH_4 + 4\ (XA)\ O_2 + 15\ (XR)\ N_2 \dashrightarrow 2CO_2 + 4H_2O + (XA)\ 15N_2 + (XR)\ O_2$$

Natural air is comprised of 21% $O_2$ and 79% $N_2$. Combustion occurs when $O_2$ reacts and combines with fuel (typically hydrocarbon). However, the temperature of combustion is not normally high enough to break all of the $N_2$ bonds, so a majority of nitrogen in the air stream passes through the combustion process and remains diatomic nitrogen ($N_2$) in the inert combustion products. Very little $N_2$ is able to reach high enough temperatures in the high intensity regions of the flame to break apart and form "free" nitrogen. Once the covalent nitrogen bond is broken, the "free" nitrogen is available to bond with other atoms. Basic chemistry dictates that free nitrogen, or nitrogen radicals will react to other atoms or molecules that can accept them to create a more stable atom. Of the possible reactions with the products of combustion, free nitrogen will most likely bond with other free nitrogen to form $N_2$. However, if a free nitrogen atom is not available, the free nitrogen will react with the oxygen atoms to form thermal $NO_x$. As the flame temperature increases, the stability of the $N_2$ covalent bond decreases, allowing the formation of free nitrogen and subsequently increasing thermal $NO_x$. Burner designers can reduce overall $NO_x$ emissions by decreasing the peak flame temperature, which can reduce the formation of free nitrogen available to form thermal $NO_x$.

The varied requirements of Power and Industrial plants entail the use of numerous types and configurations of burners. The method utilized to lower $NO_x$ emissions can differ by application. However, thermal $NO_x$ reduction is generally achieved by delaying the rate of combustion. Since the combustion process is a reaction between oxygen and fuel, the objective of delayed combustion is to reduce the rate at which the fuel and oxygen mix together and burn. The faster the oxygen and the fuel gas mix, the faster the rate of combustion and the higher the peak flame temperature.

Figure 1 plots Peak Flame Temperature against Thermal $NO_x$ created. $NO_x$ emissions increase as the adiabatic flame temperature increases. Slowing the combustion reaction reduces the flame temperature, which results in lower thermal $NO_x$ emissions. The challenge in achieving lower thermal $NO_x$ emissions is not the theory; however, it is in retrofitting the latest burner technologies into older existing boilers without adding expensive external components or processes.

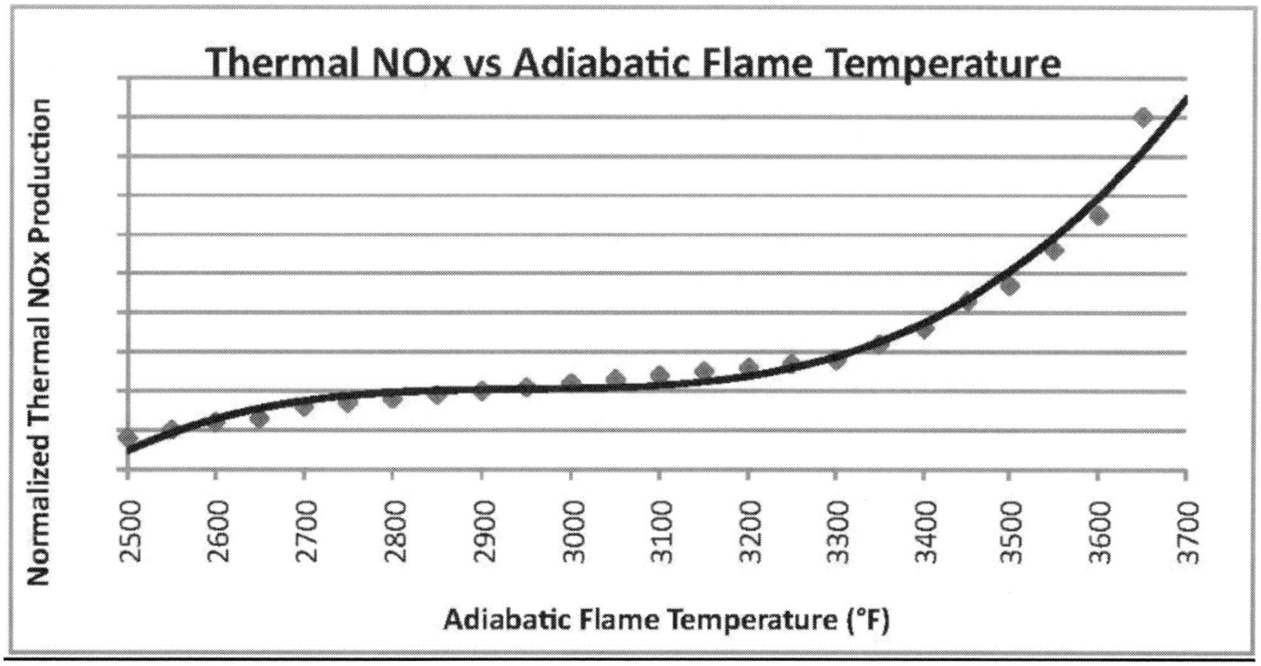


**Figure 1: Calculated Peak Flame Temperature vs. Thermal $NO_x$ Production**

The industry's standard method to reduce thermal $NO_x$ is to mix the fuel gas together with the inert products of combustion to recondition the fuel before combustion occurs. Since the reconditioned fuel is mainly comprised of inert components, the resulting composition burns at a lower peak temperature. To best utilize the inert products of combustion (flue gas) within the boiler, the fuel gas is introduced along the outside perimeter of the burner tile in an area where flue gas is present while the boiler is in operation. As the fuel gas passes through the inert products of combustion, mixing naturally occurs. This changes the composition of the fuel, and stabilization occurs at the tile exit. Since the reconditioned fuel mixture is 80 to 90% inert in most cases, the resulting flame burns at a lower peak temperature and generates less thermal $NO_x$.

The mixing of the fuel gas with flue gas prior to combustion is called Internal Flue Gas Recirculation (IFGR). When IFGR is too aggressive, it can result in an increased blower power usage, decreased burner turndown, and increased flame destabilization. Through Free-Jet Theory, maximizing IFGR while maintaining flame stability and flame length can become a challenge.

## Flame Stabilization

In order to fully utilize the Free-Jet Theory, Zeeco devised a method to stabilize the burner flame with a highly inert fuel gas/flue gas mixture. This type of combustion is achieved when the flame is stabilized in a low-pressure area created on a series of specially designed hot refractory ledges. As combustion occurs, the refractory ledge retains heat and flame stability is enhanced. Thus, to achieve improved stability and extreme Thermal $NO_x$ reduction, the Free-Jet Technology:

1. Mixes inert flue gas through free-jet methods with all of the fuel gas before combustion occurs, lowering flame temperature.
2. Stabilizes the flame on a refractory ledge, improving flame characteristics

Before combustion is initiated, a furnace is typically filled with normal air, which contains 21% oxygen. Once the burner is ignited, the oxygen content inside the furnace decreases until the burner achieves maximum duty. At this point, the oxygen content in the furnace is normally 2 to 3%. To keep the burner stable throughout the transition from start-up with 21% oxygen to maximum duty with 2 to 3% oxygen, Zeeco developed a series of stabilization ledges as shown in **Figure 2**. These ledges are a design feature of the Zeeco® Free-Jet Burner chosen for the boiler retrofit.

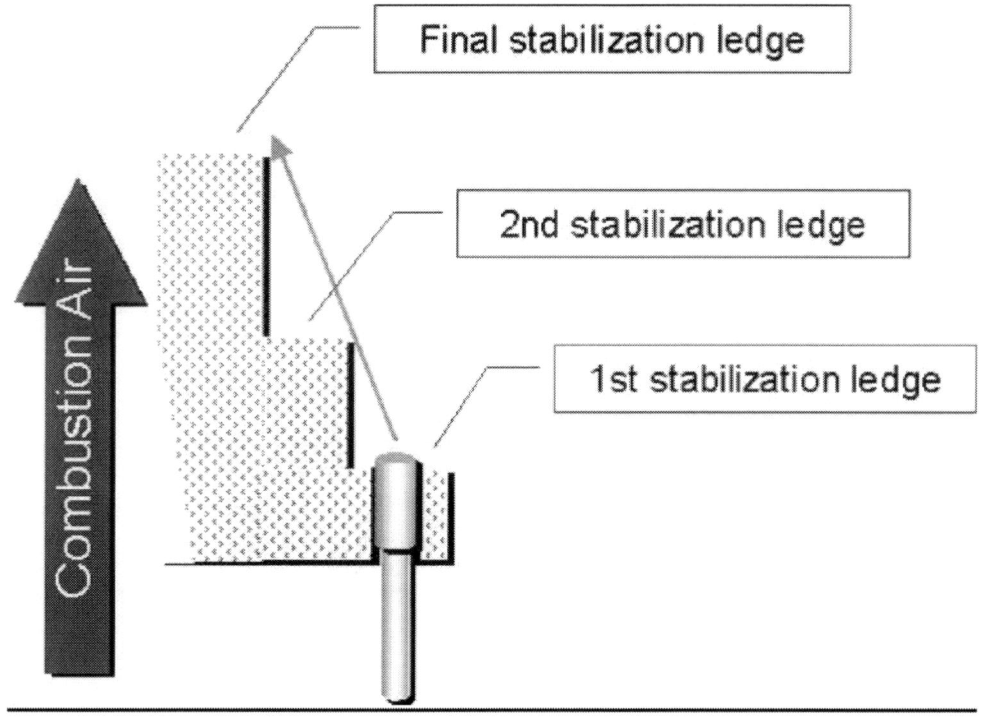


**Figure 2: Free-Jet Flame Stabilization Method Illustration**

## Burner Technology Description

### Introduction to Free-Jet Technology

The Zeeco Free-Jet burner series was designed with the specific purpose of maximizing the amount of IFGR to reduce thermal $NO_x$ emissions without sacrificing burner performance with respect to flame length, turndown, and stability. The maximization of IFGR means many of the problems associated with using high levels of EFGR to achieve low emissions can be reduced or eliminated. Specifically, Zeeco's Free-Jet design dramatically reduces or eliminates the need for EFGR by reducing blower power usage, increasing turndown, reducing maintenance and improving flame quality.

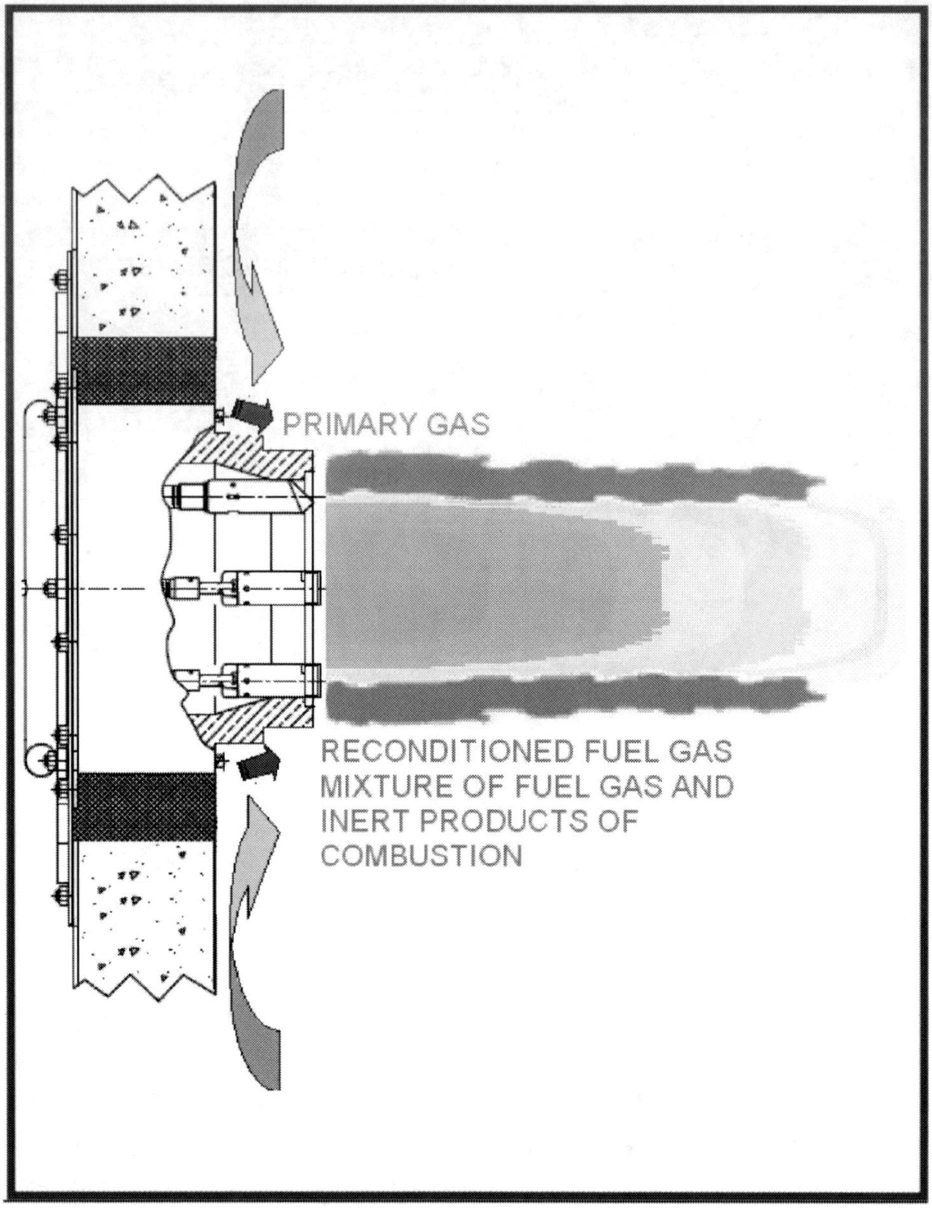


**Figure 3: Inert Products of Combustion (Flue Gas)**

## Pre-Award Engineering

Since the boiler was only in operation for a short period of time, the plant needed assistance with the selection of the new gas firing system and options that would allow them to operate the boiler at the same steam flow capacity and pressure as originally designed for petcoke firing. Zeeco designed three possible boiler conversion options for the operator to consider. Every option required removing the existing start-up burners and closing off the tube openings or retrofitting with new larger burners:

1) Remove the fluid bed bottom and vertically fire the boiler. This option was feasible for a full 100% conversion as the entire fluid bed and nozzle/air system would be removed in favor of six new burners with a corresponding windbox and air system (1500 MMBtu/hr).
2) Removing the four start-up burners and increasing the firing capacity to the necessary levels for full load steam conditions (1500 MMBtu/hr).
3.) Removing the four start-up burners with new burners, plus adding a second level of four burners above to achieve the necessary heat input for full load steam conditions (1500 MMbtu/hr).

Option One was not selected, as it would have required an entire new lower section of the boiler, plus major reconstruction of the fluid bed in order return to solid fuel firing. Option Two was not selected, as the heat release/absorption area within the boiler would have been too great for the water/steam production side.

Option Three was selected, as the owner determined that they wanted the flexibility to return to petcoke firing should the price for natural gas rise to a level higher than petcoke fuel. This allows ~80% of the system's equipment to remain available for future petcoke firing. This solution also offered the least cost to retrofit back to solid fuel firing. The plant owner also had the challenge of the water/steam side of the project, as the boiler system would now be required to operate on 100% natural gas firing instead of a high pressure recirculating fluid bed system. Zeeco assisted the plant owner and the boiler's Original Equipment Manufacturer (OEM) with the system design and selection. Several iterations were submitted by the Boiler OEM on the redesign, and Zeeco worked as a team with the three parties to select the best option for firing the boiler.

The challenge with the boiler redesign interactions was to ensure that the following conditions were met, as required by the plant owner:

- Maintain main steam flow per original design
- $NO_x$ emissions at permit levels
- Boiler able to follow steam needs per electrical generation requirements
- Ability to ramp per the original design conditions
- Turndown of 10-1
- No design impact or changes on existing water or steam circuit metallurgy (tube material changes)

Two design challenges were discovered throughout the Pre-Award Engineering Phase. The first included the air system redesign due to the fluid bed operating on primary and secondary air systems. This is further detailed in the Contract Execution Section below. Another challenge was the integration of the natural gas firing Burner Management System (BMS) and controls into the existing Distributed Control System (DCS). This is also further explained in the Contract Execution Section.

## Project Engineering

Once awarded, Zeeco immediately began the burner and BMS systems design. The plant owner required less than 26-week delivery for all components, as the need to fire natural gas was critical to the financial status of the project.

Zeeco began the contract with a full-site visit with the entire team, including the plant owner and boiler OEM. The technical requirements and time schedule were reviewed to ensure all those involved were informed of the critical nature of the project. The technical details of the project varied, as the unit was being fully converted from a solid fuel CFB to a natural gas fired utility boiler. Many items required detailed analysis during the contract phase, including:

- Burner integration into the existing waterwall and boiler structure
- Addition of one new elevation of burners and effects of flame in this furnace region
- Complete redesign of the burner combustion air system and air heater outlet to a single pressure and delivery system to the new burners
- Integration of the new BMS panel and controls logic into the existing DCS system
- Design and location of all new gas fired fuel valving and rack systems, plus local termination cabinets
- Refractory removal and modification to support new natural gas firing (lower boiler refractory was no longer needed due to the elimination of petcoke and recirculating solids from the cyclone)
- Structural modifications to support all equipment and ancillaries for the natural gas firing (combustion air ducting, burner supports, valve racks, etc.)

Burner integration posed several challenges. The new burners were required to fit into the existing waterwall opening and modified from a downfired arrangement to a horizontal firing arrangement. Zeeco advised the plant owner that the new Zeeco Free-Jet DT burner would be able to fit within the existing opening, and would allow the exact tube opening design to be used for the new upper burner elevations. Zeeco also assisted with the design of the support systems and relocation of all structural steel for the new burner locations (maintenance platforms, etc.).

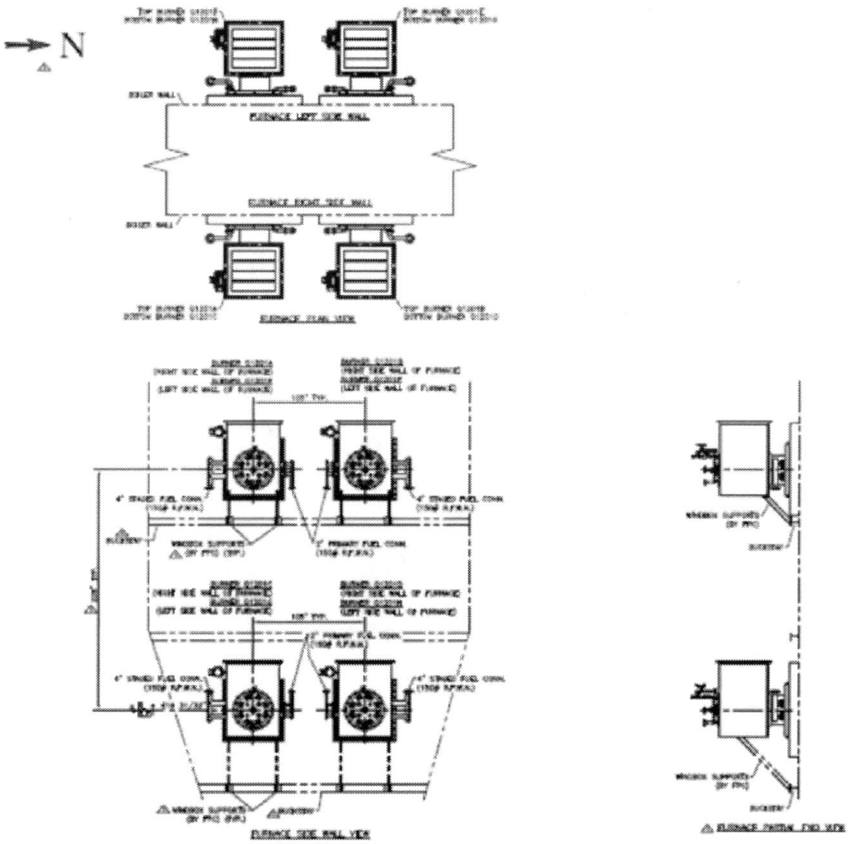


**Figure 4: Boiler Design Diagram**

It was important to ensure the new upper burner elevation had a proper flame fit within the boiler furnace and would not adversely affect heat transfer to the waterwalls or radiant furnace super heat. An elevation of ~19 feet was selected for the new burners above the existing burners as the ideal location, for not only proper flame fit but also ease the installation of the burners without major boiler structural modifications.

**Figure 5: Zeeco's DT Free-Jet Burner**

The combustion air system was completely redesigned from a two-system design (primary for the fluid bed nozzles and secondary air system for bed support), to a single pressure design. Zeeco was contracted to assist the plant owner with the entire system redesign, as well as routing to the new burner elevations. Zeeco also assisted with a new design for the air heater outlet and fan control due to the single pressure design needed for the new burners. Zeeco's Model Simulation Engineering Team spent significant time on-site with the Plant Owner's personnel designing the new system. Zeeco's engineers designed and constructed a scaled Plexiglas model of the entire combustion system and ran the Physical Flow Simulation model in our Connecticut facility. The completed model ensured +/- 3% distribution to each burner with less than 10% peripheral distribution around each burner's air entry.

**Figure 6: Free-Jet Physical Flow Simulation**

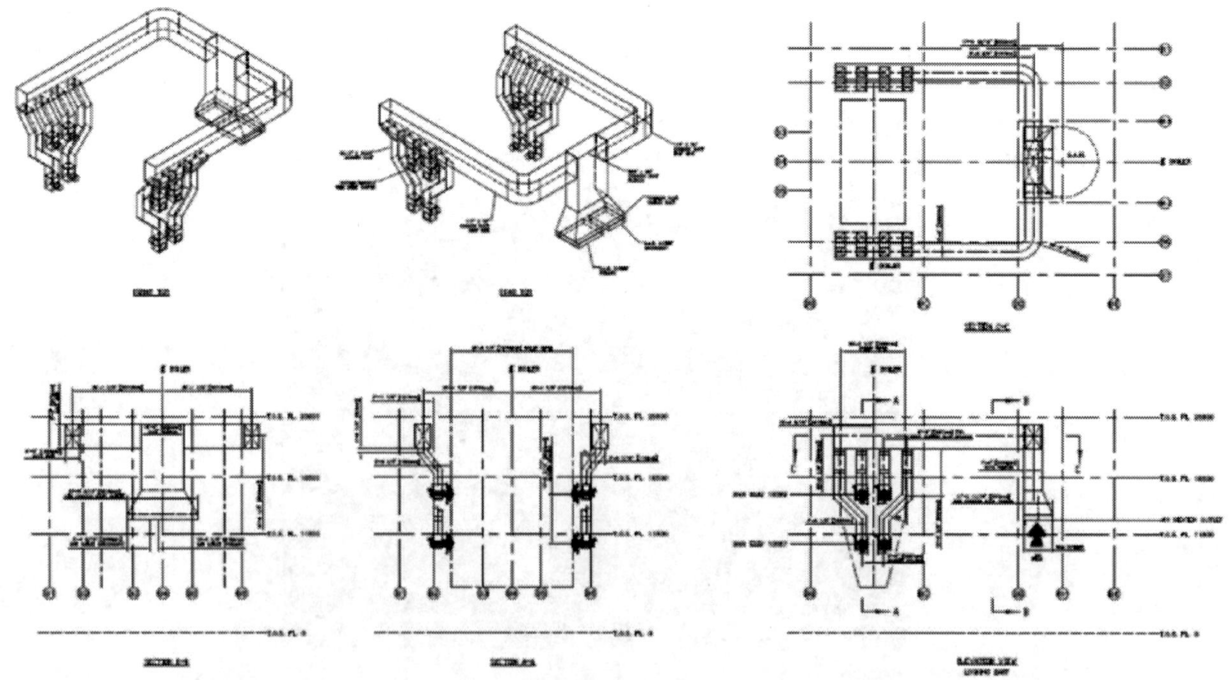

**Figure 7: Zeeco's New Design Combustion Air System**

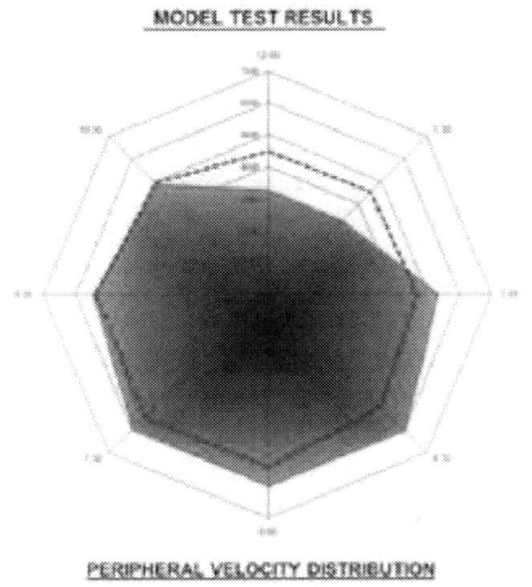


**Figure 8: Example of Model Study Peripheral Distribution Chart**

Due to the boiler's original controls and Burner Management System's design for main fuel petcoke firing and natural gas start up burners, they were not designed for full load natural gas firing. During the initial design proposal phase, it was agreed that a separate Redundant PLC BMS system would be necessary to integrate into the main DCS system. This would allow the main system to remain "as-is" in the event the unit was to return to petcoke firing in the future. Essentially, it would be a "fuel selector" switch to return the boiler's operation back to petcoke firing. This system was designed in working with the plant owner's original DCS and BMS supplier to ensure a smooth integration into the DCS system.

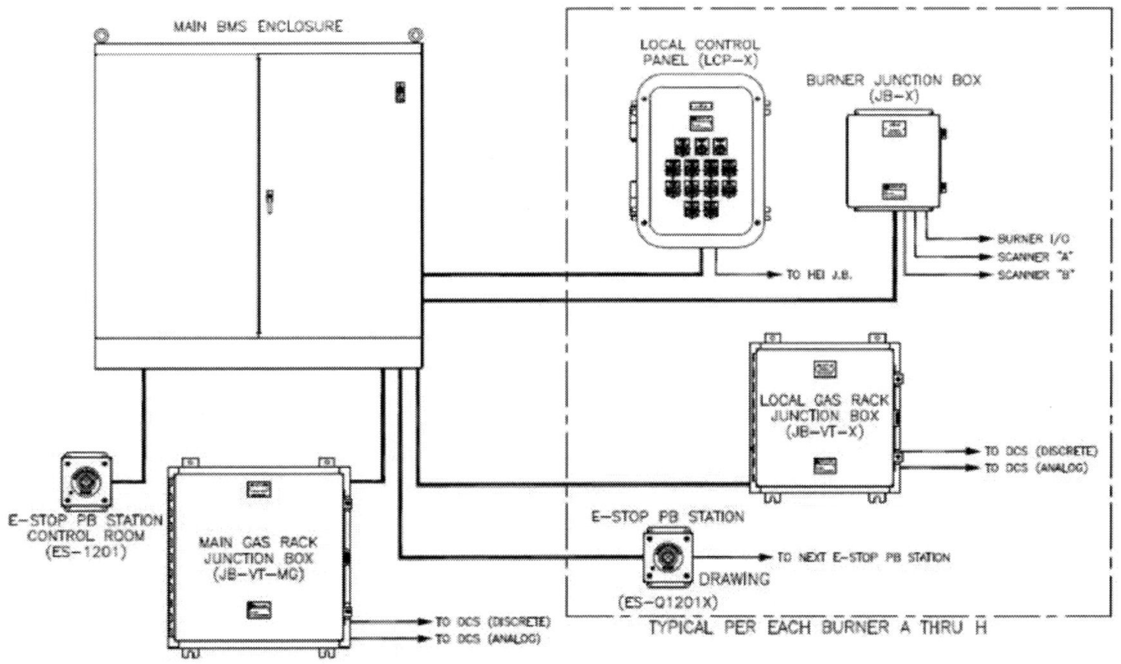

**Figure 9: Boiler's Burner Management System**

In order for the new burner system to be modified to fire on natural gas, several modifications were required for successful full load natural gas firing, including:

- Removal of furnace's lower slope refractory- During petcoke firing, the turbulent nature of the fluid bed, including the solids return from the cyclone, required the lower slope tubing to be fully protected from potential erosion. When the solid fuel is removed, this lower slope no longer required the refractory coverage to protect the tubing from solids within the cyclone, promoting further heat absorption in the furnace for natural gas firing, and to assist with lower emissions.

- As noted above, the cyclone design in a CFB is to remove solids from the furnace exit flue gas and return them to the fluid bed to promote high efficiency firing and lower carbon loss. With petcoke firing removed, the cyclone was no longer needed for this operation. The inlet to the cyclone plated off, effectively "closing" the inlet so no flue gas would pass through the cyclone. The loop seal exit was "plugged" with refractory to protect it from furnace gasses. This sealing can be deemed 'temporary", as the plate and refractory plug can be removed in the event petcoke firing is deemed cost effective over natural gas firing.

- All back-end clean up equipment are placed out of service due to natural gas firing.

Upon all final design "freeze" and plant owner approvals, fabrication and procurement proceeded and all equipment was delivered to the site by the end of March 2013. Due to the critical schedule, the plant owner was required to take the boiler out of service in late January 2013 to begin the modification process for the natural gas firing. The equipment, upon on-site arrival, was immediately transported to the boiler and the installation process began.

Due to the diligence of the entire engineering team, the integration and installation of the burners, valving and BMS proceeded very smoothly throughout the outage. All the upfront engineering and time ensured the design was cost effective and the outage was completed on time.

## Commissioning and Startup

The plant owner also contracted Zeeco's Service Engineers to provide full commissioning and installation assistance, to ensure that all mechanical and electrical equipment was installed per Zeeco's requirements. Zeeco's Service Engineering team was on-site to review and check out the electrical/mechanical/BMS/DCS installation. This proceeded very smoothly during checkout, and all minor issues were addressed in the time allotted for commissioning. Actual equipment start-up proceeded immediately after the final commissioning. Burner performance and full steam boiler operation was achieved in less than two weeks.

Final Conditions are as noted:

- Final emissions performance met the contract and permit requirements of 0.07lbs/MMBtu
- Full load steam boiler requirements were met within the two week timeframe

## Conclusion

Boiler steam load and emissions were met in a very timely fashion, and the plant owner was able to generate plant power within less than a month after the initial boiler start-up. The plant owner operated the boiler and turbine/generator near design capacity for over three months after final commissioning. The effort placed on the entire project, pre and post-award, led to the overall success of the boiler's conversion. The integration by the plant owner and the three primary engineering parties led to a fully integrated design that was reviewed by all to ensure the timely success of the project. Based upon the success of the first boiler, Zeeco was awarded the contract for the second CFB boiler conversion, and will begin with installation and commissioning to be in mid-2014.

# 351797 Vapor Intrusion - Models and Their Value

**Abstract:**

Vapor Intrusion- Models and Their Value

The problem of vapor intrusion into structures built atop contaminated properties has received increasing attention over the last decade. On the national level, new guidance from the US EPA is imminent, but in the meantime different states have adopted a wide range of policies to address the problem. Many approaches to characterization are empirical, based upon measurements at impacted locations. Since these can involve alarming and intrusive measures involving many residents, some emphasis has been placed on screening, using simple mathematical descriptions of the problem. With the vastly increased availability of powerful fluid dynamics codes, these models can guide more of the site investigation work, as well as evaluate possible possible mitigation strategies. Examples will be given of how such tools can be used to assess both petroleum-based and chlorinated solvent-based situations, based upon recent research carried out in Brown University's Superfund Research Program.

The approach involves solving the equations of vapor transport in the subsurface below the structure of interest, which is typically the soil of the vadose zone. The transport is dominated by diffusion, though entry into the structure may be strongly influenced by local advective transport near the building foundation, as a result of small pressure differences between ambient and the indoors. The indoor environment is typically modeled using a CSTR model. Subtle factors can have a very large influence on predicted indoor air exposures. These include effects related to the resistance of a capillary zone in the case of a chlorinated solvent source in the groundwater, or the form of the assumed biodegradation rate law in the case of petroleum contaminants. The models point to what might represent reliable or problematic field measurement strategies. Also, once a reliable site model is developed, the efficacy of mitigation methods such as subslab depressurization, may be more reliably examined.

*** Formatting Issue - Best Available Paper/Graphic ***

# 351798 Synthesis, Fine Structural Characterization, and $CO_2$ Adsorption Capacity of Cobalt and Nickel-Based Metal Organic Framework-74

**Abstract:**

**Abstract:** Two metal organic frameworks of MOF-74 group (cobalt and nickel-based) were synthesized, characterized, and evaluated for $CO_2$ adsorption. The BET specific surface areas of MOF-74(Co) and MOF-74(Ni) MOFs were 1,406 and 1,418 $m^2g^{-1}$, respectively with the identical pore volume of 0.84 $cm^3g^{-1}$. $CO_2$ adsorption capacities of MOF-74(Co) and MOF-74(Ni) were 3.35 and 4.07 $mmol{\cdot}g^{-1}$, respectively measured at 273 K and 1.1 bar. The oxidation state of central atoms in MOF-74(Co) was composed of Co(II) and Co(III) confirmed by XANES spectra while MOF-74(Ni) was Ni(II) central atoms. The bond distances of Co,,ŸO and Ni,,ŸO were 1.95 and 1.97 Å, respectively.

## 1. Introduction

Global population and the industrialization are increasing and hence, the energy consumption is also increasing swiftly. Till now, the main source of energy is the burning of fossil fuels. However, burning of fossil fuels releases a vast amount of $CO_2$ greenhouse gas into atmosphere which causes the raise of average surface temperature of the earth. As the present developments in CCS technology are not sufficiently effective by means of cost and energy consumption, it has prompted to improve the existing technologies and find some new alternatives [1]. Metal organic frameworks

*** Formatting Issue - Best Available Paper/Graphic ***

(MOFs) are a new class of materials that have attracted substantial attention in past few years [2]. MOFs are synthesized using organic linkers and metal joints that self-assemble to form different dimensional networks. It can generate unusually large diameter channels and cavities which is particularly attractive for $CO_2$ gas storage application. Here we report the $CO_2$ gas adsorption capacity on same type of two MOFs; such as MOF-74(Co) and MOF-74(Ni).

## 2. Experimental

In a typical experiment, 0.75 g of 2,5-dihydroxyterephthalic acid (DHTA) and 3.0 g of $Co(NO_3)_2{\cdot}6H_2O$ or $Ni(NO_3)_2{\cdot}6H_2O$ were dissolved in 150 mL of DMF with sonication in a 250 mL Duran bottle. The bottle was capped tightly and placed in a 110°C oven for 20 h. After cooling to room temperature, the mother liquor was decanted and the crystals were washed with DMF. The product was immerged into methanol for six days. The products were then evacuated and heated under vacuum to150°C over a period of 12 h.

## 3. Results and discussion

The FE-SEM images of MOF-74(Co) and MOF-74(Ni) samples are shown in Fig.1a and 1b, respectively. It shows that MOF-74(Co) has cauliflower type structure with the particle size of about 4-6 mm and MOF-74(Ni) has uniformly distributed spherical shape with the average diameter of 2-4 mm, although some particles are aggregated. The XRD patterns (Fig.1c) shows that both samples have the unique crystalline phase indexed on the basis of a face-centered hexagonal unit cell of MOF-74. Both samples indicate almost similar diffraction patterns at identical 2*fÆ* values ascribed the formation of similar crystallographic phases.

**Figure 1.** -SEM images of (a) MOF-74(Co), (b) MOF-74(Ni), and (c) XRD patterns of synthesized MOF-74 samples.

MOF-74(Co) was thermally stable up to 258°C while in case of MOF-74(Ni) the value was 344°C confirmed from the TGA analysis shown in Fig. 2. The thermogravimetric curves of both samples are consistent with prominent weight-loss steps. It can be seen that the thermal stability is not similar for these two samples although they follow the same trend of loosing weight. In case of MOF-74(Co), a fully hydrated molecular sieve contains about 27 wt.% of water and other guest molecules. and in case of MOF-74(Ni) this value is almost similar, but there is one additional step of weight loss of about 13 wt%.

**Figure 2.** - TG/DTA curves of as-synthesized MOF-74(Co) and MOF-74(Ni)

From the $N_2$ adsorption isotherms it has been evaluated that both sample follow type-I isotherms with almost no hysteresis loop which corresponding to microporous material. The BET specific surface areas of MOF-74(Co) and MOF-74(Ni) MOFs were 1,406 and 1,418 $m^2g^{-1}$, respectively with the identical pore volume of 0.84 $cm^3g^{-1}$. DFT pore size distribution of MOF-74(Zn) and MOF-74(Cu) are shown in Fig. 4a and Fig. 4b, respectively while the values of average pore size and volume are presented in Table 1. These results suggest that these samples are microporous materials.

**Figure 3.** - $N_2$ adsorption/desorption isotherms of as-synthesized MOF-74(Co) and MOF-74(Ni). Filled and open symbols represent the adsorption and desorption of nitrogen gases, respectively.

*** Formatting Issue - Best Available Paper/Graphic ***

**Figure 4.** DFT pore size distribution of (a) MOF-74(Co) and (b) MOF-74(Ni)

Fine structure and oxidation state of Zn and Cu species in MOF-74 were further investigated through X-ray absorption spectroscopy. The data were collected several times and standard deviation was also calculated. The XANES region of the X-ray absorption spectrum is very informative on both the oxidation and coordination states of metal ions. XANES spectra of MOF-74(Co) and MOF-74(Ni) samples are shown in Fig. 5a and 5b, respectively along with the standard materials. The comparison is designed to determine the oxidation states of center atoms in MOFs and understanding the effects of DHTA linkers and guest molecules on the crystal structures. It was found that the edge position (7709.46 eV) of as-synthesized MOF-74(Co) sample is very close to the Co(II). However, compared with the standard Co(II) the offset was +0.46 eV, revealed that Co ions were not entirely surrounded by oxygen atoms and hence formed hydroxides ($OH^-$) through covalent bond. This result reveals that the as-synthesized sample was not totally moisture or guest molecule free. The source of the $OH^-$ ions is either adsorbed moisture or the organic linkers that cause the little distortion in the crystal structure. In case of MOF-74(Ni) the offset was +0.57 eV. Fine structural parameters for both samples are shown in Table 2 where it can be seen that the bond distances of Co,,ŸO and Ni,,ŸO were 1.95 and 1.97Å, with the coordination number of 4.27 and 4.23, respectively.

**Figure 5.** (a) Co K-edge XANES spectra of MOF-74(Co) samples with Co(0) standard, (b) Ni K-edge XANES spectra of MOF-74(Ni) with Ni(0) standard, and (c) $CO_2$ adsorption isotherms of MOF-74 samples measured at 273 K. Filled and open circles represent the adsorption and desorption, respectively.

The $CO_2$ adsorption capacities of the synthesized samples are shown in Fig. 5c. The steady ascending slope of the $CO_2$ adsorption isotherms represents strong adsorption of $CO_2$ molecules having quadrupoles onto the open metal sites in MOF-74 samples. The $CO_2$ adsorption capacity of MOF-74(Ni) measured at 273 K and 1.1 bar was 4.07 $mmol \cdot g^{-1}$; significantly higher than MOF-74(Co) which was 3.35 $mmol \cdot g^{-1}$. However, both results are much higher than other prospective adsorbents of $CO_2$ such as Zeolite 13X [3] and ZIF-96 [4]. The elevation in $CO_2$ adsorption in MOF-74(Ni) may attribute to the higher surface area than MOF-74(Cu) and more importantly the strong affinity to $CO_2$ molecules due to different electronic configurations. It can be seen that the adsorption and desorption cycles are almost reversible for both samples. This integrity was also corroborated by the $N_2$ adsorption. The steady uphill slope of the adsorption curves indicates that higher adsorption capacity may obtain at higher pressure.

## 4. Conclusions

The MOF-74 samples were capable to store significant amount of $CO_2$ at relatively lower pressure. The difference in $CO_2$ adsorbed amount between two samples may attribute to the different affinity of cobalt and nickel ions to $CO_2$ molecules as well as the specific surface area. The moderate thermal stability confirms that these MOFs may apply to control the $CO_2$ gas from the emission of refinery or other industries.

## References

*** **Formatting Issue - Best Available Paper/Graphic** ***


1. A. Torvanger, M. T. Lund, N. Rive, Mitig. Adapt. Strat. Gl. 18 (2013) 187.

2. T. G. Glover, G. W. Peterson, B. J. Schindler, D. Britt, O. M. Yaghi, Chem. Eng. Sci. 66 (2011) 163.

3. J.-S. Lee, J.-H. Kim, J.-T. Kim, J.-K. Suh, J.-M. Lee, C.-H. Lee, J. Chem. Eng. Data 47 (2002) 1237.

4. W. Morris, B. Leung, H. Furukawa, O. K. Yaghi, N. He, H. Hayashi, Y. Houndonougbo, M. Asta, B. B. Laird, O. M. Yaghi, J. Am. Chem. Soc. 132 (2010) 11006.

# Treatment of Flowback Water from Hydraulic Fracturing with Biochar

Maoqi (Mark) Feng

Steven Cooks

Zhigang Feng

April 2, 2014

AIChE 2014 Spring Conference, New Orleans

Chemical Engineering Dept., Div. of Chemistry and Chemical Engineering, Southwest Research Institute, San Antonio, TX 78238

# Outline

- Introduction
- Technical Approach
- Experimental Work
- Summary & Conclusions

2

# Water Demand in Hydraulic Fracturing

- Water demand is huge: 100 billion gallons of water used for 39,000 wells in 2011
- Fracking accounts for 50% of total water consumption in some Texas areas

3

# Water Requirement in Shale Gas Production

- Water requirements for hydraulic fracturing, by shale plays in Texas

| Water Requirements (gallons per well) | Low Value | Median Value | High Value |
|---|---|---|---|
| Barnett Shale | <1 MM | 2.6 MM | >8 MM |
| Haynesville and Bossier Shale | < 1 MM | 5.5-6 MM | >10 MM |
| Eagle Ford Shale | 1 MM | 6-6.5 MM | 13 MM |
| Woodford, Pearsall, and Barnett-PB Shale | <1 MM | 0.75-1 MM | <5 MM |

4

## Flowback Water Recycling

- Half of fracturing wells are in water-stressed areas
- Flowback water reuse is needed

5

## Reusable Water Quality

- Parameter Range

| | |
|---|---|
| TDS at 180°C, mg/l | 9,000–16,000 |
| Turbidity, NTU | 0-5 |
| pH | 6.5-8 |
| Iron, mg/l | 1-10 |
| Chloride, mg/l | 5,000-10,000 |
| Potassium, mg/l | 100-500 |
| Calcium, mg/l | 50-250 |
| Magnesium, mg/l | 10-100 |
| Sodium, mg/l | 2,000-5,000 |
| Boron, mg/l | 0-20 |

6

## Flowback Water Treatment

- Membrane method is expensive and cannot handle solids and hydrocarbons
- Zeolites generate more solid waste
- Bioremediation is not efficient
- Nanofiltration is easily plugged
- Activated carbon adsorption is more expensive

7

## Why Biochar

- Biochar is a good sorbent
- Green
- Biodegradable
- Soil compatible and amendment
- Easy to prodcue
- cost effective
- Mass production available

8

## Biochar vs. Activated Carbon

- Biochar is produced at lower temperature
- Biochar is a precursor for activated carbon
- Activated carbon needs an activation step
- Biochar is more feedstock flexible

9

## Biochar Production

- Biochar production from wood chips by pyrolysis
- Temperature 350°C – 550°C

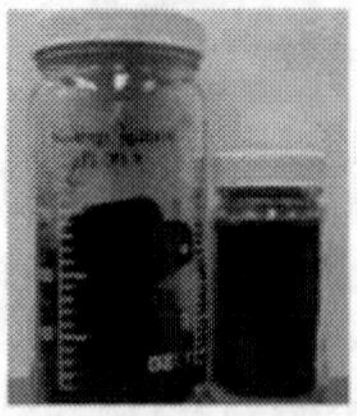

10

## Biochar Property

- Pine wood biochar carbon content vs. pyrolysis temperature

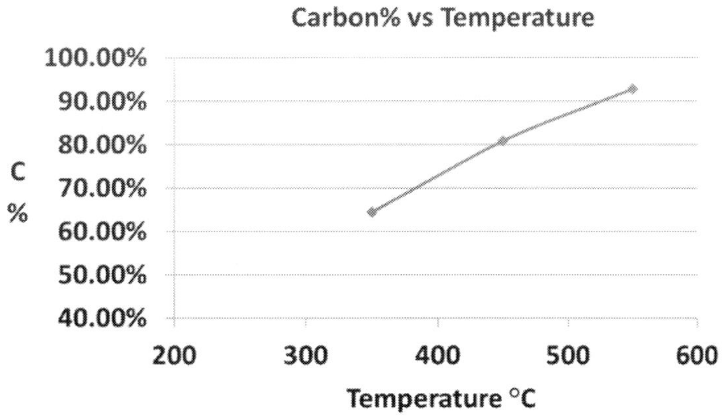


11

## Biochar Property

- Oak wood biochar carbon content vs. pyrolysis temperature

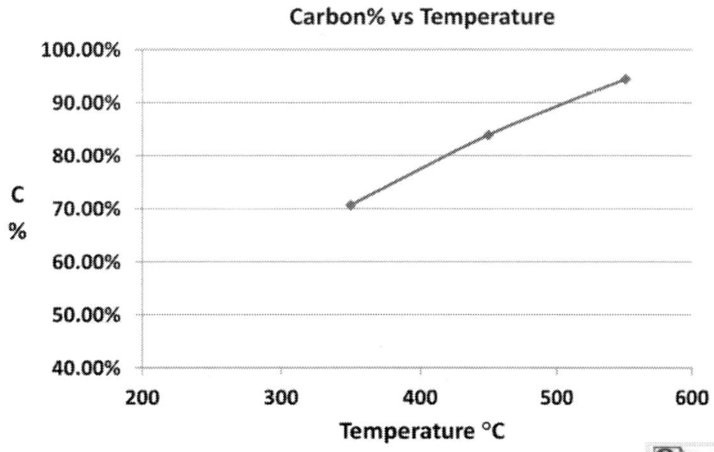


12

## Biochar Preparation

- Biochar prepared from different sources at 350°C

| Feed | Carbon% |
|---|---|
| **Cellulose** | **90.09%** |
| **Pine Wood** | **66.43%** |
| **Oat Hulls** | **58.59%** |

13

## Batch Adsorption

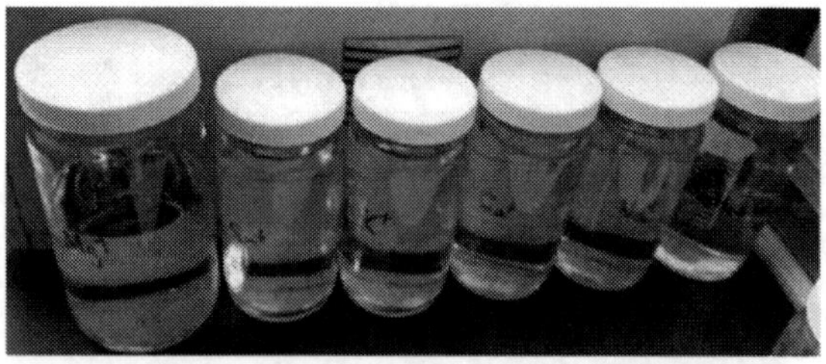

14

## Batch Adsorption

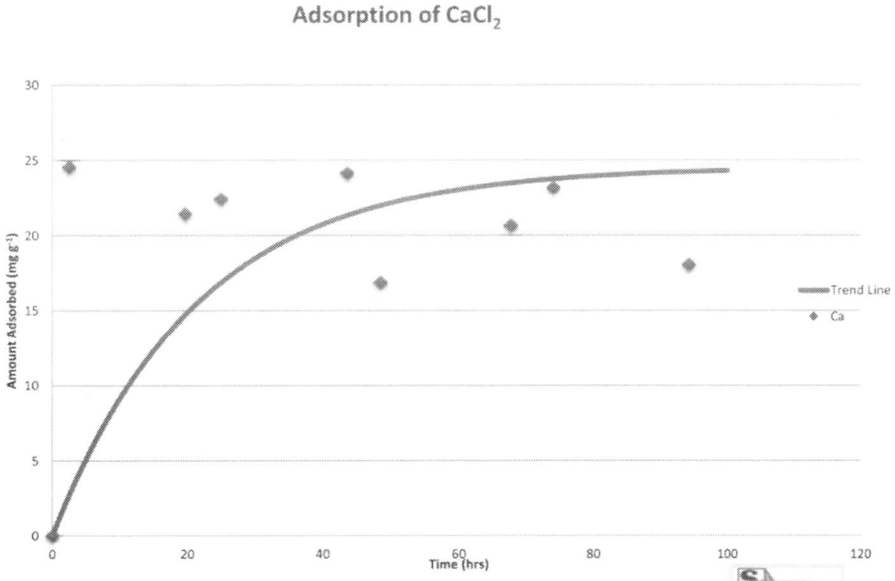


15

## Sorbent Bed Evaluation in Continuous Flow Mode

- ➢ Quick evaluation of sorbent capacity in continuous flow mode – more efficient than batch mode

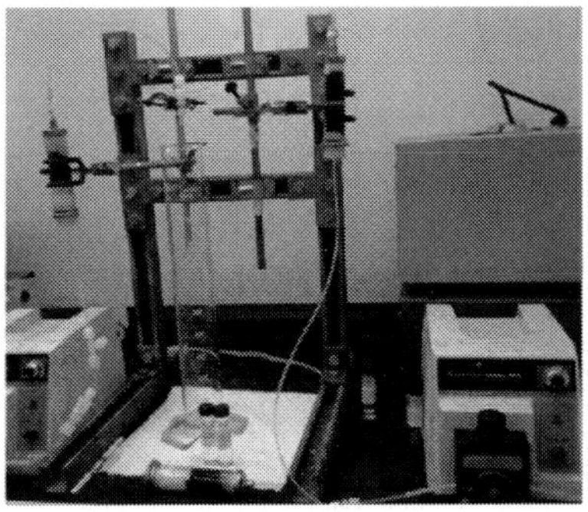

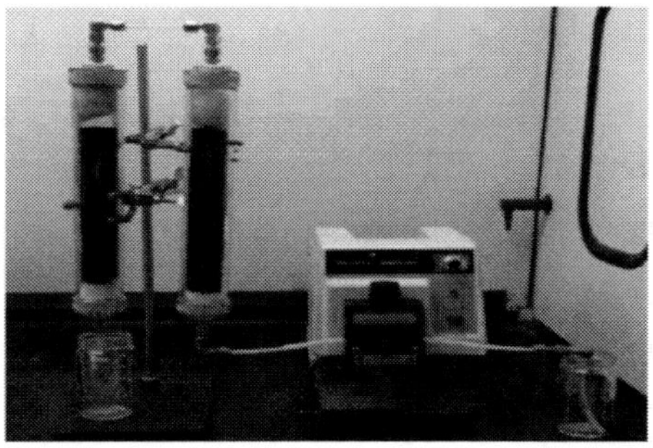

16

## NaCl Removal

- NaCl removal

| Time (hrs) | NaCl Mass (g) | Water Mass (g) | NaCl Concentration, wt% | NaCl wt% Adsorbed |
|---|---|---|---|---|
| 0.00 | 12.75 | 247.5 | 4.90% | 0.00% |
| 2.62 | 0.8921 | 18.5788 | 4.58% | 0.32% |
| 19.53 | 0.9598 | 19.3488 | 4.73% | 0.17% |
| 24.57 | 0.9886 | 19.7544 | 4.77% | 0.13% |
| 43.37 | 0.9908 | 20.1408 | 4.69% | 0.21% |
| 48.32 | 0.9065 | 18.6041 | 4.65% | 0.25% |

17

## Mg Removal

- $MgCl_2$ removal

| Time (hrs) | $MgCl_2$ Mass (g) | Water Mass (g) | $MgCl_2$ Concentration, wt% | $MgCl_2$ wt% Adsorbed |
|---|---|---|---|---|
| 0.00 | 4.8754 | 38.5514 | 11.23% | 0.00% |
| 0.5 | 4.1715 | 38.4528 | 9.79% | 1.44% |
| 1.0 | 4.3688 | 39.0205 | 10.07% | 1.16% |
| 1.5 | 4.6866 | 39.086 | 10.71% | 0.52% |
| 2.0 | 4.3229 | 38.8608 | 10.01% | 1.22% |

18

## Ca Removal

- $CaCl_2$ removal

| Time (hrs) | $CaCl_2$ Mass (g) | Water Mass (g) | $CaCl_2$ Concentration, wt% | $CaCl_2$ wt% Adsorbed |
|---|---|---|---|---|
| 0.00 | 3.975 | 38.7583 | 9.30% | 0.00% |
| 0.5 | 3.9614 | 39.2739 | 9.16% | 0.14% |
| 1.0 | 3.936 | 38.9826 | 9.17% | 0.13% |
| 1.5 | 3.9684 | 39.5214 | 9.12% | 0.18% |
| 2.0 | 3.9067 | 39.0601 | 9.09% | 0.21% |
| 2.5 | 3.9021 | 39.369 | 9.02% | 0.28% |

19

## Sr Removal

- $Sr(NO_3)_2$ removal

| Time (hrs) | $Sr(NO_3)_2$ Mass (g) | Water Mass (g) | $Sr(NO_3)_2$ Concentration, wt% | $Sr(NO_3)_2$ wt% Adsorbed |
|---|---|---|---|---|
| 0.0 | 0.4834 | 20.7043 | 2.28% | 0.00% |
| 0.5 | 0.4557 | 20.665 | 2.16% | 0.12% |
| 1.0 | 0.4491 | 21.1634 | 2.08% | 0.20% |
| 1.5 | 0.4191 | 20.3263 | 2.02% | 0.26% |
| 2.0 | 0.4114 | 20.5186 | 1.97% | 0.32% |
| 2.5 | 0.4031 | 20.1342 | 1.96% | 0.32% |

20

# Inorganics Removal

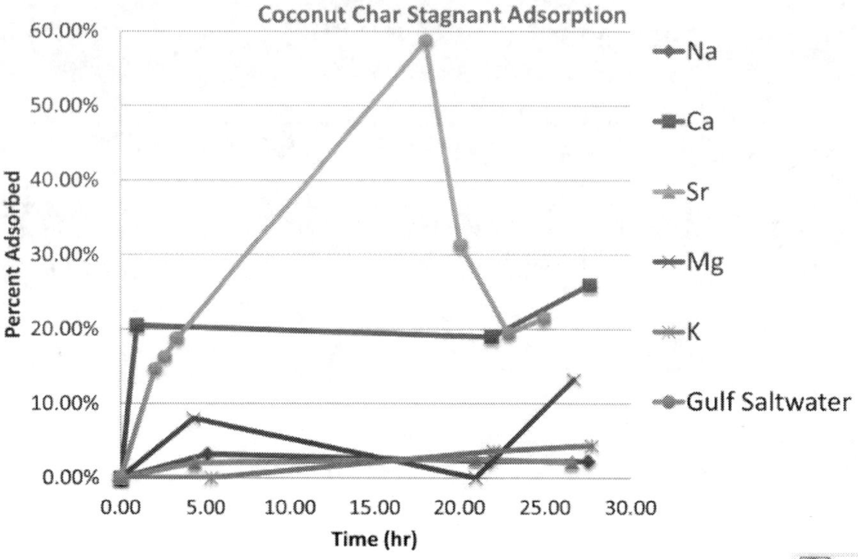


21

# Hydrocarbons Removal

- Removal of diesel hydrocarbons in water by biochar adsorption
- Per pass removal 87%
- Higher adsorption capacity for biochar produced at higher temperature

22

## Solid Particulates Removal

- Removal of solid particulates in flowback is necessary
- Pump damage
- Long settling time
- Fine solids removal is still in progress

23

## Summary

- Biochar production from wood chips
- Quick evaluation of sorbent capacity in continuous flow mode
- Successful removal of inorganic salts: $NaCl$, $K^+$, $MgCl_2$, $CaCl_2$, $Sr^{2+}$
- Successful inorganics removal from sea water
- Successful removal of hydrocarbons from simulated flowback water

24

## Some Thoughts

- Biochar adsorption bed can be used with other methods, such as ultrafiltration, reverse osmosis, and ion exchange units
- Biochar sorbent bed can be used as a pretreatment or post-polishing step for other methods

25

## Acknowledgement

Thank you!

Questions?

26

# 352927 Hazards Management for Building Locations through Air Dispersion Modeling

**Abstract:**

Oil, gas, and petrochemical industries have reached a critical era involving its timeworn infrastructure. Previously, facilities degraded with minimal effort allocated for updates that ensure personnel safety (i.e., process inventory isolation, automative control systems, etc.). More recently, industry as a whole is progressively adopting a more safety conscientious attitude about personnel and the environment. These personnel and environmental concerns are driving new codes, standards, recommended practices, and regulations to be developed and implemented. Safety studies are being performed which assess facilities to determine if these new regulations are met, while providing key insight to eliminate any deficiencies that are discovered. Available advanced air dispersion models estimate or predict the concentration of air pollutants or toxicants emitted from various sources such as industrial plants, road traffic vehicles or accidental chemical releases. Facility siting assessments are amongst the required studies used for managing the risk to on-site personnel located in new and existing buildings intended for occupancy, where air dispersion models can be used to determine the consequences of accidental releases of hazardous or toxic materials. This paper discusses case studies carried out for hazards management for building locations with different air dispersion modeling software tools including: SCREEN3, AERMOD, ALOHA, and PHAST.

# 353139 H2S Removal from Biogas Using an Ionic Liquid as Physical Absorbent

**Abstract:**

Biogas is nowadays a valuable renewable energy carrier which can be used directly as a fuel or as a raw material for the production of synthesis gas and/or hydrogen. Methane ($CH_4$) and carbon dioxide ($CO_2$) are the main constituents, but biogases also contain significant quantities of undesirable compounds (contaminants), such as hydrogen sulfide ($H_2S$) and ammonia ($NH_3$). The existence and quantities of these contaminants depend on the biogas source (i.e., landfills, anaerobic fermentation of manure). Their presence constitutes a major problem because (1) they can adversaly affect any biogas thermal or thermocatalytic conversion device (e.g., corrosion, erosion, fouling); and (2) they produce harmful environmental emissions. It is therefore important to develop a reliable and efficient biogas purification process. This work is aimed at presenting an alternative removal approach for $H_2S$ from biogas based on physical absorption using an ionic liquid as solvent.

# The Growth of Idea Ownership: Patent Infringement and Risk Management within the Chemical Process Industries

Paul LaVanway, Esq.
Fredrikson & Byron, P.A.

14 AIChE Spring Meeting, New Orleans, LA

Fredrikson & Byron, P.A.

## Overview of Presentation

I. Patent Activity Update

II. Practical Risk Management Techniques

III. New Tools

14 AIChE Spring Meeting, New Orleans, LA

Fredrikson & Byron, P.A.

# PATENT ACTIVITY UPDATE

14 AIChE Spring Meeting, New Orleans, LA

Fredrikson & Byron, P.A.

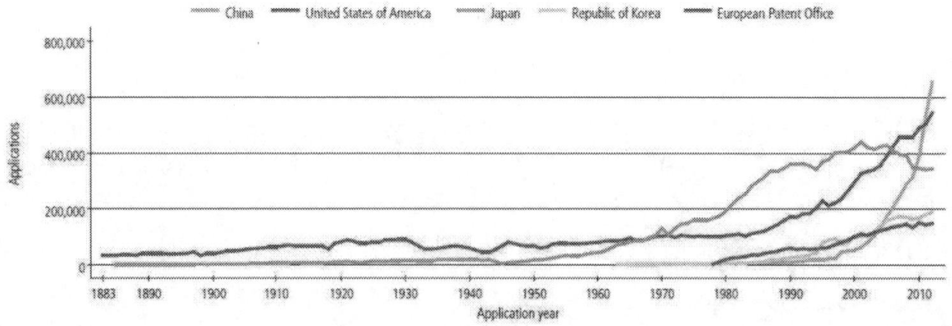


14 AIChE Spring Meeting, New Orleans, LA

Fredrikson & Byron, P.A.

# Worldwide Filing Trends

| Origin | Field of technology | | | | | | | | | | | | | | | |
|---|---|---|---|---|---|---|---|---|---|---|---|---|---|---|---|---|
| | Australia | Canada | China | Finland | France | Germany | Italy | Japan | Netherlands | Republic of Korea | Russian Federation | Sweden | Switzerland | United Kingdom | United States of America | Others |
| **Chemistry** | | | | | | | | | | | | | | | | |
| Organic fine chemistry | 1,164 | 2,798 | 18,900 | 539 | 20,317 | 38,897 | 4,436 | 38,136 | 6,148 | 10,424 | 2,185 | 5,375 | 17,365 | 12,314 | 75,204 | 30,237 |
| Biotechnology | 2,592 | 3,432 | 17,402 | 804 | 8,107 | 16,915 | 2,306 | 20,814 | 5,500 | 9,165 | 1,822 | 1,765 | 7,316 | 7,058 | 67,521 | 21,840 |
| Pharmaceuticals | 3,647 | 6,297 | 38,194 | 958 | 18,328 | 31,204 | 7,370 | 28,798 | 6,362 | 10,514 | 5,692 | 7,748 | 23,615 | 15,563 | 123,173 | 50,208 |
| Macromolecular chemistry, polymers | 382 | 816 | 11,135 | 2,173 | 4,439 | 20,009 | 2,446 | 45,323 | 4,576 | 7,906 | 935 | 325 | 3,439 | 1,714 | 31,810 | 7,538 |
| Food chemistry | 875 | 1,122 | 19,191 | 428 | 2,595 | 5,330 | 1,070 | 13,272 | 6,409 | 12,111 | 17,015 | 337 | 4,744 | 2,464 | 19,963 | 11,149 |
| Basic materials chemistry | 1,002 | 1,992 | 24,266 | 901 | 6,012 | 35,527 | 1,536 | 42,724 | 7,038 | 11,113 | 3,343 | 690 | 6,970 | 6,340 | 51,100 | 13,535 |
| Materials, metallurgy | 1,697 | 1,569 | 30,285 | 1,729 | 6,697 | 17,150 | 1,480 | 43,685 | 1,846 | 12,235 | 8,359 | 1,514 | 2,186 | 2,422 | 20,753 | 12,182 |
| Surface technology, coating | 670 | 1,386 | 11,959 | 1,185 | 4,907 | 15,954 | 1,642 | 52,591 | 1,933 | 10,246 | 1,779 | 1,200 | 2,604 | 2,498 | 36,858 | 7,336 |
| Micro-structural and nano-technology | 116 | 96 | 1,710 | 124 | 606 | 1,371 | 94 | 2,405 | 254 | 2,579 | 587 | 134 | 153 | 158 | 2,486 | 683 |
| Chemical engineering | 1,479 | 2,235 | 16,952 | 1,840 | 7,352 | 25,809 | 2,846 | 32,504 | 4,788 | 13,354 | 4,426 | 2,422 | 4,287 | 5,490 | 40,121 | 13,512 |
| Environmental technology | 762 | 1,514 | 13,620 | 855 | 5,042 | 13,594 | 1,481 | 27,259 | 2,533 | 12,888 | 2,303 | 1,116 | 1,765 | 2,719 | 20,263 | 8,002 |

14 AIChE®
Spring Meeting, New Orleans, LA

Fredrikson & BYRON, P.A.

# Worldwide Filing Trends

| Field of technology | Publication year | | | | | Growth rate |
|---|---|---|---|---|---|---|
| | 2007 | 2008 | 2009 | 2010 | 2011 | 2007-11 (%) |
| **Chemistry** | | | | | | |
| Organic fine chemistry | 52,609 | 53,873 | 53,009 | 52,559 | 51,271 | -0.6 |
| Biotechnology | 33,929 | 35,654 | 37,652 | 38,423 | 40,849 | 4.7 |
| Pharmaceuticals | 73,345 | 73,769 | 72,126 | 69,258 | 69,311 | -1.4 |
| Macromolecular chemistry, polymers | 27,627 | 28,345 | 28,800 | 28,683 | 28,748 | 1.0 |
| Food chemistry | 21,690 | 23,582 | 27,245 | 27,889 | 30,858 | 9.2 |
| Basic materials chemistry | 38,920 | 41,085 | 42,322 | 43,795 | 45,115 | 3.8 |
| Materials, metallurgy | 30,200 | 33,775 | 34,968 | 36,988 | 38,542 | 6.3 |
| Surface technology, coating | 29,348 | 30,732 | 32,792 | 33,215 | 33,954 | 3.7 |
| Micro-structural and nano-technology | 2,302 | 2,478 | 2,829 | 3,094 | 3,203 | 8.6 |
| Chemical engineering | 33,263 | 35,087 | 35,766 | 36,618 | 37,991 | 3.4 |
| Environmental technology | 21,688 | 22,605 | 24,333 | 25,572 | 26,324 | 5.0 |

14 AIChE®
Spring Meeting, New Orleans, LA

Fredrikson & BYRON, P.A.

## Comparative Trends

| Field of technology | Publication year 2007 | 2008 | 2009 | 2010 | 2011 | Growth rate 2007-11 (%) |
|---|---|---|---|---|---|---|
| **Electrical engineering** | | | | | | |
| Electrical machinery, apparatus, energy | 101,013 | 105,591 | 111,876 | 116,009 | 122,697 | 5.0 |
| Audio-visual technology | 92,978 | 91,258 | 85,411 | 80,607 | 75,881 | -5.0 |
| Telecommunications | 68,915 | 69,804 | 60,808 | 56,417 | 49,533 | -7.9 |
| Digital communication | 57,696 | 64,237 | 69,191 | 74,389 | 79,726 | 8.4 |
| Basic communication processes | 17,208 | 17,734 | 17,164 | 16,515 | 15,554 | -2.5 |
| Computer technology | 123,504 | 134,493 | 133,072 | 130,052 | 134,272 | 2.1 |
| IT methods for management | 19,370 | 21,684 | 25,022 | 23,261 | 23,532 | 5.0 |
| Semiconductors | 76,617 | 81,238 | 78,685 | 77,547 | 80,049 | 1.1 |
| **Mechanical engineering** | | | | | | |
| Handling | 42,764 | 42,783 | 42,764 | 42,414 | 44,344 | 0.9 |
| Machine tools | 36,607 | 38,315 | 40,585 | 43,242 | 46,363 | 6.1 |
| Engines, pumps, turbines | 41,891 | 43,573 | 48,156 | 48,372 | 48,423 | 3.7 |
| Textile and paper machines | 35,815 | 33,681 | 32,251 | 30,633 | 30,364 | -4.0 |
| Other special machines | 44,008 | 45,912 | 47,541 | 49,011 | 50,981 | 3.7 |
| Thermal processes and apparatus | 25,280 | 25,699 | 27,259 | 29,382 | 28,855 | 4.2 |
| Mechanical elements | 43,957 | 47,513 | 47,286 | 46,405 | 46,913 | 1.6 |
| Transport | 65,237 | 67,717 | 70,542 | 67,015 | 65,439 | 0.1 |

14AIChE Spring Meeting, New Orleans, LA

Fredrikson & BYRON, P.A.

## Change in 5 year average over last 20 Years

### Fastest Growing

| Category | Growth Rate |
|---|---|
| Computer Software | +11% |
| Data Processing | +11% |
| Semiconductor Devices | +10% |
| Video Distribution | +10% |
| Computer Hardware | +8% |
| Chemical – Crystals | +8% |
| Nanotechnology | +8% |
| Information Storage | +6% |
| Communications | +6% |
| Design | +5% |

### Slowest Growing

| Category | Growth Rate |
|---|---|
| Chemical – Purification/Evaporation /Distillation | -2% |
| Chemical – General Compounds | -3% |
| Time Measurement | -3% |
| Machine Mechanism | -3% |
| Chemical Manufacture Specific | -3% |
| Organic Compounds | -3% |
| Pipes & Joints | -4% |
| Education | -4% |
| Hazardous Waste | -4% |
| HVAC | -4% |

14AIChE Spring Meeting, New Orleans, LA

& BYRON, P.A.

# US Patent Filing Trends

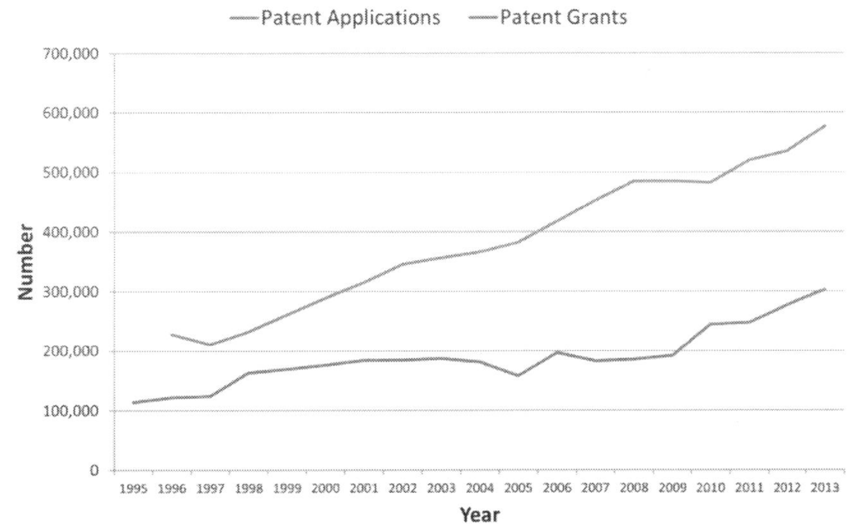


14 AIChE Spring Meeting, New Orleans, LA

Fredrikson & BYRON, P.A.

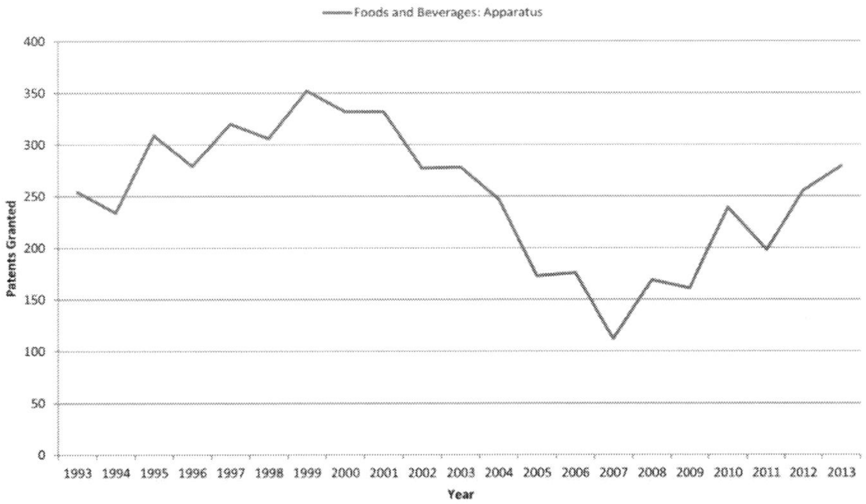


14 AIChE Spring Meeting, New Orleans, LA

Fredrikson & BYRON, P.A.

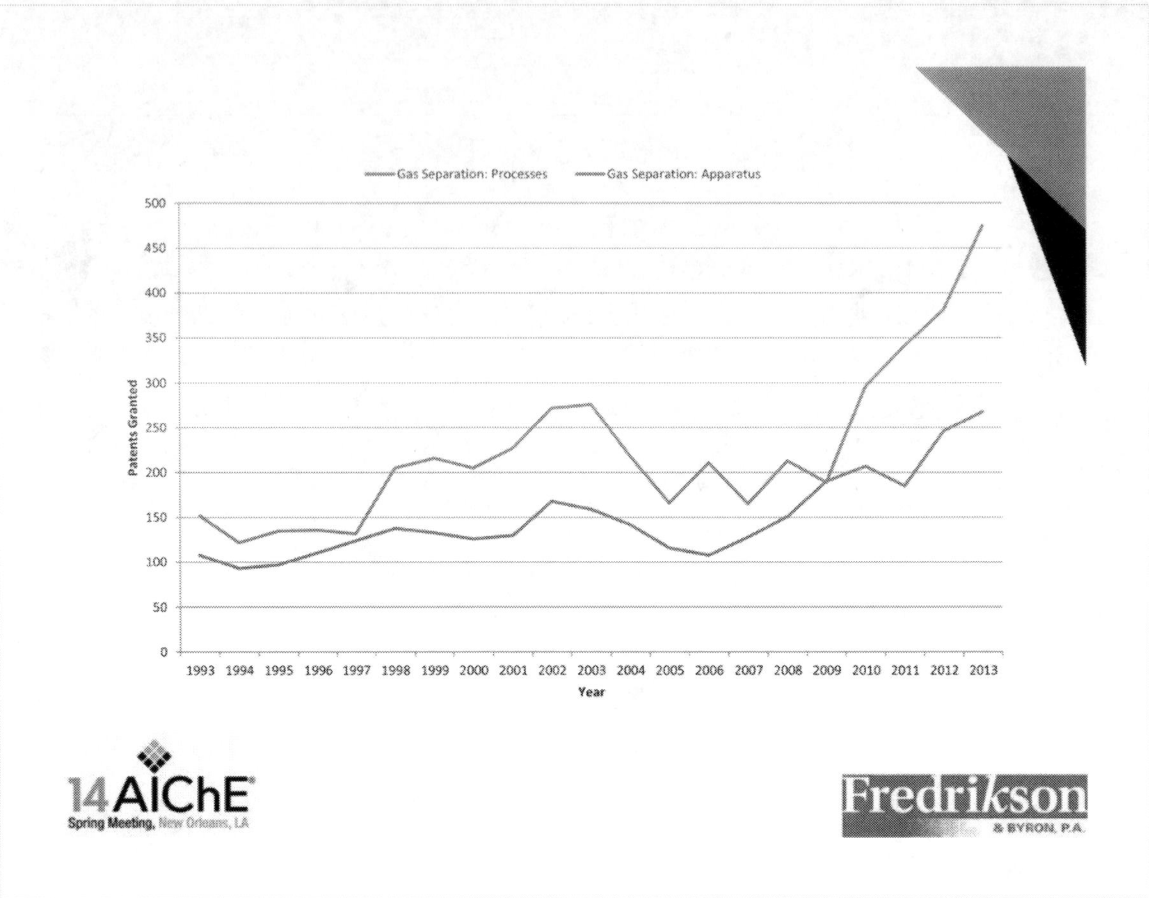

Gas Separation: Processes
Gas Separation: Apparatus
Patents Granted
0
50
100
150
200
250
300
350
400
450
500
1993 1994 1995 1996 1997 1998 1999 2000 2001 2002 2003 2004 2005 2006 2007 2008 2009 2010 2011 2012 2013
Year
14AIChE
Spring Meeting, New Orleans, LA
Fredrikson
& BYRON, P.A.


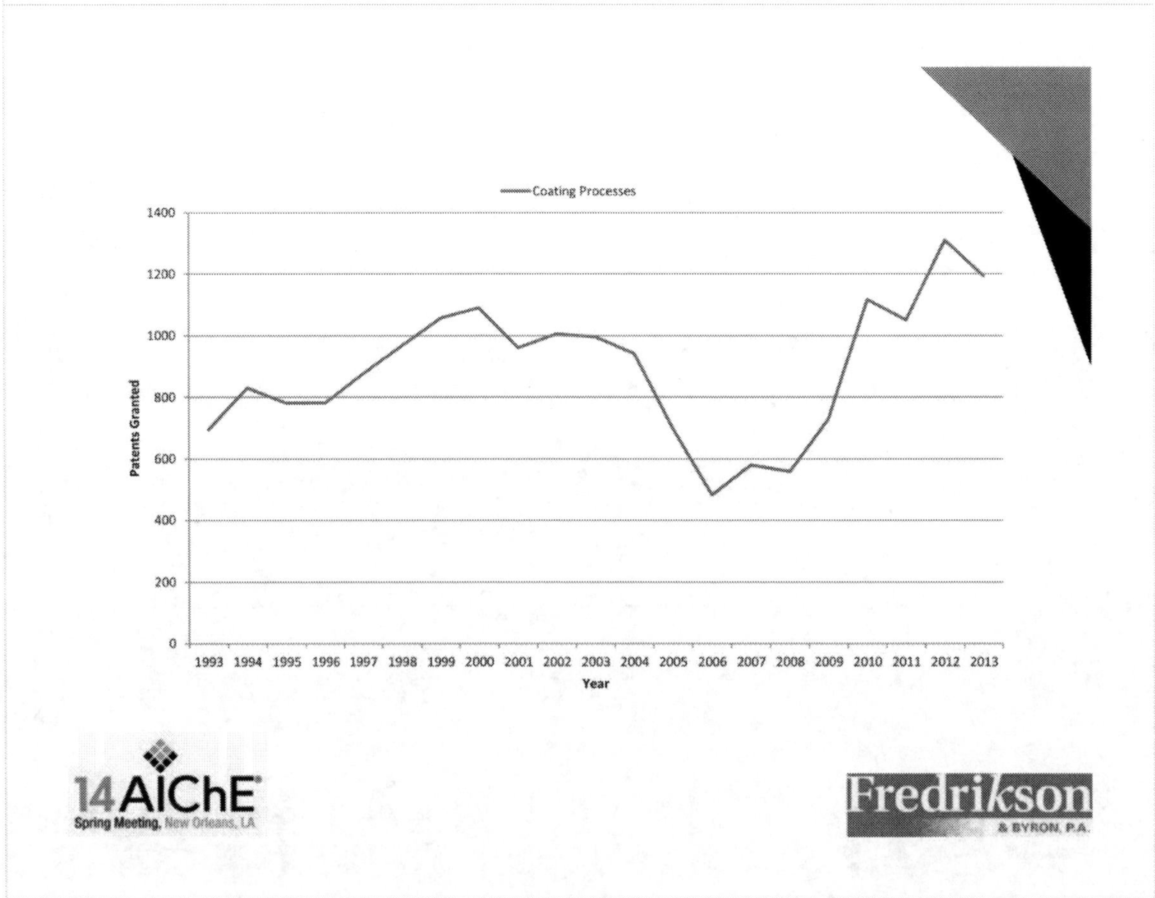

Coating Processes
Patents Granted
0
200
400
600
800
1000
1200
1400
1993 1994 1995 1996 1997 1998 1999 2000 2001 2002 2003 2004 2005 2006 2007 2008 2009 2010 2011 2012 2013
Year
14AIChE
Spring Meeting, New Orleans, LA
Fredrikson
& BYRON, P.A.

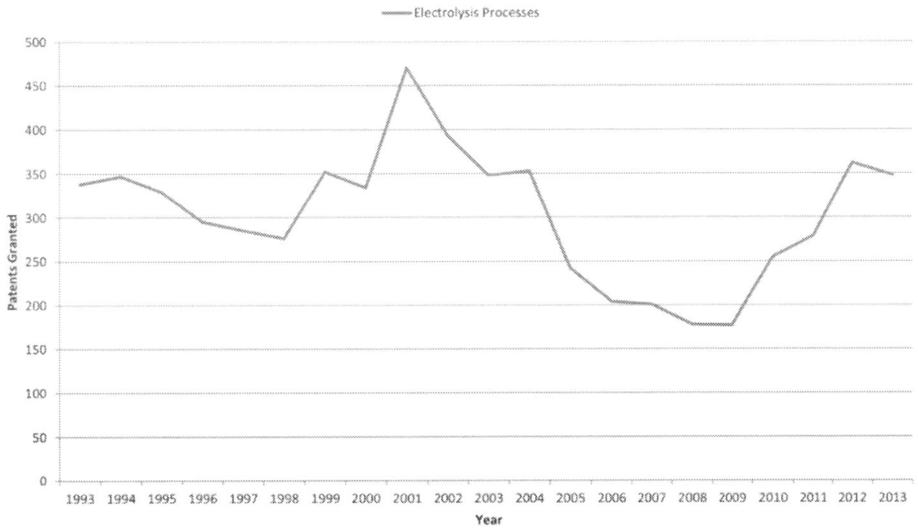


14 AIChE Spring Meeting, New Orleans, LA

Fredrikson & BYRON, P.A.

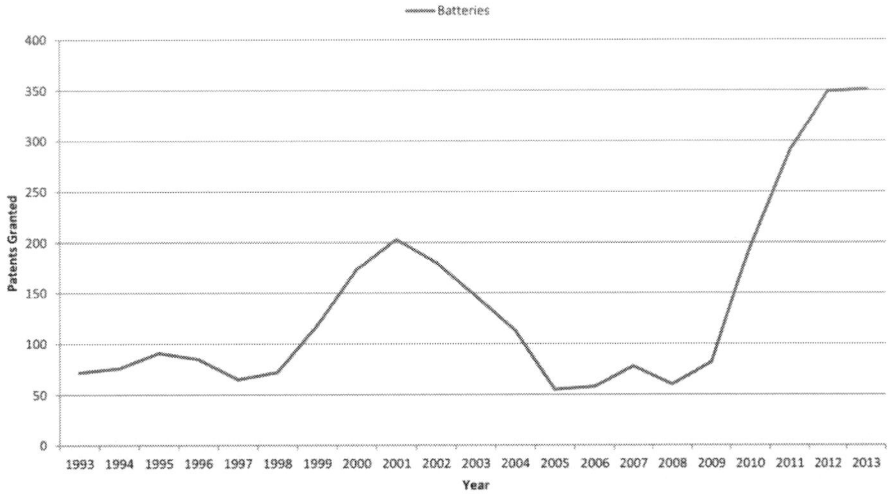


14 AIChE Spring Meeting, New Orleans, LA

Fredrikson & BYRON, P.A.

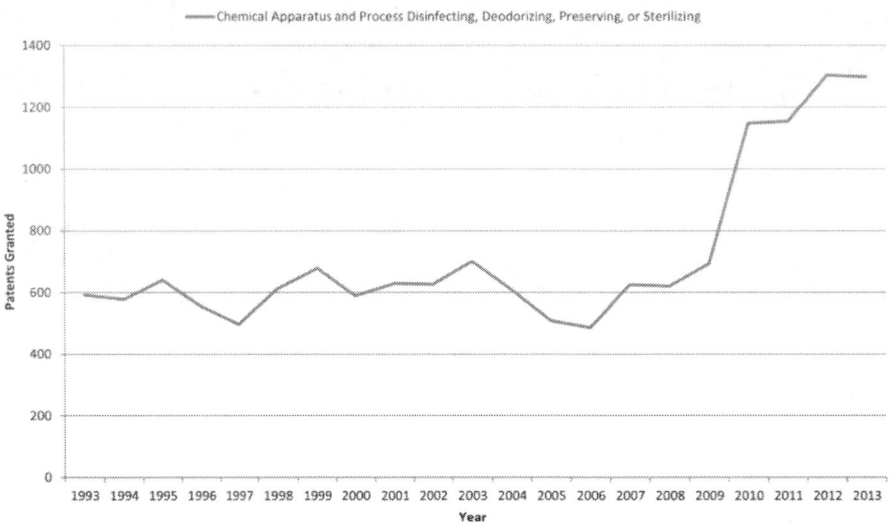


14AIChE Spring Meeting, New Orleans, LA

Fredrikson & BYRON, P.A.

# Patent Litigation Statistics

- Median damage award last 15 years all cases
  - $1.9 M → 16.5 M
- Median over last 5 years
  - $4.9M
- Chemical Industry (excluding bio/pharma)
  - Generally smaller median damages (less than $3M) but characterized by outliers
- Disruptive
  - Median time to trial 2.3 years

14AIChE Spring Meeting, New Orleans, LA

Fredrikson & BYRON, P.A.

# Cost of Patent Infringement Litigation

| Amount at Risk | Median Cost |
|---|---|
| Less than $1 Million | $700,000 |
| $1-$10 Million | $2,000,000 |
| $10-$25 Million | $3,325,000 |
| $25 Million+ | $5,500,000 |

14 AIChE
Spring Meeting, New Orleans, LA

Fredrikson & Byron, P.A.

# Distribution of Patent Cases

| Top Ten Industry | Percent of Cases |
|---|---|
| Consumer Products | 18% |
| Biotechnology / Pharma | 13% |
| Industrial / Construction | 11% |
| Computer Hardware / Electronics | 9% |
| Medical Devices | 9% |
| Business / Consumer Services | 6% |
| Software | 6% |
| Automotive / Transportation | 5% |
| Telecommunications | 5% |
| Chemical / Synthetic Materials | 5% |

14 AIChE
Spring Meeting, New Orleans, LA

Fredrikson & Byron, P.A.

# But . . . Looking at 11-15

| Rank | Industry |
|---|---|
| 11 | Food/Beverages/Tobacco |
| 12 | Metals/Mining |
| 13 | Clothing/Textiles |
| 14 | Energy |
| 15 | Agriculture |

14 AIChE Spring Meeting, New Orleans, LA

Fredrikson & Byron, P.A.

# RISK MANAGEMENT STRATEGIES

14 AIChE Spring Meeting, New Orleans, LA

Fredrikson & Byron, P.A.

## Understanding the Nature of the Risk

- "Fuzzy" boundaries Problems

**Features**

- ~~Back~~
- Seat
- ~~Four~~ Legs resting on the ground

**Revised**

- Seat
- At least one base contacting the ground

14 AIChE Spring Meeting, New Orleans, LA

Fredrikson & Byron, P.A.

## Understanding the Nature of the Risk

**Revised**

- Seat
- At least one base contacting the ground

Legal Scope Not Necessarily Co-extensive with Thrust of Description

14 AIChE Spring Meeting, New Orleans, LA

Fredrikson & Byron, P.A.

## Managing New Product Risk

- Correlate Effort to Risk
  - Repeat Use of Old Technology in Existing Field
  - Old Technology in Different Field
  - New Technology in Existing Field
  - New Technology in New Field

Increasing Risk

14 AIChE Spring Meeting, New Orleans, LA

Fredrikson & Byron, P.A.

## Other Factors to Consider

- Intensity of Competition within Field
- Past History Between Competitors
- Value of Marketplace and Opportunity
- Relative Size of Parties
- Monetization Proposition

14 AIChE Spring Meeting, New Orleans, LA

Fredrikson & Byron, P.A.

## Managing New Product Risk

- Use a Phased Approach
  - Phase I
    - Basic Understanding of Landscape and Any "Show Stopping" Patents
    - Focus on Likely Competitors – Review Patents, Literature, Products
  - Phase II
    - Revisit Base Design and Evaluate Ancillary Features
  - Phase III
    - Check Any Design Modifications or Newly Issuing Patents

14 AIChE Spring Meeting, New Orleans, LA

Fredrikson & Byron, P.A.

## Managing Ongoing Risk

- Monitor Key Competitors / Inventors
- Monitor Technology Space
- Work with Sales and Marketing Personnel to Track New Product Introduction
- Watch Partners and Consultants

14 AIChE Spring Meeting, New Orleans, LA

Fredrikson & Byron, P.A.

# NEW TECHNIQUES FOR ACTION

14 AIChE
Spring Meeting, New Orleans, LA

Fredrikson & BYRON, P.A.

# Knowing What Tools Are Available

- Pre-Issuance Submission
  - Attack the Patent Application Before it Becomes a Problem
- Simple But Limited Patent Office Attack
  - Ex Parte Reexamination
- More Involved Patent Office Attacks
  - Inter Partes Review
  - Post Grant Review

14 AIChE
Spring Meeting, New Orleans, LA

Fredrikson & BYRON, P.A.

# CONCLUSIONS

- CPI Patents Generally Trending Upward
- Great Sector Specific Variability
- Effective Risk Management Requires Being Proactive and Deploying Resources Efficiency
- New Tools to Attack Problem Patents

14 AIChE Spring Meeting, New Orleans, LA

Fredrikson & Byron, P.A.

**QUESTIONS?**

14 AIChE Spring Meeting, New Orleans, LA

Fredrikson & Byron, P.A.

# Chemical Engineering & the Law --Case Examples in Industrial Litigation

**Paper# 352190**

*Satish Almaula*
*Senior Principal Consultant*
*Baker Engineering and Risk Consultants, Inc.*
*salmaula@bakerrisk.com*

**Presented at American Institute of Chemical Engineers Spring Meeting, New Orleans LA April 2014**

## Abstract

Chemical, petroleum, and allied manufacturing sectors involved in research, technologies, engineering and project management at times encounter issues and problems in successfully meeting a project's intended financial, technical, and operational goals and expectations. Significant deviations from goals and expectations can result in lawsuits between involved parties. Alternatively, parties file insurance claims, which can also result in lawsuits.

Research and technology issues can result in patent and intellectual property legal cases. In addition, issues can arise from engineering design, licensing, product liability, safety, regulatory compliance, construction and operational deficiencies, or an inability to meet contractual and performance commitments. Accidents and incidents at manufacturing or other facilities can result in damage to property and the environment, injuries and death as well as interruption to normal operation causing business loss. Such damages and losses also can result in lawsuits and insurance claims.

The need for collaboration between lawyers, scientists, engineers and other technical specialists is obvious when conducting legal, technical and business analyses to get to the root cause of the matter to support or defend a legal complaint or an insurance claim.

This paper presents a brief introduction to lawsuits and how legal and technical teams can collaborate with their respective experience and expertise in handling a lawsuit either as a plaintiff's or a defendant's team.

The main focus of the presentation will be case examples that demonstrate how principles familiar to engineers and other technical specialists play a part in litigation involving diverse issues in the chemical, refining and allied business sectors.

# 1 Introduction

Technical professionals, whether they are engineers, scientists or technical specialists often get drawn into legal disputes. Modern day media reports of claims and settlements tend to focus heavily on events in the electronic, information technology, pharmaceutical and drug industries, especially when the issues have a direct effect on consumers and involve huge sums of money.

Petroleum, chemical processing, and allied manufacturing sectors have their share of claims and disputes involving large amounts of money but do not always get the high visibility as the above-mentioned industries. Catastrophic incidents such as fires, explosions, release of hazardous materials leading to deaths, injuries and property damage do get expanded coverage and high visibility.

It is not surprising, therefore that chemical engineers, scientists and technology specialists are often drawn into insurance claims and legal disputes. Lawyers, insurers, scientists, engineers, and technical specialists often collaborate in such matters.

This presentation provides a brief background on typical issues leading to legal disputes followed by selected case examples. Names of organizations, involved parties, and individuals are not disclosed or discussed in this presentation. The selected examples are from the author's experience in lawsuits related to chemical processing and allied industries.

# 2 Insurance Claims and Lawsuits

Insurance claims and lawsuits can arise from diverse losses and disputes such as property damage, fatalities and injuries, negligent performance or, in extreme but uncommon instances, they can arise from intentional or criminal acts.

Losses alleged in processing industry claims and disputes tend to be related to: patent and intellectual property; engineering design; licensing; product liability; safety; regulatory compliance; construction and operational deficiency; and the inability to meet contractual and performance commitments.

Technical professionals including those in chemical engineering and similar fields can be drawn into insurance claims and lawsuits as fact or expert witnesses.

Fact witnesses are generally limited to testifying about facts that they know firsthand about an issue without expressing an opinion about them. Expert witnesses are generally technical professionals with extensive experience and in-depth knowledge in a particular subject matter. They analyze facts, evidence and information and give testimony on their findings as well as their expert opinions.

Given the diversity of training and experience of chemical engineers, they can be called upon to serve as experts in correspondingly diverse fields, some of which may or may not fall under a classic view of chemical engineering.

Collaboration between insurers, lawyers and technical experts typically involves, but is not limited to, the following:

- Developing a clear understanding of the claims and allegations.
- Knowing and understanding the involved parties and the type of business they are in.
- Gathering the information, evidence and facts necessary for legal, technical and business analyses.
- Performing a screening analysis of the strengths and weaknesses of legal, technical and business factors to support or rebut the claims and allegations.
- Determining the strategies for the approach and steps to be implemented by the legal, technical and business professionals in handling the matter.

## 3 Selected Examples

### Example 1 -- Patent Related Lawsuit

A producer of asphalt shingle roof products hired a retired former senior executive as a consultant to reduce disposal costs of reject materials from production. Roofing shingles typically contain asphalt, filler material and granules evenly placed on a core material such as fiber glass or cellulose. The produced material is cut into standard sizes called "a course." The cutting produces some unusable material ("rejects" or "rejected material").

The consultant researched, developed and pilot-tested a method to recycle rejects back into production. The reject material, containing asphalt and filler material, was finely ground with limestone and then recycled back into production, saving costs of filler material and asphalt and avoiding costs associated with the disposal of the reject material. The tests were successful and easily scaled up to commercial level.

The consultant asked the company to file a patent for the new recycling method, but the company procrastinated and, ultimately, did not file for a patent.

Later, the consultant, after ending his work with the company, filed and obtained a patent for the new recycling method. Upon learning of the patent, the company sued the consultant/inventor for stealing proprietary information, which initially prevented him from offering the technology to others. The viability, cost effectiveness, and worth of the new technology as well as information protection were all issues raised in the lawsuit.

Analysis showed that the proposed retrofit of an average size recycle plant at 70 plants in the industry across the U.S. would roughly total $156,000,000 in capital costs. Estimated annual savings to the industry from reduced raw material costs and avoidance of disposal costs amounted to $190,000,000, yielding an attractive return on investment. So, the viability and cost effectiveness of the new technology were demonstrated in principle.

The worth of the new technology/licensing value was determined by financial and economic experts.

The matter went to trial, and the consultant/inventor won the case.

Principles familiar to technical specialists applied in this example: Manufacturing processes, unit operations, industry survey, capital cost estimation, annual cost estimation, return on investment, and protection of important information.

## Example 2 -- Technology Transfer, Design-Build, Project Management Lawsuit

The owner of an operating plant wanted to convert by-product waste to useful and saleable products. The owner acquired a license for new technology to be used for the conversion of waste to intermediates. The owner also acquired established technologies to convert intermediates to saleable products. The new technology licensor represented to the owner that the technology was demonstrated and ready for commercial use. The licensor also offered used equipment in good condition, which the owner agreed to purchase.

A design-build contract for the plant and process for the new technology was awarded on a fixed price ($150 million) and fixed schedule basis to an architect and engineering firm ("A&E").

The technology licensor provided a Process Information Package (PIP) for design-build work as "frozen" --- that is, ready for detailed design and procurement. Numerous and untimely changes were imposed by both the technology licensor and the owner, causing delays in the detailed design, procurement and project completion. The A&E submitted change orders to the owner, who orally approved them.

The technology sold to the owner was actually its first application on a commercial scale in the U.S. The only other reported use of the technology on a commercial scale was in Japan, where several modifications were needed to make the process work correctly. The condition of the used equipment sold to the owner was not "good," as claimed. The equipment required significant cleaning and modification to meet design and code requirements.

Project completion was delayed by three months but, when completed, the system operated satisfactorily. The A&E submitted a Request for Equitable Adjustment ("REA") to the fixed price for the additional costs. In response, the owner sued the A&E for "Bad Project Management" claiming damages amounting to seven times the REA amount. Of note is that the technology licensor was not drawn into the lawsuit and was not held accountable for changes made to the "frozen" PIP.

The matter went to mediation, but did not settle. The owner later filed for bankruptcy, and the plant was bought by another party.

Principles familiar to technical specialists applied in this example: Review of process design, PFDs, P&IDs, impact of design changes and delays on construction, project management, and contractual performance.

## Example 3 -- Process Technology Licensing Lawsuit

Two separate companies (A and B) had plants located at one site. Company A manufactured product X1 and sent its wastewater to Company B for treatment and discharge. Company A then started to manufacture another product, X2, at the same plant. Company A recognized that Company B's plant could not handle the additional wastewater created by the process to manufacture X2. Therefore, Company A decided to build its own new treatment plant on the site to handle the wastewater generated by products X1 and X2 combined.

Company A chose a wastewater treatment contractor (TC) to provide "turn-key" services, including technology, engineering, design, construction and startup of the new wastewater treatment plant. The selected technology was a "thermophyllic" biological treatment process, patented and licensed by a third party technology developer ("Licensor"). The license to operate using the technology was granted to Company A, and the technology and engineering details were given by the Licensor to TC.

The Licensor pilot-tested Company A's waste stream to confirm treatability and provided TC with process and engineering information tailored to treat the tested waste stream samples.

At startup, the new wastewater treatment plant failed to meet the treatment targets, and Company A filed a lawsuit against TC. The Licensor was drawn into the lawsuit as a third party defendant.

During the discovery phase of the lawsuit, the allegations against the Licensor were determined to be unfounded both technically and contractually. It was established that there were numerous deficiencies in Company A's design basis as well as defects in both the engineering design and installation by TC. There also were issues involving the performance warranties between Company A and TC.

The technology provided by the Licensor had been commercially proven at other facilities, and it was established that the Licensor had met its contractual obligations.

The matter was settled before trial.

Principles familiar to technical specialists applied in this example: Process evaluation, pilot testing, process design and engineering analysis, contractual commitments, review and understanding of Company A's and TC's obligations and failures.

## Example 4 -- Plant Incident and Alleged Medical Injury

A manufacturing plant was undergoing a maintenance turnaround with contractor employees working on scaffolding near the top of a large carbon steel tank containing concentrated sulfuric acid. An undetected water leak into the tank diluted the acid and a relief valve on top of tank momentarily lifted and released vapors with a sulfurous odor. The contractor employees immediately evacuated the area and reported the incident. The turnaround work was shut down for the day, and all workers were sent home.

Reportedly, the contractor employee who was closest to the top of the tank went to a restaurant with his co-workers, where he collapsed. He was attended to and recovered. Later, the contractor employee sued the plant for negligence, alleging that the gas he was exposed to was hydrogen sulfide ($H_2S$). During the litigation, medical statements were produced stating that the contractor employee's nervous system was permanently damaged from exposure to $H_2S$.

The defense argued that the released gas was more likely to be sulfur dioxide ($SO_2$) and not $H_2S$. Reported public information showed that exposure to $SO_2$ could have bronchial effects but not permanent nerve damage.

A literature search as well as an examination of the conditions of the carbon steel tank (pressure, temperature, and acid concentration) on the date and time of the incident suggested that the gas released was as a result of corrosion, and it likely contained primarily $SO_2$ with trace amounts of $H_2S$.

Defense lawyers engaged an independent third party laboratory to test evolution of gases from reaction of concentrated and diluted sulfuric acid on carbon steel, thus simulating conditions in the tank on the day and time of the release. The tests showed that the evolved gas content was mainly $SO_2$ with likely trace amounts of $H_2S$.

A chemistry professor for the plaintiff testified in his deposition that the gas released at the time of the incident was primarily $H_2S$, but his testimony was strongly challenged by defense lawyers. A second professor from another university testified in his deposition that the plant violated OSHA's Process Safety Management (PSM) regulation. The second professor's testimony was challenged on the basis of his lack of experience with PSM.

Later, it was learned that the contractor employee who filed the lawsuit had been previously employed as a sheet rock worker and had developed indications of dust-related issues involving his bronchial system.

The matter was settled between the parties before trial.

Principles familiar to technical specialists applied in this example: Chemistry of reaction of dilute acid with carbon steel; examination of the incident timeline and possible cause(s) of water backflow into the acid tank; basic metallurgy of carbon steel; review of third party test results; and review of the plant's PSM program.

## 4 Observations and Conclusion

1. Insurance claims and lawsuits can arise from a variety of reasons and situations, and outcomes are not easily predictable.
2. Lawsuits and insurance claims can arise from unwritten "perceived expectations" by the involved parties.
3. Besides the terms and conditions defined in a contract, other factors, such as standard of care and generally accepted good practice in a given industry, are often encountered in allegations.
4. In addition to directly involved parties, all other entities that are in any way associated with the matter may be drawn into the lawsuit - especially those perceived to have "deep pockets."
5. Claims of advantages and success using new technology on a commercial scale should be thoroughly evaluated before acquiring the license for its use.
6. Clear agreements between a company, its consultants or involved third parties regarding ownership and rights regarding intellectual property need to be executed in advance.
7. Applying fixed price and fixed schedule terms to design-build contracts in capital projects (or any) in which a firm and definitive scope cannot be described clearly in the contractual documents will most likely result in cost overruns and schedule delays, prompting lawsuits.
8. A technical professional can take steps to minimize liability, although it is doubtful one can completely avoid it. Some exemplar suggestions are:
    - Try to clearly understand the technical and (to the extent possible) commercial needs of an assignment.
    - Take on only what you are capable of performing on a sound technical basis. Do not make unsupportable assumptions or try to "wing it."
    - Speak up if you are not the best person to take on the assignment. Have it passed on to someone else who can best perform the work.
    - Keep good written factual records of your work and your communications about the work you are doing.
    - Most of all, be factual and truthful in all you do.

# Determining The Liability Of A Fire Starting Product

C. J. Abraham
Scientific Advisory Services Ltd
Great Neck, New York

**Abstract**

**Natural Fire Starter is an ethanol based fuel gel which was designed specifically for wood and wood pellet stoves. It is sold in a plastic bottle with a child proof closure. This product had never been cited in any reported accident until a 7-year old was badly burned when he was allowed, unsupervised, to toast marshmallows in a wood burning fireplace. The subsequent events reported by the child, his parents and brother are discussed as well as the experts and lawyers claiming the liability of the defendants. Said defendants include the manufacturer of the formulation, the bottle manufacturer, manufacturer of the closure and the packager.**
**In order to reconstruct the accident and determine the cause and liability involved, this investigator had to consider the testimony of the injured child, his parents, brother , the ENT personnel, medical personnel, plus testing and testimony of all of the experts as well as the statements made by the plaintiff's attorney.**

## Determining The Liability Of A Fire Starting Product

**As simple as the product appears, over 50 characteristics and variables of the bottle of fire starter were considered in reconstructing the accident.**

**After extensive testing by this investigator and all of the experts on both sides, analysis of all of the documents and videotaped depositions, the conclusion of this investigator indicates that the case, as presented by the plaintiff's attorney, is highly suspect and the opinions of the plaintiff's experts are unreliable and without scientific substance.**

### Background to the accident

**On or about February 2010, Plaintiff "A" received burn injuries at his home located in West Virginia. According to the Complaint, he was using a product called Natural Fire Starter Gel, which is an ethanol-based fuel gel.**

**According to testimony of "A" and his parents, a fire had been burning in the fireplace on the lower level of the house for much of the day. The Natural Fire Starter Gel was near the hearth. At some point in the evening, A was given permission by his mother to go to the lower level of the house unsupervised to roast a marshmallow prior to taking a bath He was naked,.**

**According to the complaint, A applied the Natural Fire Starter Gel to kindling wood in the fireplace and the Natural Fire Starter Gel container suddenly failed near the lower portion of the bottle, projecting flaming gel onto "A", resulting in injury.**

**Natural Fire Starter Gel is designed specifically for pellet and wood burning stoves. It is manufactured and assembled by Packaging Service Company (PSC) using a plastic bottle manufactured by C.K.S. Packaging, Inc. and a cap manufactured by Stull Technologies, Inc. The product is distributed by Hearthmark, LLC. The sole purpose of the gelled product was that the product was to be used for only wood pellet and wood burning stoves.**

**The subject bottle was purchased by Plaintiffs at a local Walmart store. According to the Complaint, Plaintiffs' allege the Natural Fire Starter Gel was not reasonably safe for its intended use, was defectively designed and manufactured, and carried inadequate warnings.**

**The undersigned reviewed over 10,000 pages of documents obtained through discovery in addition to performing, examining, or analyzing the following:**

1.**the subject bottle and exemplars**
2. **Testimony of injured party**
3. **View of fireplace and living room after the accident**
4. **Analysis of the PVC container**
5. **Analysis of the stopper for the PVC container**
6. **Proving or disproving the testimony of the plaintiff, his parents and brother**
7. **Testing and testimony of plaintiff's experts**
8. **Testing and testimony of the defense experts**
9. **Deficiencies in testing by defendant's experts**
10. **Deficiencies in testing by plaintiff's experts**
11. **What are the scientific basis, from a mechanical engineering stand point why either the plaintiff or defense experts were incorrect in their analysis?**
12. **Testimony of all experts.**
13. **Testing by all experts**
14. **Medical documents**

**In addition to the above, it was important to evaluate all of the chemical, physical, thermal, and flammable properties of the product in addition to evaluating the warnings and instructions.**

1. **Accident Reconstruction**
2. **Annealing**
3. **Basics of blow molding**
4. **Basics o f ethanol based fires**

5. **Basics of gelled fire starters**
6. **Bottle closure**
7. **Brittle versus ductile fracture**
8. **Compatibility of a solvent and chemicals with P VC containers**
9. **Compatibility of ethanol and PVC**
10. **Consumer Product Safety Commission Reports**
11. **CT scans (Computerized Tomography Scan)**
12. **Curvature of bottle at shoulder**
13. **Deflagration, detonation, explosion**
14. **Destructive testing**
15. **Design of PVC container**
16. **Expansion o f PVC plastic**
17. **Explosion limits**
18. **Flame arrestor technology**
19. **Flame speeds**
20. **Flashback**
21. **Flashover**
22. **Heat damage**
23. **Heat exposure and PVC**
24. **Laminar flame speeds of ethanol-air mixtures**
25. **Material Safety Data Sheets (MSDS)**
26. **Napalm**
27. **Non destructive testing**
28. **Origin of fracture**
29. **Pictorials**
30. **Plastic bottle rupturing**
31. **Plasticizers**
32. **Pressure release valves**
33. **Prior reported accidents**
34. **Recalls on subject product and similar products**
35. **Scanning electron microscope**
36. **Seams in plastic container**
37. **Similar products**
38. **Single holes versus multiple holes**
39. **Stains in plastic bottle**
40. **Stress in plastic containers**
41. **Stress cracking**
42. **Stress whitening**
43. **Specification of PVC container**
44. **Supersonic versus subsonic**
45. **T ensile strength of PVC**
46. **The effect of dye in ethanol in PVC bottle**
47. **The effects of gelling ethanol**
48. **Thermal gravimetric analysis**
49. **Viscosity**
50. **Volatility**

**51. Warnings and instructions**

**The Plaintiff, 7 year old boy "A"\ (Deposition summary with comments)**

**At the time of the incident, "A" was 7 years old and claims in his deposition, that he shoots his own gun. He started shooting at the age of 5. He stated in his deposition that he has never started a fire without an adult present. His mother told him not to use the gel on a burning fire because it was a waste .** ***He stated his mother told him that he could start the fire that day. Contrary to that statement*** **"A"'s mother testified that she started the fire prior to taking the family to church.**

**"A" stated that he** ***"used the gel to start a fire and also to restart a fire."*** **This** ***admission is important and*** **more than documents the fact that "A" squirted a lot more of the Diamond product into the fire than he testified to.** ***On the day of the incident his "mom's back was out and he was sick with pneumonia."***

**He stated that he was getting ready to roast marshmallows. However, his mother told him to take a bath. He asked her if he could roast a marshmallow. She said yes. He went to the kitchen, opened up the cabinet** ***and got a bag of marshmallows.*** **The alleged bag of marshmallows was not located or identified in any of the documents reviewed or photographs taken at the scene of the accident.**

**According to "A", the fireplace was in the basement and his mother, father and brother were upstairs. He walked down the stairs, went to the fireplace and stated that there was no fire . "A" testified that the coals were black with white powder over them, and he thought the fire had gone out.** ***If there was white powder all over the coals, how could he see the color of the coals?***

**He then stated that he set the marshmallows down and then put one on the stick and laid the stick right by him. He then went and obtained the fire gel.** ***He moved the screen and put it on the side.*** **Moving the screen left him without any protection whatsoever. He put some wood up in a teepee and started squirting gel on it. He testified that the gel landed on a coal and started lighting. The heat emanating from the coals is over 400 degrees Fahrenheit and impossible for anyone to expose any part of their body especially an individual's hands. He then stated that he had the bottle up about a foot and the fire went up 6 or 7 inches. This was not accurate based upon the testing performed by the undersigned.**

**"A" stated that he held the gel bottle about a foot from the teepee made of wood. He was on his knees andwas holding the bottle and squeezing it angled downward . He then stated thatas he tilted it up and and backward, the gel expelled out of the bottom.He stated it exploded and his brother heard the noise upstairs while he was watching television.**

**He then stated that he was forced backwards onto the couch and got up and ran into the bathroomon the same floor. He grabbed a cup of water and threw it on theback**

**of his neck in an attempt to put the fire out. He then ran upstairs to his mother. He stated, contrary to what "T" testified to, that his father grabbed him) and threw him in the bath tub and started running the water.**

**"A" testified that his mom and dad taught him how to start a fire using a teepee. Contrary to that statement, his father stated under oath that it was "A"'s idea to use the teepee.** ***Regardless, starting that type of fire did not include using the gel.***

**Although he had observed his mom's friend start a fire using lighter fluid, he has never observed any adult pour lighter fluid on a pre-existing fire.**

**On the day of the fire, his mother started the fire prior to going to church. The fire was burning most of the day and went out approximately 1 or 2 o'clock in the afternoon. He stated the accident occurred at approximately 8:30 p.m. If that is true and no one replenished the fire with additional logs, it would be questionable if any embers still existed.** ***Note - How would*** **"A" would not be able to build a teepee if the embers were still hot** ***and not all ash?*** **He stated that there were no coals in the fireplace, there were just ashes. He also stated that matches were "right there." He also stated that if the coals were hot, he would have just put paper on the coals and not used the fire starter gel.** ***("A"'s testimony does not make sense. If he didn't use paper and he just observed ashes how could he have built a teepee if there was sufficient heat to ignite the gel? There was a reply by "A" that he did not feel any heat while he made the teepee.***

**At the time he was squirting the gel into the fire, he had to stand on the ledge (hearth extension) and lean in to pour the gel onto the wood. He was able to get up on the hearth using his hands and knees. He set the gel down, climbed up on the ledge and started squirting all around the teepee. He** ***stated he squirted all around before there was a fire.. He also stated that he was still on his hands and knees when the fire came at him.***

**Comment: Reconstruction of the accident by the undersigned with an exemplar fireplace documents that the fire does not flash back** ***but goes directly in the vertical direction.*** **Fire travels directly in the vertical direction.** ***At this time, while on his hands and knees he tilted the bottle back and the bottom came out with a noise.*** **Contrary to the analysis of the plaintiffs' experts, this more than documents that their analysis is without merit and without a scientific basis.**

**The amount of squeezes the plaintiffs' expert had to complete prior to meeting within the explosive limits of the ethanol does not coincide with "A"'s testimony or the real-world.**

**"A" stated that when he first picked up the gel bottle,** ***it was not warm. As he fell backwards, he stated that he threw the gel bottle onto the couch. (If there was any leak or opening in the bottom of the subject container, the gel would have***

***squirted out the leaking end as well as the opening at the top. "A" never testified that as he was squeezing the gel into the fireplace that any other part of the container was leaking.)***

**He had no clothes on when his mother told him to take a bath. However, his older brother, "T", stated that his mother used the wet towel from "A"'s bath that he took prior to roasting the marshmallows in the fireplace.**

**Further in his testimony he stated that he used a lighter to start fires and would get gel and squirt it on the wood and light it. When he was finished lighting the fire, he would close the cap and set it down about 10 feet from the fireplace. He would then put more wood on the fire. , "A" also stated he was told to be careful of the gel. *He knew that the gel "could light on fire" and it was dangerous.***

**"A" stated that he thought the fireplace was still lit. He had observed that fireplace with the wood burning hours before that time. *(If he thought the fireplace was still burning, why didn't "A" first try paper on the coals as he was taught?)***

***He stated that he used the gel even though the ashes were still warm because he wanted to get it started.* He described the ashes as whitish, black and grey. He states that he did not see any red.. *However, he previously testified that the coals were not warm.***

**Further on in his testimony on "A" stated that the fire was out about 1 p.m. and he went down to roast marshmallows around 8:30 p.m. (the incident occurred at 7 p.m.). He did not poke a stick in the ashes to see if there were any embers. *(His parents stated that they taught him how to start a fire and use the gel which is not consistent with "A"'s behavior and actions.)* "A" was allegedly taught how to start a fire and use the gel, which is an unreasonable misuse when it involves a child.**

**In addition, based on "A"'s stated awareness and knowledge, and based on the fact his mother taught him not to put the gel on an existing fire or hot embers, "A"'s actions were not reasonable. Moreover, it is not reasonably foreseeable that a 7-year-old child will use the fire starter gel to start or accelerate a fire.**

**He reiterated the statement he made previously. "A" stated that he squirted the fire starter gel on the hot coal and it flamed up only about 6 or 7 inches. *(However, he previously stated that the coals were not warm.)* As he tilted the container it came closer to the fire and then "it exploded." This statement cannot be accurate, because the air space contains close to 100% ethanol vapor. The explosive limits within the container are not reached until "A" squirts the gel into the fireplace a number of times so that air can replace the ethanol vapor and bring the concentration of the ethanol to within the explosive limits. *Hypothetically, assuming, for example, that there was a leak in the bottom seam, deflagration could not***

***occur because the opening in the bottom would have released all pressure created by the reaction taking place.***

**Even though the bottom came closer to the fire, he never saw the flame come in contact with the bottle. "A" reiterated again that once he was on the hearth, he was just on his knees with his hands on the bottle.** ***(Note - It would be impossible for the bottom of the container to come near the fire because "A" was squirting into the fire with the nozzle facing the fire and not the bottom of the plastic container. In addition, if "A" was squirting the bottle as he testified to, the gel would have leaked out through the small opening at the bottom if it did exist. There was no testimony by "A" that, as he was squeezing the gel into the fire, that any of the gel came onto any part of his body from the opening in the base of the container.)***

***Comment - If "A" was on his knees on the hearth, there should not be any burns below his knees. It would have been impossible for him to be thrown back onto the couch. The burn diagrams from Shriners Hospital show that he is not burned a few inches above both knees and down to his toes in the front of his body except for the inside of his right calf.***

**The history as recorded by Shriners Hospital for Children states that "patient is a 7 year old white male with recent past history sustained on 08-28-2010.** ***The patient had been roasting marshmallows inside when an accelerant for the fire caught on fire and exploded onto the patient. The patient then fell backwards into the fire before being recovered by his father."*** **The areas that were not burned were his scalp, his back and left side. There is no mention of a couch. If he fell backwards into the fire, his back would have been severely burned. It was not. However, it was possible that he fell forward into the fire. That would account for the intense burns on the front part of the body.**

**The fact that "A"'s scalp and face were not burned refutes his testimony that he raised the container above his head and it exploded onto his body. If that were accurate, he would have received second and third degree burns on his scalp and his eyes and eye site would have been considerably damaged.** ***(This also refutes the plaintiffs' contention that there was a small opening in the bottom of the container prior to the incident.)***

***"A"'s mother, in her deposition, stated that she had problems recalling what the warnings stated on the subject container. However, she did remember that the product was flammable based on her experience. If that is true, how could she expect "A" to recall or fully understand any part of what she read and understand the meaning and intent of the warning?***

## **"Father"**

**"A"'s "Father" also stated that he taught "A" at the age of 5 or 6 to make a fire. He taught him to place a piece of towel or tissue under the small kindling and light it. "A", he stated, came up with the teepee approach.** ***(This is contrary to "A"'s testimony.)***

**"Father" stated that he used the gel less than 10 times to start a fire and "A" watched him do it. He never saw "A" start a fire with the gel. More importantly, he told "A" that he was never to use the product because it was dangerous. Also, he never put the gel on the mantel and never saw it there. "Father" started the log fires with starter logs.**

**He purchased only one bottle of the Diamond product at Wal-Mart. That was the only time he ever purchased the product. This comment is contrary to "K"'s comment stating that she purchased it. He stated that he read the label and it said "keep out of the reach of children." That is why he told "A" not to use it. He does recall that it said to "store away from flame or heat sources." That is why he kept it on the floor which, in his opinion, was the coldest point in the room.**

***It should be noted that, contrary to the plaintiff's allegations, if it was kept on the floor, that would mean that it was well below the hearth and not be exposed to any heat from the fireplace which, the plaintiff's claim caused a leak in the container that ended up causing the fire and the deflagration.***

**The statement by the father also contradicts "T" who stated that the gel was kept on the mantel. "Father" always let the fire die down in the late afternoon and early evening. He again contradicted "T" when he stated that he tried to smother the fire with a towel.**

**When he went down to the basement, he does not remember seeing the gel bottle that day or after the fire. He did hear that "A"'s grandmother found the bottle of gel outside. I was under the couch. After the accident, his wife obtained a restraining order preventing him from seeing the children.**

**"Father"' stated in his deposition where, he quoted an entry in "K"'s (his wife) diary. It is alleged that after she got home on the evening of March 10 and checked the fireplace where the incident occurred, she** ***discovered that "A" did not squirt the lighter fluid*** **(The gelled product)** ***into the fire as she first thought. The bottle was sitting close to the fire and appeared to have a melted crack near the bottom of the bottle allowing it to spray into the fire causing "A"'s burns. By that time, she had already retained a lawyer to sue.***

***Question: "K"'s entry into her diary contradicts everyone else's testimony under oath. Her entry infers that "A" was not on his knees on the hearth holding the container as he testified to under oath. It also contradicts "Father"'s testimony. However, it does fit her attorney's theory and that of***

***the plaintiff's experts. This author, having over 40 years of experience in evaluating similar cases, the statements made by the family members are incredulous.***

**"Father" was not aware that "K", his wife at the time, allowed "A" to use the gel under her supervision. He did not expect "A" to start a fire on the evening of the incident. He also stated that before the incident, he thought he heard a popping noise. The question is whether you can actually hear a popping noise in the basement while he is on another floor watching and listening to a TV program.**

**"K" ( mother of "A")**
**"K" stated n her deposition that "A" should not start fires by himself. She also stated that she never** ***caught him starting a fire.***

**Contrary to "Father"'s testimony, she stated that** ***she bought the fire starter*** **because it was thick and it was right beside or right near the fire logs in the Wal-Mart store. <u>Without reading the warnings,</u>** ***she felt that the product was safe and could be used to continue the fire.*** **She also knew to keep it out of the reach of children and that the product could explode if it got too close to the heat source.(This contradicts "A"' testimony where he stated that his mother read the warnings to him in the store.)**

***It should be noted that there is not one word about explosions on the Diamond product label.*** **When "A" and his mother returned from Shriners Hospital in May he told her that** ***he was squirting the gel on the fire.*** **She also told aa registered nurse in the emergency department in the hospital that** ***"The patient squeezed the accelerant into the flames and exploded on the patient."*** **At her deposition, she stated that she didn't recall telling anyone that statement. However, the statement made and recorded in the hospital notes fit the reconstruction of the cause of "A"'s burns and contradict the statements made by "A" and "K" under oath and also contradicts the opinions set forth by all of the plaintiff's experts.**

**She also claimed that she saw a hole in the bottom of the bottle and assumed that it must have sat near the fire because it is only that it could have gotten there, because it got hot and melted.**

**She does recall seeing the marshmallow on the end of the fork .A small residue of what might have been the remains of a marshmallow was noted in one of the photographs. However, that evidence was never retained for examination. The "bag" of marshmallows were never found.**

**"K"'s mother found the bottle of the gel outside of the front yard under the couch. There is no explanation how her mother was able to remove the plastic container from under the couch.** ***It should be noted that some of the deformation on the subject container was due to the weight of part of the couch on the container caused deformation after the incident.***

**She does not recall reading part of the instructions where it stated that one should "Close bottle and remove from immediate area as vapors are flammable." However, she knew that it was flammable.**

**When asked if she ever read the instructions on the Fire Starter Gel to "A", she replied "I do not recall." However, a question was then asked again of the reading of the label and she stated that she read it to both "A" and "T" and told "A" that he cannot use the product .**

**"K" stated in an answer to interrogatories that "A" had previously used the Fire Gel) under her supervision). When "K" allowed him to use it under her supervision, there was no fire. His father objected to allowing "A" to use the product. "K" does not recall how many times she allowed "A" to use the product. This contradicts her prior statements under oath. In fact, "A"'s father told "A" not to use the product.**

- **"D" (registered nurse)**
  **Donielle is a registered nurse working full-time for the emergency department at the Memorial Hospital and the County EMS.**

**Her intake states that the patient was roasting marshmallows in the fireplace. *Patient squeezed an accelerant onto flames exploded on patient*. Patient had severe burns over entire body except for his back. Estimated 70% of second and third degree burns observed. No other injuries from explosion.**

***This recorded statement contradicts "A"'s testimony, his mother's, his father's and "T"'s.***

- **"J" (registered nurse)**
  **"J" is an EMT-B. He also testified that he was told that "A" went down to the fireplace and *sprayed some type of aerosol or something on the fire and it blew back on him.***

**Shriners Hospital (dictated by "R" MD in April of 1020)**
***The patient had been roasting marshmallows inside when an accelerant for the fire caught on fire and exploded onto the patient. T<u>he patient then fell backwards into the fire before being recovered by his father.</u>*** **This more than documents the fact that "A" sprayed an excessive amount of the Diamond product on to the existing fire. When he fell back, he was exposed to close to 1800 °F. It should also be noted the screen was not in place when this occurred.**

**Critique of Dr. "G" (Expert for plaintiff)**
**Dr. "G"'s academic background and experience is excellent. However, his background in storage, transport and processing vessels for chemicals, gasoline and other flammable products in addition to the implementation of various flame suppression and arresting technologies for the safe handling of combustible products**

**in general and gasoline tank design and leak detection systems for gasoline stations and gasoline storage systems does not relate to the issues and the cause of the accident in the subject case. The "simple screen flame arrester for the inside of a gasoline tank" is not the solution in this specific case. The product in this case is thixotropic and the design of the opening in the cap must be at a size that would allow an individual to squeeze out a desired amount with minimal pressure on the container.**

**Dr. "G" conveniently forgets about the fact that the plaintiffs testified that "A" was going to roast marshmallows. There was no evidence found of any marshmallows in the area of the fireplace or the living room except for the small amount of white residue on the fireplace fork. There was some white color or remnant on the end of a fork in one of the photographs. However, that was never examined by any of the investigators.**

**He goes on to state in his report that that "A" "holding the bottle with both hands with the nozzle pointed down and pored the kindling wood when the fire gel hit the kindling/embers and then the heat source ignited the bottle due to the flashback."**

**Dr. "G" never attempted to reproduce the subject accident with an exemplar fireplace and the subject Diamond formulation. He accepted the questionable testimony of "A" and his mother when designing and performing his testing.**

***How could*** **an explosion or flashback cannot occur with one squeeze when the head space contained almost 100% ethanol vapor and testing by the plaintiffs experts documented that a large number of squeezes were necessary to reduce that concentration of ethanol vapor and have sufficient enough air in the head space so that the vapors could be ignited. The experiments performed using an exemplar fireplace by the undersigned disproved all allegations made by the plaintiff and the plaintiff's experts.**

**Dr. "G" neglected to state that "A" testified he was on his knees on the hearth and that he allegedly made a teepee out of kindling and squeezed the bottle on the teepee. He also neglected to state what the rest of the family testified as to what occurred as well as other recorded evidence. He also did not mention the fact that there was no evidence of any kindling in the area of the fireplace or the fire.** ***If "A" constructed a teepee out of kindling, where was the kindling wood located? Not one of the photographs contained any evidence of kindling wood.***

**There is no evidence that the container caught on fire and exploded. If it did, it is more likely than not, that all of the contents would have been burnt up and there would have been extensive deformation of the subject container. His explanation of the deformation of the container is incorrect. The plastic container can be deformed, as it was in this case, when a weight of sufficient size is placed on the plastic for a period of time. The deformation was mechanical and had nothing to do with the incident.**

**More likely than not, the fire test performed by Dr. "D", another plaintiff's expert , documented that the container had to be placed in a specific position for an extended period of time before it would catch on fire and deflagrate. The amount of time that Dr. "P" had to squeeze the container before anything was notable was well over two dozen times.**

**Analysis, methodology and opinions (Part of Dr. "G"'s report)**

**A. "G" Opinion: The rupture to the subject bottle is consistent with an internal explosion.**

**Comment: The rupture to the subject container is not consistent with an internal explosion.** ***The rupture was more consistent with the deflagration. It also required a very large number of squeezes at a specific angle over a long period of time before the fire was spread inside the container before the rupture occurred, and this was not from a "flash back" occurrence. Tests performed by the undersigned with the Diamond container filled with the thixotrophic formulation and refute the experts theory and opinions.***

**B. "G" Opinion: The burns on "A" are consistent with burning gel being expelled from the bottom of the ruptured bottle and spraying on his body.**

**Comment: The burns on "A" are also consistent with "A" spraying an excessive amount of gel in many areas** ***(into the fireplace and onto his body).*** **It is also consistent with "A" falling** ***back*** **into the fireplace** ***(see medical documentation).*** **This can expose "A" up to temperatures reaching up to 1,800 degrees F. No one knows exactly how many times "A" squeezed the plastic container.** ***For example, it is obvious that he squeezed a large amount of the gel directly in front of the hearth. There is also evidence that there were many squirts all around the fireplace, on areas surrounding the fireplace, the walls and ceilings.*** **As proven by the undersigned, in all of the experiments performed, no flashback occurs at anytime no matter how much gel is squirted into an open flame.**

**C. "G" Opinion: Explosion tests were conducted with 5 exemplar bottles of Natural Fire Starter Gel. Ten tests were conducted with single hole caps by squeezing the bottles of gel over a candle flame. In each of these tests, the bottle had an internal explosion. In one of these tests, the bottle ruptured. The ambient temperatures of these tests were between 55 and 65 °F.**

**Comment: The explosion tests that Dr. "G" conducted on five exemplar bottles did not involve methodology that was scientific. The tests conducted on 8/29 self extinguished. No explosion occurred. The container deflagrated.**

**The test conducted on 9/11 by "G" where the base of the container blew out did not, in anyway, reproduce the results produced in the exemplar container. The container was held in a horizontal position for an extended period of time which is unrealistic. The same is for the second test that was held on 9/11.. In addition, the initial ignition of the external vapor was snuffed out because the internal concentration of the ethanol was not within the explosive limits.**

**There is no documentation that the subject container was only partially filled containing a lot of head space. Flammability tests performed by the undersigned using an exemplar fireplace disproved the theory and opinions presented by Dr. "G". The thixotropic Diamond product was squirted directly in front of the fire exposing the spout, vapor and thixotropic liquid to over 200 degrees Fahrenheit and through a hole on the top of the fireplace exposing the spout, vapor and thixotropic liquid to over 300 degrees Fahrenheit. No flashback or internal deflagration occurred while the undersigned personally squeezed the plastic container. The spout was so hot that the undersigned was not able to close it until it cooled down**

**The additional test *performed by Dr. "G"* with a different cap are also non scientific and do not, in any way simulate the conditions that existed at the time "A" squeezed and squirted the gel all over the pre-existing fire, carpeting and walls.**

**Using the same formulation with a different cap documented what probably occurred on the original cap on the subject bottle. The subject bottle did not explode but the squirting of the gel all over the fireplace by "A", including his body on a pre-existing fire significantly contributed to "A"'s burns when he was ignited.**

***If "A" was really going to toast marshmallows, there would have been evidence of marshmallows that were roasted on the pre-existing fire and glowing embers. There was no evidence in the room where the fireplace was located that there was even a supply of marshmallows anywhere else in the house. "A" nor his mother never testified where the marshmallows were originally located. "A" never testified that he took the fireplace fork into any other part of the house and placed a marshmallow or marshmallows on the fork.***

**D. "G" Opinion: The concentration of the vapor space in the gel bottle was measured with an O2 sensor. This concentration was found to be in the explosive range for temperatures above 50 °F.**

**Comment:** When the cap is first removed the headspace contains well over 19% ethanol vapor. When the container is constantly squeezed, external air constantly dilutes the concentration of ethanol vapor in the headspace to a point where it can fall within the explosive range of 3 to 19%. This can only occur if one squeezes the container a large number of times. That is the only way that the plaintiff's experts were able to create sufficient oxygen in the headspace when the container was constantly squeezed and held in a horizontal angle. When the concentration of ethanol vapor is above 19%, the flame front will be snuffed out.

E. "G" Opinion: The cap on the Natural Fire Starter Gel bottle had on 0.113 inches (2.87 mm) hole and with this cap all of the tests showed internal combustion.

**Comment:** The statement is not true. Most of the ignition tests were self extinguished *because the concentration of the ethanol vapor was above 19%.* It is only when the container was continually held in the horizontal position on a continuous basis where air was introduced into the air space and the concentration of the ethanol vapor falls within the explosive limits did Dr. "G" obtain internal combustion and not an explosion. If he recalls, "A" testified that he was holding the container in two hands and aiming downward. If that is the case, then there would be no head space and no internal combustion. In the tests that Dr. "G" conducted he should have directed the cap in the downward direction rather than horizontal. In addition, had he placed more gel in the container, it is more likely than not it would have no internal combustion taken place at all.

F. "G" Opinion: Four tests were performed with Kingsford charcoal lighter fluid caps (3 smaller holes) on a Natural Fire Starter Gel bottle. There were no internal combustion for these tests.

**Comment:** The tests conducted with three small holes were not conclusive and the methodology was not scientific. In addition, the opening in the subject cap was designed to expel a thixotropic formulation. Three holes could have been used for a liquid but not the thixotropic formulation. The use of a three-hole cap would not prevent the underlying incident and would not have acted as a flame arrestor. Moreover, a three-hole cap could expose users to more dangers because if they misuse the product near a flame or hot ember, the consumer has to position the bottle closer than the one with the one-hole cap. In addition, the end of the cap was held much further away from the candle than all of the prior tests with the standard closure.

**G. "G" Opinion: The Consumer Product safety committee induced Anderson Chemical to recall 168,000 bottles of Eco-Lite Charcoal starter in 1991. The CPSC learned of five reports of the bottles exploding in consumers' hands when this charcoal lighter fluid was sprayed on hot charcoal. Four of these explosions involved serious injuries, including hospitalizations for burns.**

**Comment: The recall of the Eco-Lite Charcoal Starter in 1991 had nothing to do with the subject formulation. That product was a liquid formulation not a thixotropic and was designed for a different intended use than the product at issue. A red herring.**

**H. "G" Opinion: Eco-Lite (82% ethanol, 8% isopranol, 3.5% methanol) is made of a mixture of alcohols and has a flash point of 56 °F. The other more familiar charcoal lighter fluids are made of petroleum products and have a flash point above 100 °F. It should be noted that the Diamond Natural Fire Starter Gel (85-95% ethanol) is made of ethanol and Bitrix and it has a comparable flashpoint to the Eco-Lite mixture.**

**Comment: What Dr. "G" does not understand is whether the flashpoint was 56 °F or 150 °F, it wouldn't make any difference when the vapor is exposed to an ignition source of 1,500 to 1,800 °F. In addition, the flash point reported was incorrect. The closed cup flash point was significantly higher.**

**I. "G" Opinion: Caps with 3 small holes are commonly used in the industry and well known. A standard 3 hole cap, manufactured by Stull Technologies, would have prevented this incident as the smaller holes act as a flame arrestor. An igniting flame cannot pass through the small holes and into the container to cause internal combustion. Thus, an explosion cannot occur.**

**Comment: Dr. "G", in all his tests, did not reproduce the alleged blow (deflagration) out of the subject container in any of his tests. His statement regarding flame arresters is utter nonsense.**

**The undersigned has been involved with exploding batteries with micro porous flame arresters for over 30 years. The flame front goes through the flame arrester into ullage area much too fast for even the plastic to melt and snuff out the flame front before the battery explodes. The purpose of the flame arrester is to snuff out the flame before it reaches the internal head space of the flammable liquid. When it comes to polymeric materials, the flame front does not melt the opening fast enough to snuff out the flame. It is not foreseeable when one performs a risk analysis on the subject formulation that a foreseeable consumer would constantly squeeze the container when only a half a tablespoon would be used to start a fire. The product was not designed for a 7 year old to use the product as a toy. That is not a foreseeable misuse of the Diamond product.**

**J. "G" Opinion: Warnings do not alleviate the product manufacturer and retailer of the duty to design and sell a safe product. It is a basic, generally accepted mechanical engineering principle that performance of a risk assessment (e.g. failure modes and effects analysis) is necessary in product design. It is also a basic, generally accepted mechanical engineering principle to preferentially design out risks as opposed to warn against them. Manufacturers, distributors and resellers have a duty to consider anticipatable uses and misuses of a product.**

**Comment: The warnings and instructions on the subject container were well done for a pictorial of a wood pellet stove.** ***The product is not designed for 7 year olds. It is designed for a small segment of our population of adults who own wood pellet stoves.*** **It is not designed for fireplaces otherwise a pictorial of a fireplace would have been on the front panel of the product. The wording on the labeling on the container covers that specific area.**

**Dr. "G" fails to point out that there were starter logs on the side of the fireplace and that the parents knew that there was a safe method of starting a log fire in their fireplace. He also fails to point out that the parents knew that the subject fire starter was highly flammable and dangerous, must be kept away from children and was designed for wood pellet stoves as shown on the pictorial on the label of the product. The warnings were written for a person with a fifth grade education and would be obvious to any consumer that the small container, like gasoline, should be kept in the garage and out of reach of young children. A manufacturer is not responsible to follow a consumer to their residence and hold their hand while they use their product.**

**K. "G" Opinion: It is apparent that none of the defendant's *experts* in this case performed any tests of any kind to determine if the container and cap could be safely used by a consumer in the presence of a flame. Had the management or engineers for these defendants conducted appropriate testing of this product, they would have determined that the Diamond Natural Fire Starter Gel and its container pose a significant and unsafe explosion risk.**

**Comment: The defendants did not have to perform any tests because the properties of the gel are known as well as the use and application for pelletized wood stoves. The properties of the one hole cap and its integrity and use for thixotropic formulations is well known for many applications including harmless products such as soap. The same design is also used for squirting insect repelling products inside and outside residential and commercial buildings.**

The public also knows that gasoline in any container poses a significant and unsafe explosion risk. The same goes for Drano which is composed of almost 100% of sulfuric acid. *These* products are not designed for 7 year olds.

What Dr. "G" fails to point out is that there have been no reported accidents with fireplaces and the subject product with young children. In addition, there have been no reported accidents with wood pellet heating stoves especially when only a tablespoon of the Diamond product was used to start the fire in a wood or wood pellet stove.

This author measured the force required to release the snap cap of the subject bottle. The tests indicate that it takes over 8-pounds of force to pull the cap off. The cap would be considered child proof. Therefore, it is questionable whether a 7-year opld would be able to easily open the cap with one finger.

L. "G" Opinion: It is also apparent that none of the defendants in this case performed a reasonable risk assessment of this product kind to determine if the container and cap could be safely used by a consumer in the presence of a flame. Had the management or engineers for these defendants conducted an appropriate risk assessment of this product, they would have determined that the Diamond Natural Fire Starter Gel, its cap, and its container pose a significant and unsafe explosion risk. Further they would have discovered the problem had been solved centuries ago and the solution implemented.

Comment: The problem that was allegedly solved centuries ago did not involve 7 year olds. The flammable tests using an exemplar fireplace performed by the undersigned refute all of Dr. "G"'s opinions. In addition, he is not in any position to come to that conclusion without performing his own tests with an exemplar fireplace and the subject Diamond formulation.

M. "G" Opinion: This product as packaged is unreasonably dangerous to be sold as a consumer product.

Comment: The product as sold to the consumer was not unreasonably dangerous because it was not designed or sold with the intent that a 7 year old would be squirting the contents of the bottle on a fire. *In fact, there is no evidence that "A" squirted the contents of the bottle on embers. There is evidence to the contrary that he actually squirted the contents of the bottle onto the pre-existing fire.* In addition, no one knows how many times or how much of the gel that "A" squirted onto the fire once it got started. He also squirted additional amounts to restart the fire. *We also know that he squirted the liquid in a large number of other areas. This is illustrative of the fact that "A" was playing with the Diamond product and did not use any of the warnings and instructions his mother stated she gave him previously.*

N. "G" Opinion: An internal explosion can occur in a Diamond Natural Fire Starter Gel bottle as it is being squeezed onto a flame or ember.

The statement that Dr. "G" made has a low probability *and is totally false*. Experiments performed by the undersigned continuously documented that no explosion or deflagration can occur internally no matter how much gel was expelled into the fire from the front or directly on top of the fire. This was documented in the videotape test performed by the undersigned. It is more probable than not that "A" had to squeeze excessive amounts of the gel on the fire in front of him. *What no one can answer at this point is when the bottom seam separated and melted. It is more probable than not that the bottom separated well after "A" was on fire.*

In addition, not one of the experts for the plaintiff was able to reproduce the appearance of the subject container in anyway whatsoever. Similarities resulted by squirting the container excessively until sufficient air entered into the head space and reduced the concentration of the ethanol vapor to within the explosive limits. *This illustration by the plaintiff's experts did not, in any way, simulate "A"'s testimony.*

Squirting the contents excessively into a burning fireplace caused a large flashback onto "A"'s exposed body, *which had been previously squirted on by "A" himself,* to the elevated temperatures emanating from the fireplace while he was *allegedly* on his knees facing the unscreened fire. *The high temperatures emanating from the fireplace as measured by the undersigned in the tests performed, was more likely than not, much too high for anyone to withstand.*

Contrary to Dr. "G"'s statement, the subject product was not designed to be squeezed on glowing embers or pre-existing fire *in a fireplace* on a continuous basis by an adult or by a 7 year old. *The product was created and sold for pellet and wood*

***stoves only. This was open and obvious to any adult looking at the label on the container.*** **If several tablespoons were placed on the glowing embers or pre-existing fire, "A" would not have been injured.**

**O. "G" Opinion: This internal explosion can occur when the ambient temperature is above 50 °F.**

**Comment: Dr. "G"'s tests were not supported by scientific testing and documents readily available in the subject case.**

**P. "G" Opinion: While the outside temperature was 24 °F, the accident happened in a finished basement of a house where a fire had already been burning to roast marshmallows so the temperature at the time of the incident was likely over 50 °F.**

**Comment:** ***There was no testimony or comment by "A" that, prior to the subject incident, he was able to safely roast and consume a marshmallow.*** **Not one of the experts reported marshmallows being roasted or present in the area. If the experts continually believe the mother and "A", they will continue to make errors and produce results that are not scientifically acceptable. In addition, there was no evidence of a bag of marshmallows available anywhere.**

**Q. "G" Opinion: The 5 previous explosions of the Eco-Lite Charcoal starter in 1991 are very similar to the explosion of the present case. The flash point for this previous Eco-Lite product and the present gel are both approximately 56 °F and the chemical compositions of the two products are substantially similar.**

**Comment: The formulation for the Eco-lite Charcoal Starter in 1991 was a pourable formulation and is not related to the gelled thixotropic formulation. These two formulations should not be intermixed.**

**R. "G" Opinion: The flashpoint for alcohol is particularly hazardous since at room temperatures, alcohol vapors are in the explosive range. Petroleum distillates like diesel fuel have a higher flashpoint and their vapors tend to be too lean at room temperatures to be explosive. Fuels like gasoline with a very low flashpoint have vapors that tend to be too rich at room temperatures to be explosive.**

**Comment: When alcohol is placed in a thixotropic formulation, it is much less**

**dangerous than the pourable liquid. The tests performed by the experts document this statement. Furthermore, the flammable tests performed by the undersigned using an exemplar fireplace and the thixotrophic formulation demonstrated the integrity and the formulation of the container.**

**S. "G" Opinion: A flame arrestor could prevent internal explosions of this alcohol gel. Flame arrestors have been proven to prevent the propagation of flame into a container. Even a cap with smaller holes, such as the Kingsford cap tested by BEAR, can act as a flame arrestor.**

**Comment: Dr. "G" failed to test the squirting capabilities of the smaller holes versus the one in the subject container. What he also is demonstrating is that the subject gel does not flow in comparison to a non-gelled liquid. If the product is so highly flammable, it would have emitted sufficient vapors and ignited when first exposed to the radiated temperature emanating from the fireplace.**

**What Dr. "G" fails to demonstrate is that the open holes have to be small enough to snuff out the flame front entering through the hole openings in the cap, or that the openings will melt and close fast enough to snuff out the flame front before it enters the head space area in the container. That is impossible.**

***In summary, I am sure that the billing for Dr. "G"'s work conformed to the inadequate testing over a long period of time.***

**There were many areas that proved that the subject product, when used in accordance with the pictorial that was on the front of the label was relatively safe and not designed to be used by 7 year olds who were suppose to be roasting marshmallows.**

***More probable than not, based upon over 50 years of experience in the area of flammable materials, fires and the reconstruction of accidents, "A" did a lot more than just squirt the thixotropic product on the fire. There is no doubt that he squirted the product into a number of areas ranging from the carpeting near the hearth, the hearth, the walls and other areas of the living room in addition to squirting a large amount onto a pre-existing fire. The pre-existing fire was a result of "A" starting the fire and then adding the thixotropic gel in large quantities in the area around the fire, the walls, the back of the fireplace and over parts of his body. Dr. "G" ignored the reality of what actually occurred.***

**Critique "D" (Another PhD plaintiff's expert)**
**"P" describes the accident stating that ""A" Nelson, a 7 year old boy, was attempting to ignite a fire in a home fireplace when the bottle of fire gel ruptured causing the napalm gel contained in the bottle to spray back onto him resulting in severe burns."**

**A. "P" states "that the bottle was an unsafe container for a napalm or napalm-like gel."**

**Comment - Napalm (naphthenic and palmitic acids) is a thickening/gelling agent generally mixed with gasoline or a similar fuel used in military operations. It is a tactical weapon to remove vegetative cover.** ***The product at issue is not considered a napalm gel.***

**"P"'s conclusion that "the bottle was unsafe" has no scientific or technical basis."**

**B. "P" stated "that the bottle was manufactured which is not *fully* compatible with the hazardous contents (i.e., ethanol based napalm gel)."**

**Comment - PVC containers have been safely used for ethanol and many other solvent and flammable products for generations. The use of the word "fully", does not, in any way indicate that the contents of the subject bottle are not safely contained for its foreseeable use and application. At room temperature, the container is completely compatible with ethanol if one performs a risk analysis of the foreseeable use and foreseeable misuse of the subject product, the container will not, foreseeably, be exposed to elevated temperatures. The undersigned has personally used PVC plastic containers for ethanol from 1955 to the present time. All of the research performed by the undersigned in graduate school that incorporated pure ethanol was always transferred from large containers from the supplier to small PVC containers. In undergraduate school, the undersigned worked in the chemical supply room for 4 years transferring 80 to 100% ethanol from large containers to a small PVC containers for students to be used for laboratory experiments. The undersigned had the same experience in industry. There was never a question as to the integrity of PVC and its use as a container for ethanol liquid.**

**Furthermore, this investigator had, in the 90's patented an ethanol based product in a PVC bottle that contained a non-toxic repellent. Not one accident in the years it was sold throughout the United stated, Europe and China.**

**C. "P" states "that the bottle wall did not have sufficient wall strength to accommodate the internal forces of an accidental internal vapor ignition."**

**Comment - What "P" does not admit was that there is no polymer container that can withstand an internal the deflagration or explosion. Why would he single out this specific polymer and not include every other type of polymer? The flammability experiments performed by the undersigned with an exemplar fireplace and exemplar Diamond containers of the thixotropic formulation refute "P"'s statement.**

**D.** "P" states "that the bottle wall had too broad a wall thickness variation leading to stress concentration in the thin areas near the bottom."

**Comment - As stated previously, there is no polymer container that can be designed with respect to the subject and similar ones that can withstand internal deflagration and/explosion. However, the flammability experiments performed by the undersigned using an exemplar fireplace and exemplar Diamond containers of the thixotropic formulation refute "P"'s statement.**

**E.** **"P" states "that the thinnest area (weakest and most likely point of rupture) in the wall was near the bottom portion of the bottle which was the area closest to the person holding the bottle."**

**Comment - What "P" fails to state or consider is that the head space above the ethanol gel contains close to 100% ethanol vapor or, in the alternative, well over the upper explosive limit of 19%. When the container is turned upside down and a small portion is poured out (a spoonful or two), no change takes place because the thixotropic material moves towards the spout and the head space moves to the rear or bottom of the container. It is only when an individual, such as "A", continually squeezes and squirts the contents many times causing air to then mix with the head space and reduce the concentration of alcohol vapor to within the explosive limits can the vapor ignite and eventually deflagrate internally.** ***This was proven and well documented by "P"'s tests. (See reply in I.)***

**Furthermore, the differencews in thickness amounted the the thickness of an average human hair.**

**F.** **"P" states "that the appearance of the subject bottle is consistent with the testimony of "A" concerning how the explosion occurred."**

**Comment - "P"'s reliance on "A"'s deposition testimony versus everyone else's documented records is incredulous.** ***He never attempted to reproduce the subject incident using an exemplar fireplace with the actual product.*** **His testing does not in any way reproduce even "A"'s testimony assuming even part of what he testified to was true. His testimony contradicts the scientific analysis of the ignition and the injuries documented.**

**G.** **"P" states "there is no way to design and manufacture a squeezable blow-molded plastic container that can safely contain in the detonation of a napalm gel. However, it is possible to design a bottle to handle a deflagration/slow burn without blowing napalm back onto the consumer using the product."**

**Comment - The subject container and similar thixotropic products have been safely used by tens of millions of consumers for decades. He continually forgets that the**

**product was not in any way designed for 7 year olds to play with *or a fireplace*. The product was designed for a small segment of our population that used wood pellet or wood stoves.**

**Contrary to "P"'s opinion, if there was a practical and inexpensive design for the same container to handle a deflagration, slow burn without blowing "napalm" back onto the consumer it would have been produced and used in lieu of the subject one. If CKS was responsible, and they are not, in manufacturing a plastic container filled with thixotropic ethanol, that would be safe for a 7 year old, it would be supplied in a steel safe with no combination to open it.**

**If "P" performed a risk analysis relating to the foreseeable use and the foreseeable misuse of the subject product, he would be enlightened to know that the product was designed specifically for pelletized wood burning stoves and not fireplaces. He would also agree that the instructions state that only a tablespoon of the material must be placed within the pelletized wood stove burn pot, that an additional half tablespoon of gel is to be placed on top of the wood pellets, and that the product should not be squirted by a 7 year old all over a room including the fireplace where an excessive amount was squirted continually by "A". In fact, "P" reproduced how many times one has to squeeze the subject container before enough oxygen was drawn into the head space to bring the ethanol vapor concentration down to within the explosive limits. *That is not a foreseeable misuse of the product. It is not foreseeable that parents would allow a 7 year old child to use the product as a toy.***

**H. "P" states "Based upon my experience I believe that bottle manufacturers should communicate with their customers to learn the intended use of their bottle. The bottle manufacturer and/or the customer should test the bottle to determine if the bottle is safe for the intended use. If the bottle is found to be unsafe for the intended use, the bottle manufacturer should refuse to market the bottle for that application."**

**Comment - This more than documents the fact that "P"'s experience is limited in reference to the real world of involving the manufacturing of plastic containers. The sealed area is usually the weakest part of both plastic and glass containers. "P" should also know that if he was capable of performing a risk analysis on the subject container, it would have been easy to *prove that it was not foreseeable that a 7 year old would be playing with the container as a toy*; holding it in a horizontal position adjacent to a flame for an extended period of time and then raising it above his head where he testified it exploded. It should be noted that there is not one other reported incident with the subject product where the incident was described similar to "A"'s.**

**I. "P" states "a PVC bottle is not appropriate as a container for napalm gel but manufacturing the bottle with the bottom being the weakest point (closest to the user) in the container was exceptionally reckless."**

**Comment - "P"'s comments again demonstrates his lack of expertise in the area of plastic containers. The seals of all plastic containers are the weakest part of the container. It is also the weakest part of a metal and glass containers. The subject container had seals at the base and on each side of the container. The seal that will be based on how the container was being held while the internal vapors inside the container was on fire. The seals could have failed at any of locations where the seals were located.**

**The subject plastic container in this case was designed to satisfy load drop-test and also satisfy the internal pressure produced by ethanol vapor. A drop test, in this connection, tests the ability of the bottle to withstand impact forces when the bottle is dropped on its mouth or neck portion because the mouth and/or neck portions are generally dimensionally the smallest parts of the bottle, any impact forces applied to those regions create maximum stresses in the walls of the container that tend to cause the container to burst or rupture at the weakest areas, typically the mold seams. Existing bottles, such as the one in the subject case, used by the chemical industry, have satisfied both requirements or specifications established by the various government authorities and transportation laws. It is important to note that a spherical surface where the seams are located in the subject container is the optimum design for the most resistance to failure.**

**PVC (Polyvinyl Chloride) is a softer plastic, which is relatively inexpensive. It is very durable and can be dropped without much of a problem. It also takes a lot of deliberate strength to tear PVC apart. The variation of the hardness of PVC is based upon how much of the plastic is put into a specific area. PVC can also be recycled. In addition, PVC has been used to contain ethanol and other alcohol for decades.**

**"P" also states in his opinion, that PVC is "marginally compatible" with ethanol and that is not true. The rest of his opinion on page 12 of his 28 page report is purely subjective without scientific documentation.**

**"P" also states that when he was with Dow Chemical, Dow was obligated to work with their customers to insure the safety of the consumer. In this case, CKS was not responsible to perform any risk analysis studies because they supplied the PVC containers as requested. However, he neglects to state that liquid ethanol and other alcohols such as isopropanol have been sold in PVC containers for decades. If that is true, there is no other obligation or responsibility on the part of CKS with reference to the end use application of the PVC container in this case.**

**If one examines "P"s litigation experience on page 16 and his scientific presentations listed on pages 17-26 as well as the patents listed on pages 28-29, there isn't one reference that relates to PVC, ethanol, fires, explosions and deflagration of containers.**

*Lastly,* the undersigned assumes by choice, that "P" chose not to address the subject of marshmallows even though he had an opportunity to review the depositions of "A" and other members of the family as well as an opportunity to examine all of the photographs associated with the accident. He also failed to comment on the fact that if there was an opening at the base of the container that any squeezing of the container would have released the thixotropic ethanol through the opening in the top closure as well as the alleged opening in the base of the container.

Furthermore, tests performed by the undersigned using an exemplar fireplace and exposing a half filled Diamond product container to well over 200 degrees Fahrenheit documented that the bottom seam would not rupture or melt and that the container actually expanded as shown in the videotapes and photographs taken by the undersigned during and after the tests.

Critique by "H" (Expert for the plaintiff)
Dr. "H" stated in her report that "The product is marketed as a fire-starting accessory, it is suggested for use with wood or pellet stoves."

From Section 3.0 (Background) to page 7 of Dr. "H"'s report, she re-iterates what is already known through prior publications and/or company literature. What is important on page 8 is that she measured head space of the volatiles in the PVC container and found that it was essentially ethanol. There was no report that air and/or oxygen was present in the head space. That is important because no explosion can occur if the concentration of the ethanol vapor is greater than the upper explosion limit.

Dr. "H" refers to the CPSC's 1991 voluntary recall of Eco-Lite charcoal starter. This is a product composed of ethanol and individuals were warned not to spray the material over a burning fire and/or charcoal because, like gasoline it may result in a flashback.

*Comment*: The objective of the Natural Fire Starter Gel was exactly what the name refers. The name and pictorial on the label designate that the product is designed to be used for wood and pellet stoves and not fireplaces. The warnings Eco-Lite product was not thixotropic and designed to start coal fires for barbeques, as an example. The two products are not designed for the same application.

It is open and obvious to a consumer that one does not squirt gasoline into any type of fire. Therefore, when the consumer is properly warned to keep the product away from children and a parent, as in this case, does not place the subject product out of the reach of their children, including "A", and claims they allow "A" to use the product. Manufacturers cannot be responsible for their actions when they violate the manufactures instructions. Manufacturers cannot be responsible for the parents'

**actions when they decide, on their own, to use a product designed for wood and pellet stoves to be used in a fireplace. Dr. "H" ignores that issue.**

***When a product involves a fire starter of any type, there are no liquid or gelled fire starters that are inherently safe. However, in the subject case, even "A"'s father stated that he used the Safe Fire Starter logs that were on the side of the fireplace. Why any parent would use any other product when they have a safe way of starting a fire is questionable.***

**What Dr. "H" fails to also point out is that all of the other products do not designate that their products should be used for wood and pellet stoves with a pictorial on the front. *In the opinion of the undersigned, that significantly differentiates the subject product in this case versus all of the others*. The CPSC staff are not warning experts and never addressed that issue with reference to the fact that the Natural Fire Starter Gel with the pictorial and the fact that it was designed only for wood and pellet stoves distinguishes that product from all of the others.**

**Dr. "H" continually relates the Natural Fire Starter Gel to the Eco-Start Natural Lighter Fuel. They are two separate products designed for two separate objectives. She continually fails to recognize or even discuss the differences in the two and the pictorial of one product versus the other. In fact, on page 12 of her report, she does not feel that there are any differences in the two.**

**Dr. "H" relates the two by stating that they would act identically with a hot coal or a piece of burning wood "regardless of whether the products are used in wood stoves, barbeque grills, or fireplaces." Dr. "H" is not a warning specialist and, from that stand point alone, *there is a significant difference*.**

**The Natural Fire Starter Gel is specifically designed for wood stoves and pellets and not barbeque grills or fireplaces. In addition, from a risk analysis standpoint, it is not foreseeable that a parent would use a product designed for wood and pellet stoves or fireplaces when they already had purchased and are experienced in using the Safe Starter Logs. Furthermore, it is not foreseeable that a parent, under supervision or not, would allow a 7 year old to use the product with or without their permission.**

**"H"'s reports her comments relating to the determination of the subject fuel gel versus what was reported by the manufacturer and the CPSC may vary slightly but it is not an issue in this case. The temperature of the flame of a fire is in the range of 1800 °F and that of burning coals is over 900 °F. What difference would it make whether the flash point of the gel is 54, 73 or 130 degrees? Once the vapor is ignited in the fireplace, it would still ignite and flash back whether the flash point is 52 or 130 degrees.**

**Therefore, the comments Dr. "H"'s report do not make any sense, and in the opinion of the undersigned, ,the fire starter is safe when it is used in accordance with its instructions. for wood and pellet stoves.**

**Every consumer has the right to purchase any product they want in Wal-Mart or any other store. For example, many stores sell tanks filled with propane. Propane is highly explosive and has explosive limits between 2.1 and 10.1 percent. Gasoline is also explosive and has explosive limits between 1.4 and 7.6 percent.**

**When gasoline is used for gasoline lawn mowers, gasoline hedge cutters and emergency electrical power generator, the gasoline is usually dispersed from small container ranging from 1 to 10 gallons. There has never been a recall for any of those products even though they are more dangerous than the thixotropic ethyl alcohol. When the thixotropic ethanol, propane gas and gasoline are used for their intended purpose by adults, the risk of foreseeable injuries is reduced significantly. Dr. "H" fails to continuously point out that all of these products are not designed for parents allowing 7 year olds to use the product under any conditions, *especially when "A" used the product as a toy.***

**Lastly, Dr. "H" spent a lot a time reproducing data that is readily available by the manufacturers of the products and documents readily available on line. There was no need to perform any laboratory tests proving what is already known and attempting to reinvent another wheel.**

**1 "W"J.D.(Expert for the plaintiff)**

**Mr. "W" defines his area of product safety management as a "system that a reasonable prudent manufacture puts in place before the first product is conceived and ensures that the final product along with its warnings, packaging and marketing materials is reasonably safe."**

**What Mr. "W" neglected to discuss is that every consumer has the right to purchase any product they want in Wal-Mart or any other store. For example, many stores sell tanks filled with propane. Propane is highly explosive and has explosive limits between 2.1 and 10.1 percent. Gasoline is also explosive and has explosive limits between 1.4 and 7.6 percent.**

**When gasoline is used for gasoline lawn mowers, gasoline hedge cutters and emergency electrical power generator, the gasoline is usually dispersed from small container ranging from 1 to 10 gallons. There has never been a recall for any of those products even though they are more dangerous than the thixotropic ethyl alcohol. When the thixotropic ethanol, propane gas and gasoline are used for their intended purpose by adults, the risk of foreseeable injuries is reduced significantly. Mr. "W" fails to point out that all of these products are not designed for parents allowing 7 year olds to use the product under any conditions.**

**The hazards were identified in each of those products; the risks were assessed and adequate safety measures were applied to minimize both hazards. The same approach was made by the manufacturers and distributors of the Natural Fire Starter Gel which is much less dangerous than propone and gasoline. *As stated previously,***

***the undersigned has been involved with the use and application of 80 to 100% ethanol and PVC containers for close to 58 years.***

**It should be noted that Mr. "W" is not a warning specialist especially in the area of hazardous products and chemicals. He points out on page 7 of his report that he is developed on-products warnings and instructions for a number of manufacturers and distributors. Not one of them involved flammable chemicals with physical and chemical characteristics similar to the Natural Fire Starter Gel.**

**Vending machines, ATVs and pool products are not at the level of warnings and instructions required for products such as the Natural Fire Starter Gel. In contrast, the undersigned has created warnings and instructions for a number of corporations, presented OSHA safety meetings at a large number of construction sites, created Material Safety Data Sheets for flammable products, evaluated polymers from room temperature to hyperthermal temperatures, consulted to NIOSH (the research arm of OSHA) and performed many explosive tests on a variety of flammable products.**

**Mr. "W" also claims that he created warnings and instructions for fitness equipment in Dick's Sporting Goods. The undersigned has evaluated many accident cases involving fitness equipment. I have also invented sports products that were sold in Dick's. Contrary to Mr. Kitzes' claims, the warnings and instructions for fitness equipment are created by the manufacturers of the equipment and not the management of Dick's Sporting Goods.**

**Mr. "W"'s' report he points out the use of a pictogram but fails to point out the pictorial of a wood stove on Natural Fire Starter Gels label versus that of a fireplace. He also fails to point out that the words in the warning are at the level of a fifth-sixth grader. In addition, he failed to point out that the product must be kept away from children and is not designed to be used by a 7 year old.**

**The label of the Diamond product states in block letters "PERFECT FOR ALL WOOD AND PELLET STOVES." There is also a pictorial of a wood stove - NO FIREPLACE is shown or mentioned on the front or back labels of the Diamond product. Mr. "W" failed to address that issue which would be considered self-serving on his part by not commenting on that information on the label.**

**Mr. "W" describes "A"'s behavior prior to the accident without noting all of the different statements made to the EMTs and medical specialists at the hospital. He even failed to mention "A"'s brother's statement under oath where he stated that "A" had already taken a bath and his wet towel was upstairs in the bathroom. In fact, the description of the accident described by Mr. "W" is totally inaccurate and technically incorrect.**

**Mr. "W" makes a statement regarding "A" being an "avid outdoorsman, hunting, fishing and trapping." What he doesn't point out that "A" at the age of 7 was able to read and did well in school and was sophisticated for his age.** *is immature and*

***irresponsible, based upon his behavior prior to and up to the time of the incident. In addition, contrary to Mr. "W"'s statement, it should be obvious to him that "K" did not read the Diamond Natural Fire Starter Gel label to "A" on prior occasions. That conclusion is obvious to the undersigned.***

**Although Mr. "W" had been associated in some way with the CPSC (not true), he is not a chemist, chemical engineer, safety specialist, safety engineer, professional engineer or certified in warnings and instructions. In addition, he is not qualified to perform a risk analysis, based on his background and experience on the product, Diamond Natural Fire Starter Gel.**

**The fact that "K", "A"'s mother,** ***could not read certain important parts of the label and, on top of that,*** **disregarded the warnings concerning children, negates all of his opinions in regarding the subject warning. When you have a consumer such as "K" dismissing the dangers which she was aware of and the warnings concerning her own child, how could Mr. "W" criticize anything on the label of the subject bottle.**

**Mr. "W" knew, or should have known, especially with his law degree, that when you have a recalcitrant worker on a construction job, who refuses to wear protective equipment and is injured, that worker is responsible for his own injuries. In this case, "K" openly refused, in spite of her husband's opinion, to follow the simple instructions and warning on the back label of the Diamond Natural Fire Starter Gel. Now that the horse has run out into the pasture, both Mr. "W" and "K" want to place a lock on the barn door. The undersigned handles similar situations like this each week when the son or daughter of an individual receives a brain injury in soccer and can't go back and play unless they wear his headgear (see www.forcefieldheadbands.com). By that time, it is much too late** ***to help the child reduce the risk of a brain injury.***

**Although the warnings and instructions on a subject label were accurate based upon the manufacturer's intended use of the product in wood and wood pellet stoves, the instructions regarding children were completely disregarded by "K". If she just obeyed that one warning, "A" would never have been injured.**

**In spite of "K"'s behavior, Mr. "W" continues to quote various sections of the CPSC without addressing the real issues in the case. For example, on page 12 of his report he describes the vapor "may burn with near invisible flame." What does that mean to a 7 year old? The fact is that the warnings and instructions were never designed for a 7 year old to read and use the product. What Mr. "W" fails to recognize is that "K", the mother of "A", had complete control of the product and completely ignored the warnings.**

**There was nothing complex about the simple warning concerning children. There is nothing that a manufacturer can do if a consumer, such as "K" refuses to follow and ignores important instructions concerning the safety and welfare of her own child.**

***On top of all of the instructions that should have been followed if "K" used the product, no one used a tablespoon of the product at any time. If a tablespoon amount was placed on any fire, no injuries would have occurred.***

**The undersigned has been involved with warnings and instructions regarding all types of chemicals from the early fifties to the present time, from the warnings and instructions published by the Manufacturing Chemical Association to the current ANSI Z 129.1 which is the labeling standard for Hazardous Industrial Chemicals-Precautionary Labeling. It is important to note that this is a voluntary standard and not statutory. All of the other references Mr. "W" makes from do not relate to the real issues in the subject case.**

**Most important, the manufacturers of the product knew that the product will meet the needs of a small segment of the market which is wood and wood pellet stoves.**

**"W" states that "the highly flammable ethanol based gel with a nearly invisible flame, could flashback to the bottle, causing it to explode in the hands of the consumer and spread the flaming contents over the body of the user." This statement is totally inaccurate and false. It contradicts the tests performed by "P" and also documents the fact that Mr. "W" is limited with his lack of background in the areas of chemistry, chemical engineering, fires and explosions, deflagrations, flashbacks and why and how the container failed in the instant case. In addition, although Mr. "W" never addressed the subject, it is more likely than not, that no polymeric, glass or metal container could have withstood the deflagration that occurred.**

**What "W" is attempting to do is validate the defective scientific methodology used by the other plaintiff's experts that did not, in anyway, attempt to reproduce "A"'s description of his accident using an exemplar fireplace and the Diamond thixotropic product. The same group of "experts" chosen by the plaintiff are comparing the same inaccurate models and opinions with one another-but not versus the real world.**

**Furthermore, Mr. "W" never addressed the subject in which "A" irresponsible actions resulted in him squirting the contents of the Diamond Natural Fire Starter Gel in many areas on the carpet, the hearth, the walls and the firewood while it was burning. In addition he held the container for an extended period of time at a horizontal angle just like "P"'s test demonstrated before the vapor, emanating from the open container, caught on fire resulting in the deflagration of the container.** ***This did not happen.***

**Mr. "W" deficient background relating to the issues in this case is well beyond his capability of explaining what occurred in this case posing the subject incident and resulting injuries to "A".** ***His reliance on the other experts inaccurate and inept attempt to reproduce the questionable testimony of "A" and his family rather than independently prove or disprove his case, defines his lack of***

***experience and background in the technical issues associated in the subject case.***

**Fire and burning tests**

**Based upon over 50 years of experience in testing the flammability of a variety of solid and liquid chemical formulations in addition to testing the subject formulation in a simulated wood fireplace, it is the opinion of the undersigned, with a reasonable degree of safety engineering and scientific certainty, that "A" did not testify in his deposition as to his actual behavior prior to and up to the time he was injured. There is no question, based upon the testing by the undersigned and the plaintiff's expert's testing, that "A" did more than just squirt a little of the thixotropic alcohol into a burning fire. It is more likely than not that he squirted an excessive amount of the contents not only on the carpeting and the areas surrounding the fireplace such as the hearth, the back wall of the fireplace, the riser to the hearth, the walls, etc., but the straw that broke the camel's back was the excessive amount of formulation squirted on himself and the fire itself for an extended period of time until the vapors emanating from his body ignited due to the fire and heat radiating from the fireplace ending up igniting the vapors.**

**Experiments by the plaintiff's experts document the fact that an excessively large amount of squirts. The only way the vapors inside the container could be ignited was if "A" constantly squirted the contents out and air continued to enter the head space until there was sufficient air in the container to lower the concentration of the ethanol vapor to within the explosive limit so it could ignite. That was also verified in the video tape submitted by the plaintiff's expert.**

**In addition, the fact that "A" did not receive burns on the top of his head and back documents the fact that he did not raise the container above his head while the gel was coming out the bottom. That had to occur elsewhere such as on the or around the couch. (The resistance of the container from igniting and burning was well demonstrated in the fire test performed by the undersigned.)**

***Furthermore, this also verifies the fact that, in the opinion of the undersigned, "A"'s mother, "K", contrary to her sworn testimony under oath, never read the instructions to any of her children and never demonstrated that only a tablespoon amount of the contents should be placed on an unlit piece of wood in a wood stove. If she had, this would also document that "A" did not understand anything she attempted to warn "A" about and, as a 7 year old, he was irresponsible and disobedient.***

. ***In addition to the above,*** **the undersigned squirted excessive amounts of the Diamond product into the burning fireplace from the front and from the top exposing the spout to well over 200 degrees Fahrenheit in the front and over 300 degrees Fahrenheit one foot from the flame from the top. All of these tests were videotaped and photographed. In both instances, the undersigned could not touch the closure on**

**the top to close it. It was much too hot. In all cases, there was no indication of any flame front entering the container because it would be impossible for that to occur. Therefore, those tests disprove "A"'s account and testimony of what occurred prior to the subject incident.**

***The undersigned has consulted on several hundred burn cases to both children and adults over the last 44 years. Based on that experience and studies, I am able to convey that experience and observations to the following:***

***a. Explosions of natural gas, propane, gasoline and other flammable liquids, including ethanol, isopropanol, cause intense heat for a very brief time. Clothing, unless it ignites, is protective in flash burns. Hence, flash burns generally have a distribution involving all exposed skin, with the deepest areas facing the source of ignition. For the most part, flash burns reach progressive layers of the dermis in proportion to the amount and kind fuel that explodes.***

***b. In contrast to flash injuries, flame burns are invariably deep dermal if not full-thickness because of more prolonged exposure to intense heat.***

***c. Adding and squirting a voluminous amount of Diamond gel into the fireplace in all directions did not result in any flashback whatsoever. The heat emanating from the fireplace was too high in temperature for any human to withstand. If that is true, where was "A" actually positioned at the time his body caught on fire because of the large amount of gel all over his body? How did he really react once the gel all over his body caught on fire?***

***Based upon the extensive experiments performed by the undersigned on an exemplar fireplace squirting extensive amounts of gel on a burning fire, there was never a flashback experienced under any conditions. Any testimony regarding flashbacks squirting the Diamond product into a fire cannot be reproduced and would be incorrect.***

***d. With reference to the subject incident and "A"'s burns, it is highly probable based upon the amount of contents that he squirted into the fire, that he could have easily received a significant amount of his burn injuries from the gel that "A" squirted all over his body prior to getting on his knees in front of the fire.***

. **Marshmallows**

**There is some evidence of a white residue on a large metal fork in one of the photographs. *If "A" did roast one or more marshmallows and placed the fork down, this would then lend itself to "A" finishing the marshmallow(s) and then continuing to play with the Diamond product.* Squirting of the gel onto the carpet, hearth and other areas around the room, including his own body prior to kneeling before the unscreened fireplace and fire contributed to all of "A"'s burn injuries. All of the vapors emanating from his body was exposed to the heat from the fireplace. The temperature was more than enough to ignite the vapors. In fact, the temperature was more than enough to ignite most other solvents. No one ever found the bag of marshmallows "A" claimed he took downstairs to the fireplace.**

**Reconstruction of Subject Accident (Flammability Testing)**

**The undersigned conducted extensive fire tests in Boynton Beach, Florida by creating a exemplar fireplace using hickory wood chunks, paper and the Diamond Fire Starter Gel to start the fire. The following was observed:**

a. **As the fire developed, the undersigned squirted large volumes of the gel onto the fire. No flash fire was observed. What was observed was a fire of higher intensity.**

b. **Large volumes of the Diamond product were squirted into the fire directly from the front. No flash fire was observed. What was observed was a fire of higher intensity.**

c. **A metal surface was covered with the gel and placed near the front of the fireplace. This was done to stimulate the gel covering "A"'s body that he previously squirted on himself. The surface burned with high intensity within a short period of time.**

d. **The surface of the 2" X 6" piece of wood, approximately 16" long was coated with the gel and placed along the floor surface adjacent to the front of the fireplace. Within a short period of time the vapors from the surface ignited and the gel on the surface burned.**

e. **Large volumes of the gel were squirted onto a blazing intense fire in the fireplace from a position directly above the fire and in front of the fire. No flashbacks occurred at any time. *The spout was exposed to temperatures ranging from 200 to 300 degrees Fahrenheit. The spout, containing the gel never caught on fire.* This contradicts all statements made by the plaintiff's experts and "A" in his sworn testimony.**

f. **The Diamond Fire Starter Gel container was placed directly in front of the fireplace. Nothing initially happened to the container. However, additional tests on the integrity of the container proved that it held up exceptionally well when** exposed to temperatures over 200 degrees Fahrenheit as documented in the videotapes and photographs taken by the undersigned during the flammability and integrity testing of the container. *The container was half filled and the only observation made while the container was exposed to over 200 degrees Fahrenheit was that the body of the container facing the fire expanded as shown in the videotapes and photographs.*

g. The Diamond Fire Starter Gel container with approximately 40 cc's of gel then thrown into the fireplace. There was no explosion and it took an extensive amount of time before the container started to burn documenting the fact that the integrity of the design of the container was more than satisfactory.

h. The undersigned squirted large quantities of the Diamond Natural Fire Starter Gel from the opening in the top of the fireplace shown in the photographs and videotapes. The temperature opening of the top of the fireplace was well over *300* degrees Fahrenheit. It was so hot that the undersigned could not touch the top of the closure and seal the top with my bear hands. The temperature was well over ignition temperature of the ethanol vapor. In spite of that, no internal ignition was observed and no explosion occurred contrary to the testing and comments by Dr. "G" and "P". The latter two plaintiff experts did not, in any way, experimentally demonstrate their incorrect opinions.

Testing the Integrity of the Diamond Natural Fire Starter Gel Plastic Container
The undersigned tested the integrity of a half filled container of the Diamond Natural Fire Starter Gel. The fire was started using the Diamond Natural Fire Starter Gel as the accelerant. Instead of a tablespoon of the material, approximately 1/3 of the container was sprayed on the wood nuggets and the vapors were ignited with a Scripto flame lighter. A continuous videotape was taken and the experiment was also documented with a series of photographs. A half filled container of the Diamond Natural Fire Starter Gel was placed approximately 14 inches away from the fire. The temperature of the heat emanating from the fireplace, at the point where the container was placed, was approximately 200 degrees Fahrenheit. No human being could withstand the temperature.

A half filled container of the Diamond Natural Fire Starter Gel was placed in front of the fireplace for over 30 minutes. After approximately 20 minutes, the area below the headspace that was facing the heat emanating from the fireplace swelled after approximately 25 minutes. There was no failure whatsoever and the bottom seam maintained its integrity contrary to the claims made by the plaintiff's experts. It is the opinion of the undersigned, with a reasonable degree of safety engineering and scientific certainty, that the plastic container is more than suitable for the formulation for the Diamond Natural Fire Starter Gel. In addition, unlike many formulations containing alcohol does not deform while in storage.

An additional important note was that the temperature emanating through the top hole in the cover of the garbage can where the undersigned squirted streams of the gel

**into the fire below, was well over *300* degrees Fahrenheit. It was so hot, that the undersigned could not handle the closure for the top of the bottle after squirting the gel through the hole. Contrary to the claims made by the plaintiff's experts, there was no flame front that went through the opening in the spout that entered the internal headspace of the container. The reason was that the concentration of the ethanol vapor was above the upper explosion limit *and that the headspace was at the bottom of the container and not the top. When the container was placed in the upright position, and the spout left open until it cooled down, there was no ignition of the ethanol vapor escaping from the opening in the spout.***

**The headspace above the half filled container contained mostly ethanol vapor. When exposed to the excessive heat in front of the exemplar fireplace, the gas expanded and the container responded by expanding as well. Contrary to the *defendants* allegations regarding the seam on the container, it would be predicted to anyone with knowledge of plastic containers, liquids and vapors within the container, that the most vulnerable area of the container when exposed to elevated temperatures would be the headspace within the container and not the bottom seam. *Even though this was foreseeable, it was also proven during the experiment by the undersigned.***

**Furthermore, based on the demonstration by the undersigned, the plaintiff's claim that a leaked developed at the bottom seam of the container when the container was exposed in some unexplained way to their fireplace, is an impossibility. If it was exposed to excessive heat from the fireplace, the container would demonstrate an area of expansion above the headspace. The evidence in possession of the plaintiff documented that there was no expansion whatsoever. Therefore, it would be impossible for the accident to occur as explained under oath by "A", members of his family and the reports and experiments released by the plaintiff's experts.**

- **The retail package provided sufficient barrier to prevent migration of the product or volatiles from the product.**

- **The Material Safety Data Sheet relating to this product was prepared in a reasonable manner and within all applicable standards.**

- **The delivery of this product, as delivered by PSC to Jarden/Hearthmark, was new and merchantable, free from any defects in workmanship or material, and conformed to applicable specifications, state of technology and the standard of care in the industry. Specifically, it is my opinion to a reasonable degree of engineering and scientific certainty, that the product met the specifications set forth by Jarden Home Brands / Hearthmark in the Performance Sepecifications for the Diamond Fire Starter Gel.**

**Tests were conducted by the undersigned to determine the force required to open the closure that snapped in place. All measurements were over 8-pounds of force. That**

**would make the closure child resistant and very difficult, if not impossible for young children to open it up.**

**Conclusions**

**It is the opinion of the undersigned, with a reasonable degree of safety engineering and scientific certainty, that contrary to the allegations set forth by the plaintiffs and the plaintiffs expert's, that the cause of all of "A"'s injuries were directly related to his mother's *willful and intentional* disregard of the safety and welfare of her own son and had nothing to do with the formulation of the Diamond Natural Fire Gel or the integrity of the container. *The fact, at this time she is attempting to expiate her guilt by blaming her negligence on anyone of the defendants is a sham.* The conflicting statements by each member of her family, the EMTs and the medical specialist in the hospital more than document this opinion.**

**The flammable testing performed by the undersigned refuted and contradicted all of the allegations made by the plaintiff and by all of the plaintiff's experts. All of the testing performed by the plaintiff's experts was not realistic and did not relate, in any way, to the real world and what actually occurred at the time "A" was burned.**

**Summarily, the undersigned further opines that the Diamond Natural Fire Starter gel product is reasonably safe for its intended uses and reasonably foreseeable misuses, that the actions of PSC, Inc. in communicating with Jarden/Hearthmark and blending/packaging product were within all applicable standards of care, and PSC, Inc. did not cause or contribute to "A"s incident.**

**The experiments performed and opinions made by the plaintiff's experts did not represent the actual behavior of "A" prior to and up to the time of the actual incident. All of the experts' opinions were *"validated" by the plaintiff's handpicked group of alleged group of "experts" that compared* based on the results of inaccurate testing and questionable testimony of "A" and his family, *but* not versus the real world and real science. Even so, the Plaintiff's experts still could not replicate the version of events provided by the family.**

7. **Opinion**
   a. **With over forty years of experience in the area of flammable fabrics, explosions, fires, children's burn injuries, warnings and instructions I cannot find liability in Diamond's Natural Fire Starter Gel.**
   b. **"A"'s parents have a non-delegable duty to safeguard the safety welfare of "A.". They blatantly ignored that responsibility and allowed him to play unsupervised with a dangerous instrumentality. They are responsible for the consequences of their irresponsible behavior.**
   c. **The undersigned has compassion for "A" and his injuries. However, the liability and the cause of his burn injuries are directly related to the disregard of "A"'s's safety and welfare by his parents.**

**C. J. Abraham,, PhD, JD, PE, CPC, FRSC, DEE, FTI, BCFE, DCEE, CChE, CChem has a varied background that can be found at: www.scientificadvisory.com. He also has over 45 patents in a variety of arenas as well. Many of the inventions were commercialized. The latest invention can be found on www.forcefieldheadbands.com**

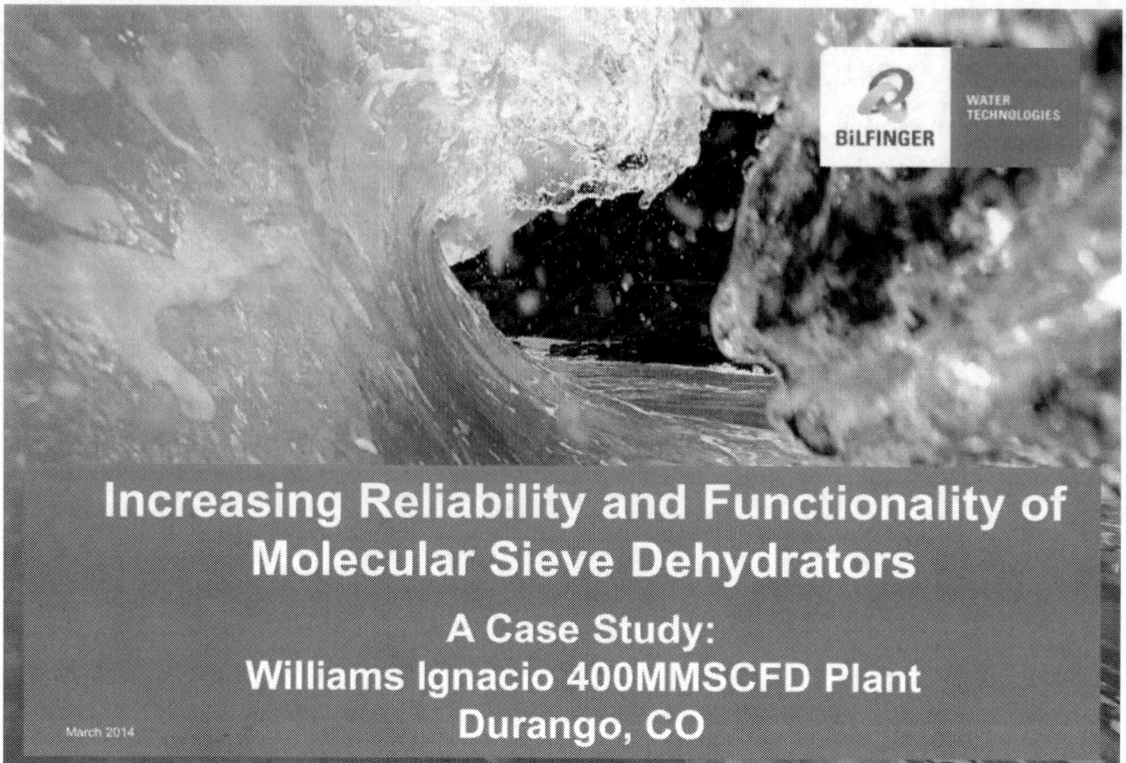




## CONTACT INFORMATION


**Benjamin Schmitt, PE**

**Senior Applications Engineer**

Phone +1 651 638 3274
Mobile +1 651 253 1258
Benjamin.schmitt@johnsonscreens.com


**Hydrocarbon Processing**
**GLOBAL BUSINESS UNIT**

# PRESENTATION OVERVIEW

BILFINGER WATER TECHNOLOGIES

- Review Pre-Installation Operation
- Options Available to Address Known Problems
- Known Engineering Challenges
- Engineering Configuration Alternatives
- Installation Review
- Review Post-Installation Processing Operation



# PRE-INSTALLATION: OVERVIEW DATA

BILFINGER WATER TECHNOLOGIES

**Dryer Overview:**

- 400MMscfd
- Fish Engineering (1956)
- (4) Adsorber Train
- 8' Diameter x 23' Tan-Tan
- Multitude of Conversions to Achieve Final Loading
- 34,200lb Mole Sieve / Vessel

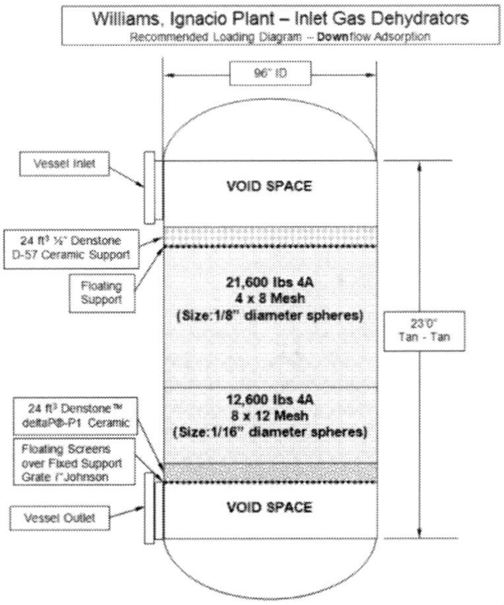


©2014 All rights reserved. Bilfinger Water Technologies, Inc., New Brighton, MN USA Information contained in this presentation cannot be copied or distributed without the express written consent of Bilfinger Water Technologies, Inc.

# PRE-INSTALLATION: SUPPORT GRID RELIABILITY

BILFINGER WATER TECHNOLOGIES

- Multiple Unplanned Shut-downs Between 2010-2013
- Extensive Unplanned Down Time & Maintenance Costs
- Loss Of Production
- Loss Of Sieve Containment

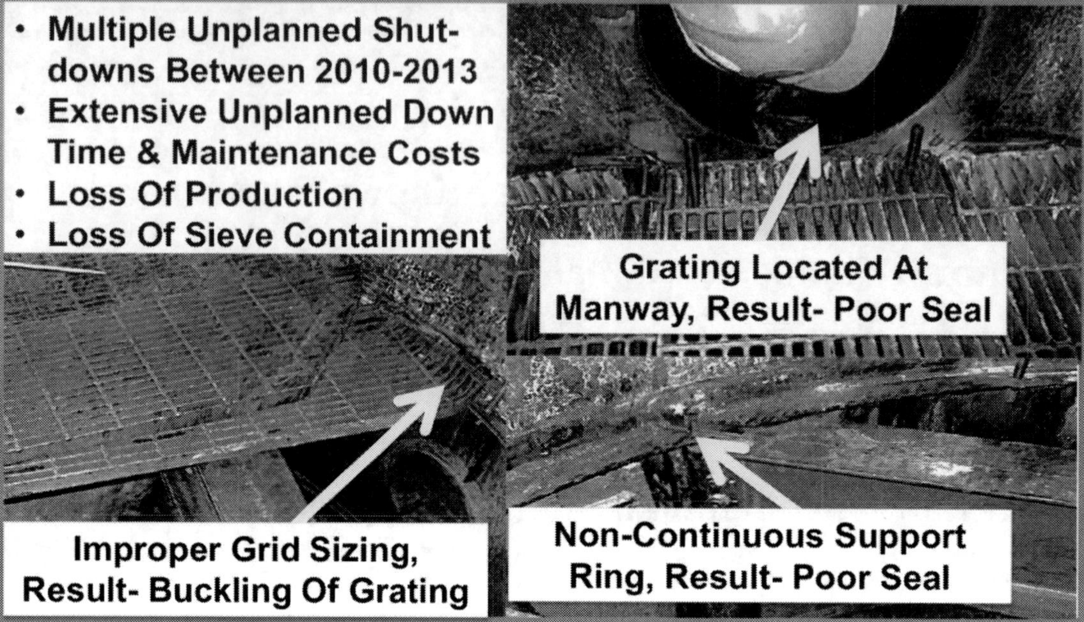




# PRE-INSTALLATION: SUPPORT GRID CONFIGURATION

BILFINGER WATER TECHNOLOGIES

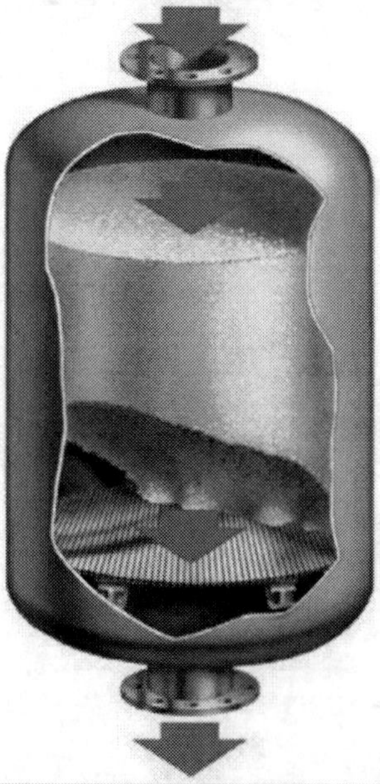

- Applicable in Mole Sieve Dryers
- Full Cross Sectional Catalyst Support
- Versatility of Design
- Ability To Support Catalyst In Direct Contact

©2014 All rights reserved. Bilfinger Water Technologies, Inc., New Brighton, MN USA Information contained in this presentation cannot be copied or distributed without the express written consent of Bilfinger Water Technologies, Inc.

## JOHNSON SCREENS SSG™

BILFINGER WATER TECHNOLOGIES

- Applicable in Mole Sieve Dryers
- Versatility of Design
- Ability To Support Mole Sieve In Direct Contact
- Increased Mole Sieve Weight



## KNOWN ENGINEERING CHALLENGES

BILFINGER WATER TECHNOLOGIES

- Mechanical Reliability
- Mole Sieve Containment
- Thermal Expansion
- 16 Inch Manway
- Space Limitations
- Late 1950's Vessels
- No Welding To Vessel

## KNOWN ENGINEERING CHALLENGES

- Side Outlet / Inlet
- Unknown Head Condition
- Increased Bed Weight
- TA Schedule
- Internal Piping Connections



## CONFIGURATION ANALYSIS:

| | SUPPORT GRID | SSG™ |
|---|---|---|
| Mole Sieve Weight | | ✓ |
| Mole Sieve Usage | ✓ | ✓ |
| Reduced Regeneration Cycling | | ✓ |
| Fluid Distribution/ Hydraulic Efficiency | ✓ | ✓ |
| Support Media Volume | ✓ | ✓ |
| Bed Containment Mechanical Reliability | ✓ | ✓ |

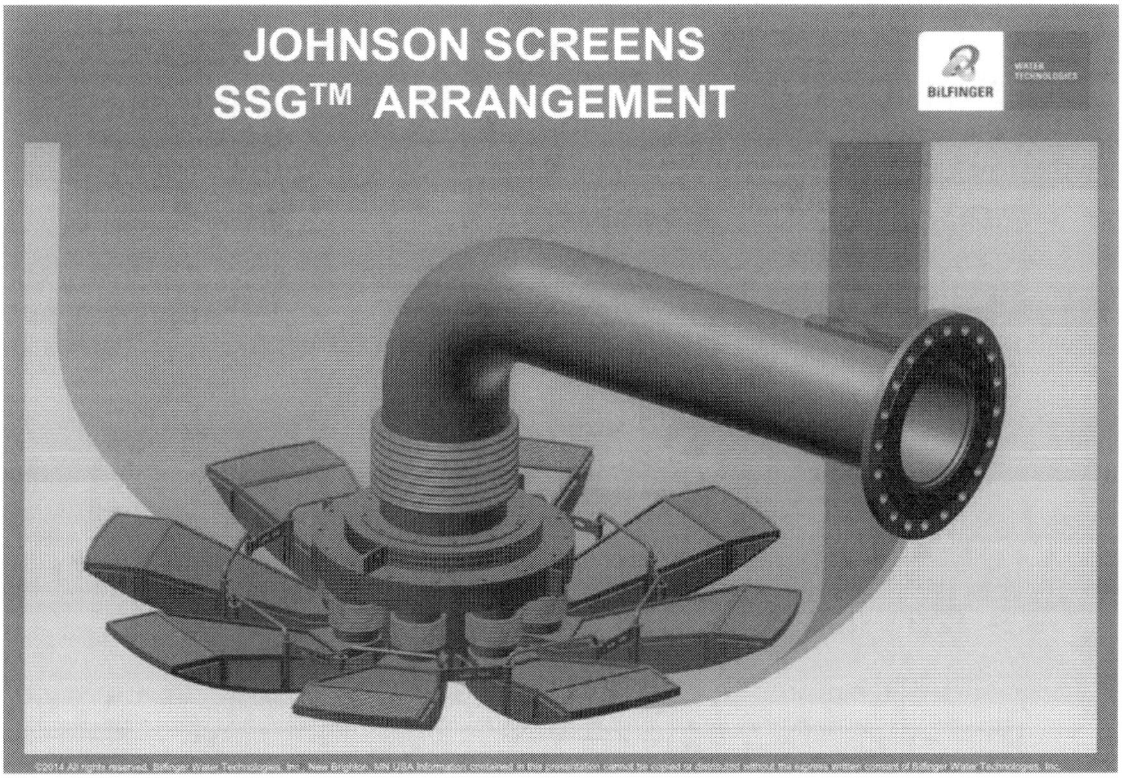
JOHNSON SCREENS
SSG™ ARRANGEMENT
BILFINGER
WATER TECHNOLOGIES
©2014 All rights reserved. Bilfinger Water Technologies, Inc., New Brighton, MN USA Information contained in this presentation cannot be copied or distributed without the express written consent of Bilfinger Water Technologies, Inc.

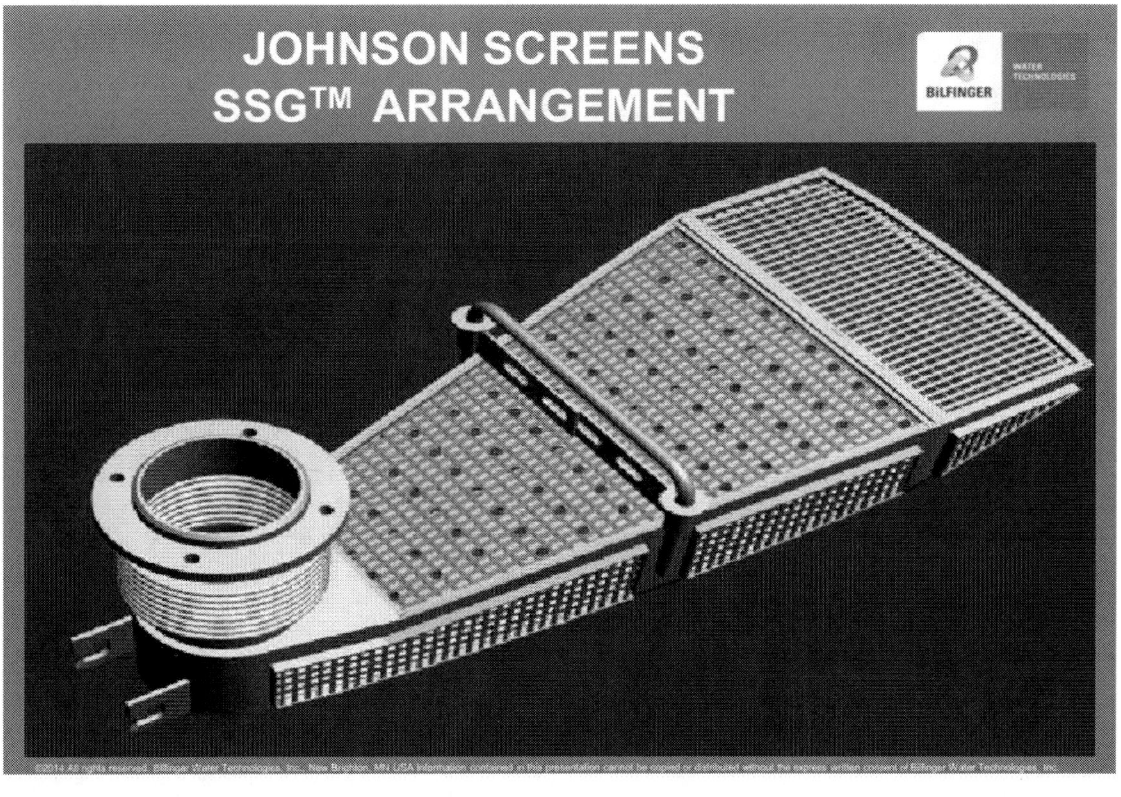
JOHNSON SCREENS
SSG™ ARRANGEMENT
BILFINGER
WATER TECHNOLOGIES
©2014 All rights reserved. Bilfinger Water Technologies, Inc., New Brighton, MN USA Information contained in this presentation cannot be copied or distributed without the express written consent of Bilfinger Water Technologies, Inc.

## PRE-INSTALLATION: OVERVIEW DATA

- 400MMSCFD (4 Vessels)
- 8' Diameter X 23' Tan-Tan
- Installed Weight: (Each Vessel)
    - 43,200lb Mole Sieve
- 2' Unused Void Space In Head
    - 4,500lb Additional Sieve Potential
    - 42.3% Increase In Sieve Volume

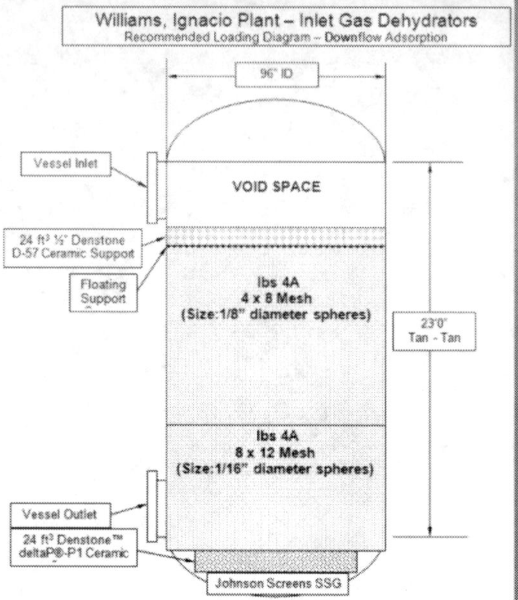




## GENERAL CONFIGURATION LOADING DIAGRAM

| Diameter: 8'<br>Tan-Tan: 23' | Support Grid | SSG™<br>As Loaded | SSG™<br>Potential |
|---|---|---|---|
| Sieve Weight | 34,200 lb | 43,200 lb | 48,700 lb |
| Inert Material Volume | 48 $ft^3$ | 48 $ft^3$ | 48 $ft^3$ |
| Un-Used Volume | 397 $ft^3$ | 148 $ft^3$ | 48 $ft^3$ |
| Sieve Weight Increase | | 26.3% | 42.3% |

# CFD REVIEW: SUPPORT GRID

BILFINGER WATER TECHNOLOGIES

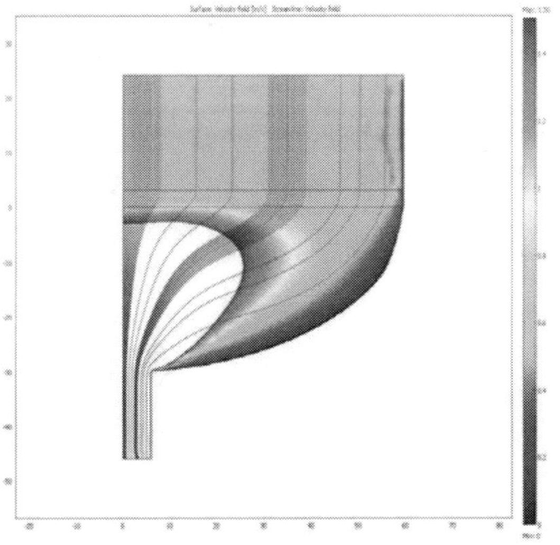

**Modeling Assumptions:**

- Down Flow
- Homogeneous Catalyst Bed
- Uniform Inlet Flow
- 2:1 Elliptical Head Configuration
- Support Grid Located At Tangent Line Of The Vessel
- Shadowing Effect Of Support Beams Not Included In Analysis



# CFD REVIEW: SSG™

BILFINGER WATER TECHNOLOGIES

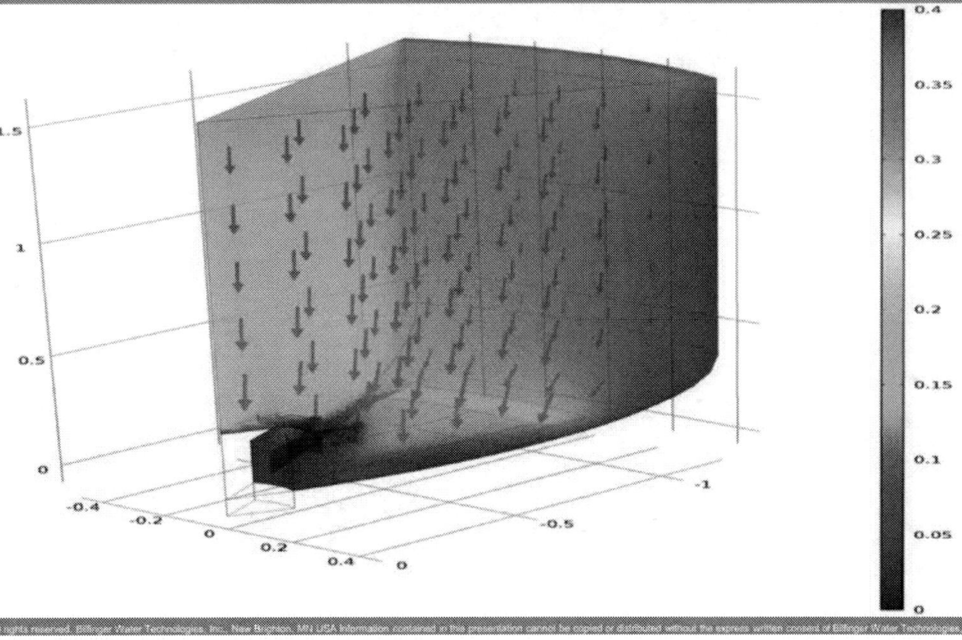

# INSTALLATION REVIEW

# INSTALLATION REVIEW

## INSTALLATION REVIEW

BiLFINGER WATER TECHNOLOGIES

## INSTALLATION REVIEW

BiLFINGER WATER TECHNOLOGIES

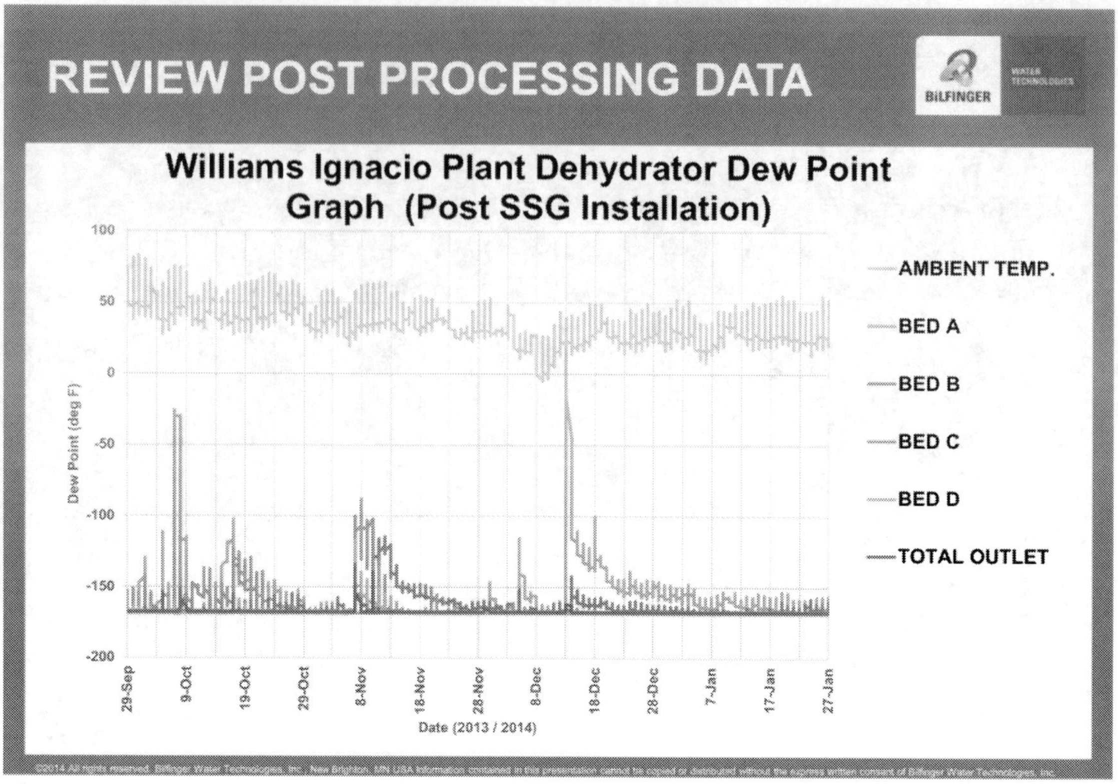
REVIEW POST PROCESSING DATA
BILFINGER
Williams Ignacio Plant Dehydrator Dew Point Graph (Post SSG Installation)
100
50
0
-50
-100
-150
-200
Dew Point (deg F)
29-Sep
9-Oct
19-Oct
29-Oct
8-Nov
18-Nov
28-Nov
8-Dec
18-Dec
28-Dec
7-Jan
17-Jan
27-Jan
Date (2013 / 2014)
AMBIENT TEMP.
BED A
BED B
BED C
BED D
TOTAL OUTLET

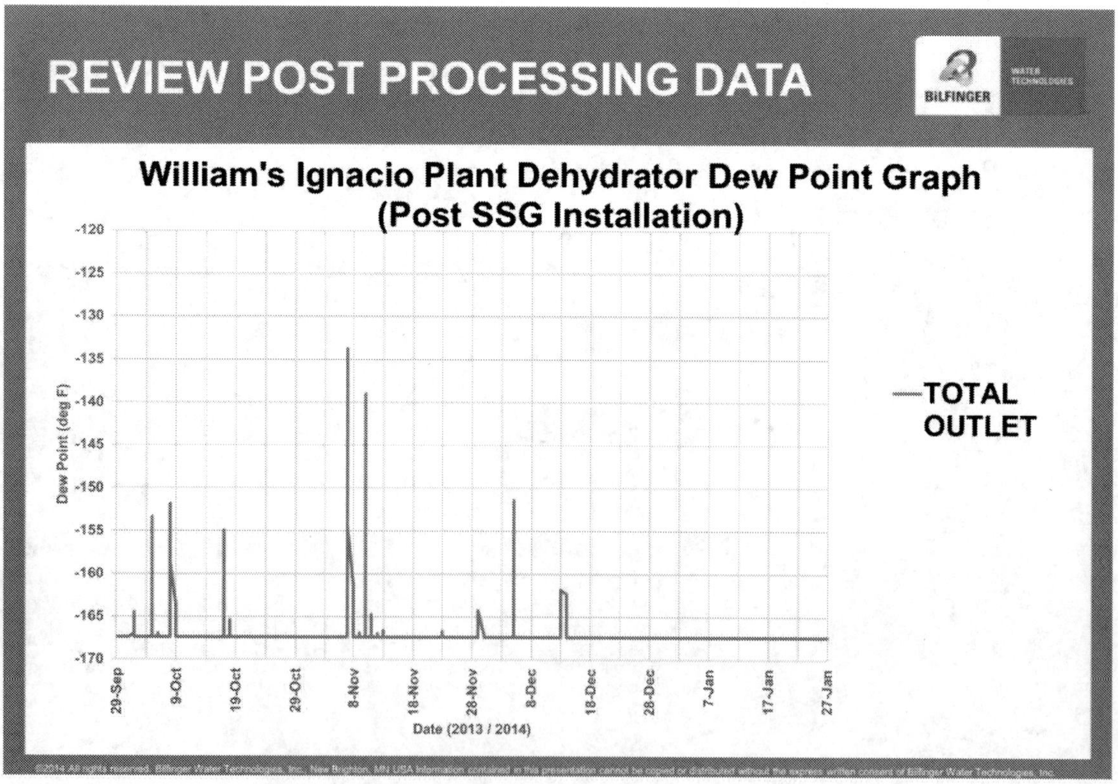
REVIEW POST PROCESSING DATA
BILFINGER
William's Ignacio Plant Dehydrator Dew Point Graph (Post SSG Installation)
-120
-125
-130
-135
-140
-145
-150
-155
-160
-165
-170
Dew Point (deg F)
29-Sep
9-Oct
19-Oct
29-Oct
8-Nov
18-Nov
28-Nov
8-Dec
18-Dec
28-Dec
7-Jan
17-Jan
27-Jan
Date (2013 / 2014)
TOTAL OUTLET

## UNPLANNED ADDITIONAL BENEFITS

BILFINGER WATER TECHNOLOGIES

### INSIDE THE VESSEL ADVANTAGES

- Bed Migration Eliminated - All Segments Are Sealed
- Increased Bed Capacity 26.3%, with 42.3% Possible
- Maximize Mole Sieve Utilization
- Fewer Regeneration Cycles Required
- Lower Fuel Consumption



## UNPLANNED ADDITIONAL BENEFITS

BILFINGER WATER TECHNOLOGIES

### OUTSIDE THE VESSEL ADVANTAGES

- Bed Migration Eliminated - All Segments Are Sealed Removes The Need For Cleaning Down Stream Equipment
  - Lower Fuel Consumption
    - 400,000 Cubic Feet Gas Used Per Bed For Regeneration
    - Fuel Cost $3.50 Per Million Cubic Foot
    - 400,000 X $3.50 X 29 (Regen) X 4 (Beds) = $162,400 / 2 = $81,200 Savings Per Year

## UNPLANNED ADDITIONAL BENEFITS

### OUTSIDE THE VESSEL ADVANTAGES

- Fewer Regeneration Cycles Required
  - Less Valve Actuations Required
    - 19 Valves Actuated Per Regeneration
    - 29 Fewer Regenerations
    - 19 X 29 = 551 Fewer Valve Cycles / Year (Potentially Doubles Valve Life)



## JOHNSON SCREENS SSG™ Arrangement

## CLOSING REMARKS

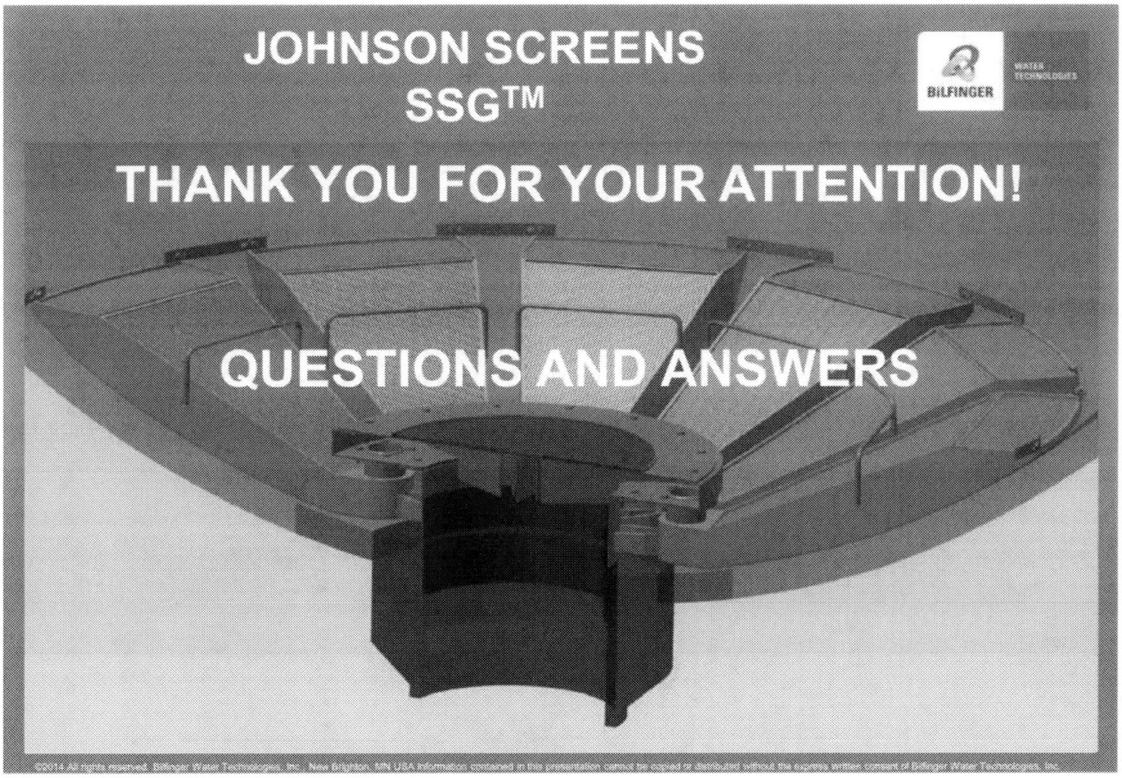
JOHNSON SCREENS
SSGTM
BILFINGER
WATER TECHNOLOGIES
THANK YOU FOR YOUR ATTENTION!
QUESTIONS AND ANSWERS
©2014 All rights reserved. Bilfinger Water Technologies, Inc., New Brighton, MN USA Information contained in this presentation cannot be copied or distributed without the express written consent of Bilfinger Water Technologies, Inc.

# ZEOLITE ADSORPTION STUDIES FOR CONDITIONING OF HIGH-PRESSURE NATURAL GAS FLUIDS


Behnaz Hojjati
Robert A. Marriott
*University of Calgary*
*Calgary, AB*


*AIChE Spring Meeting, New Orleans, LA, March 30- April 3, 2014*

# Dr. Robert Marriott's research group

Optimization of high-pressure conditioning requires advanced understanding of solid-gas phase behavior.

How much water can we tolerate and still avoid hydrate issues?

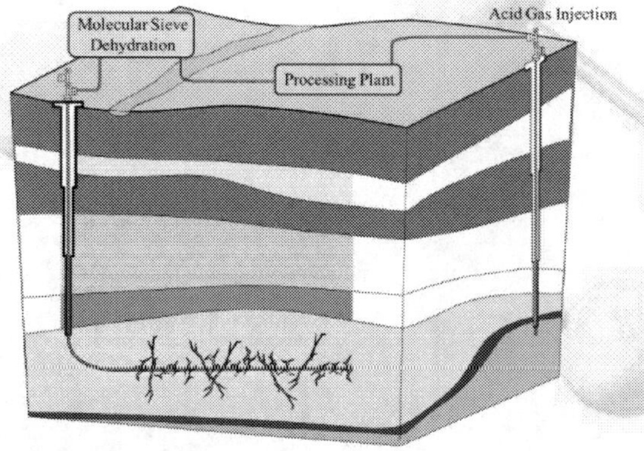


Can we inject sulfur or sulfur with viscosity modifying additives?

Why is there unexpected $H_2S$ in some shale reservoir fluids?

2

# NATURAL GAS

- In order to upgrade produced natural gas to sale gas, some unwanted components must be removed $H_2O$, $H_2S$, $CO_2$ , and COS

- Dehydration of natural gas
  - *Absorption (glycol)*
  - *Adsorption (molecular sieves)*

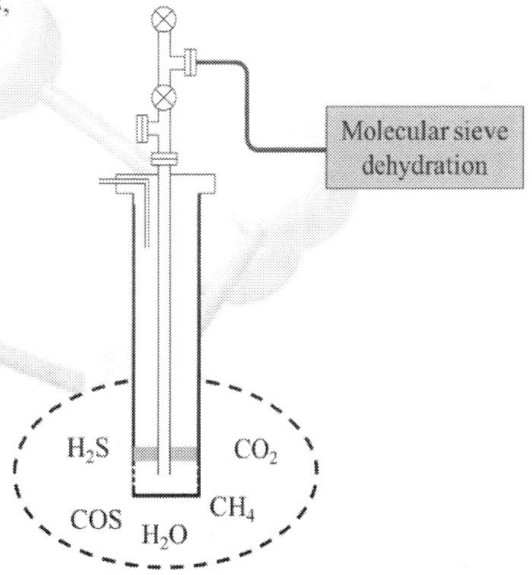


- Selective high-pressure adsorption : *Molecular sieves*
  - can reduce water to very low dew-point concentration
  - small pressure drop
  - dehydration of natural gas at low cost (depending on location)
  - relatively free from problems of corrosion

3

# ADSORPTION OF $H_2O$ AND $CO_2$ ON THREE MOLECULAR SIEVES

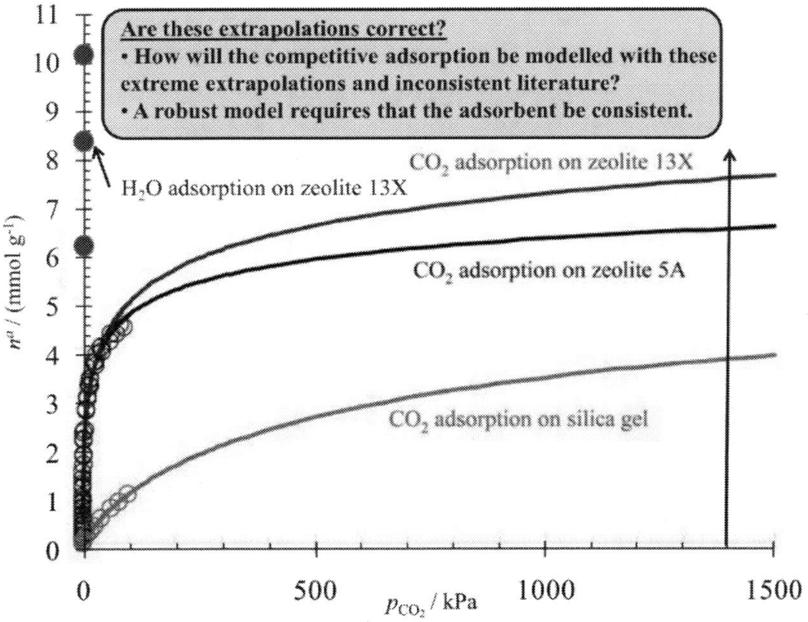


Data from Wang and LeVan *J. Chem. Eng. Data* **2009**, 54, 2839-2844.
Fugacities calculated using Span and Wagner, J. Phys. Chem. Ref. Data **1996**, 25, 1509-1596.

4

# OPTIMIZATION HIGH-PRESSURE SOUR GAS ADSORPTION

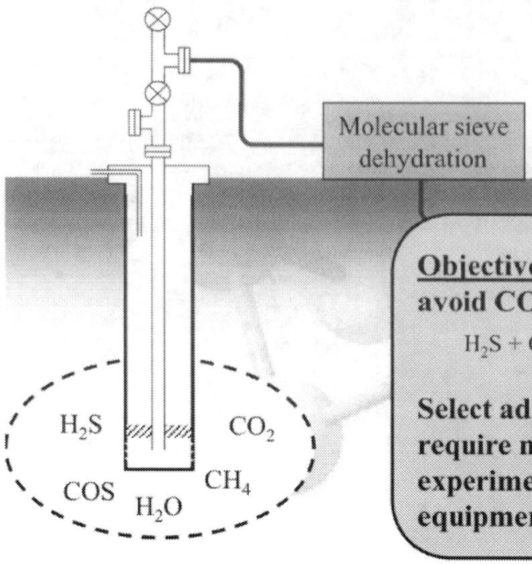


**<u>Objective:</u> optimize regeneration conditions to avoid COS production.**

$$H_2S + CO_2 \Leftrightarrow COS + H_2O$$

**Select adsorption relationships/equations require multi-component high-pressure experimental data from wet sour serviceable equipment (method development)**

5

# WHY BEGIN WITH ZEOLITES?

$M_{2/n}O{\bullet}Al_2O_3 \bullet ySiO_2 \bullet wH_2O$

- Zeolites are inorganic microporous materials having a **highly regular** pore structure.
- Aluminosilicate zeolites are easily synthesized under hydrothermal conditions from reactive silica-alumina gels in alkaline media at temperatures between $T = 80$ °C and 200 °C.
  - Low silica zeolites (Si/Al:~1–1.5) A, X
  - Intermediate silica zeolites (Si/Al: ~2–5) Y, L
  - High silica zeolites (Si/Al: ~10–100) ZSM-5, Silicate

Zeolite A

$M_{2/n}O{\bullet}Al_2O_3{\bullet}2SiO_2$

Zeolite 4A: M= Na
Zeolite 3A: M= K
Zeolite 5A: M= Ca

6

www.zeoliteguide.com

# SCANNING ELECTRON MICROSCOPY AND ENERGY DISPERSIVE X-RAY SPECTROSCOPY

Electron Dispersive X-Ray Spectroscopy: O 50.79
Na 17.78
Al 15.48
Si 15.95

Si/Al : 1.14-1.15

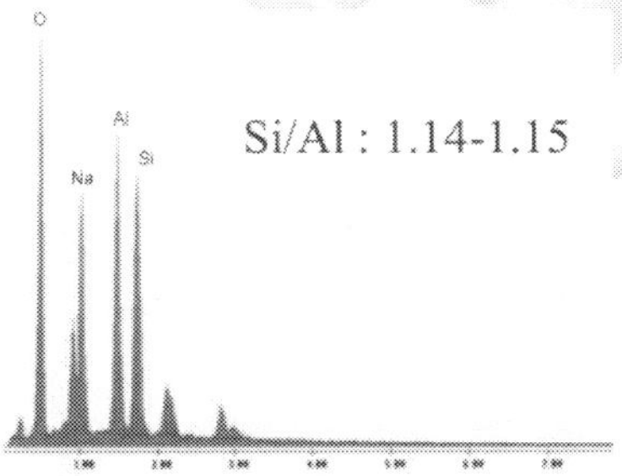


This Zeolite 4A Commercial Zeolite 4A

Dynamic Light Scattering, $\bar{d} = 4.45$ μm

7

# POWDER X-RAY DIFFRACTION

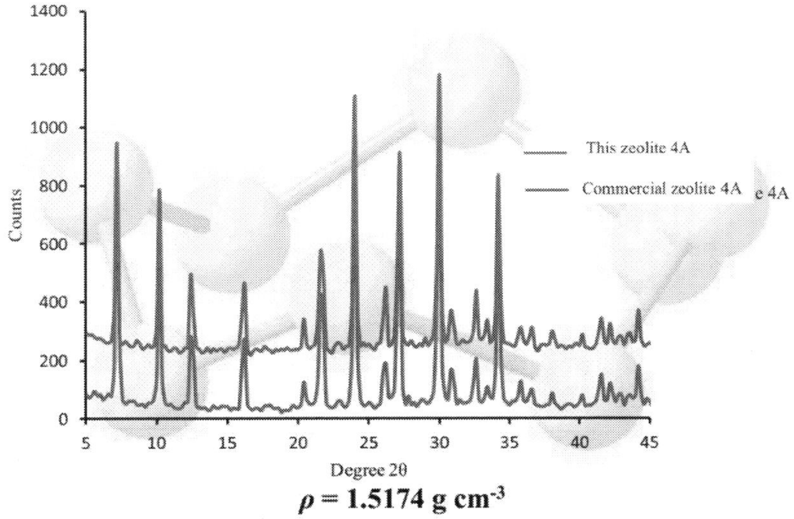


$\boldsymbol{\rho}$ **= 1.5174 g cm$^{-3}$**

**IZA reports $\rho$ = 1.5182 g cm$^{-3}$ for fully dehydrated Zeolite 4A**

8

# *In-situ* ATR-FTIR &FT-RAMAN ADSORPTION OF $CO_2$ONTO ZEOLITE 4A AT $T$ = 25°C

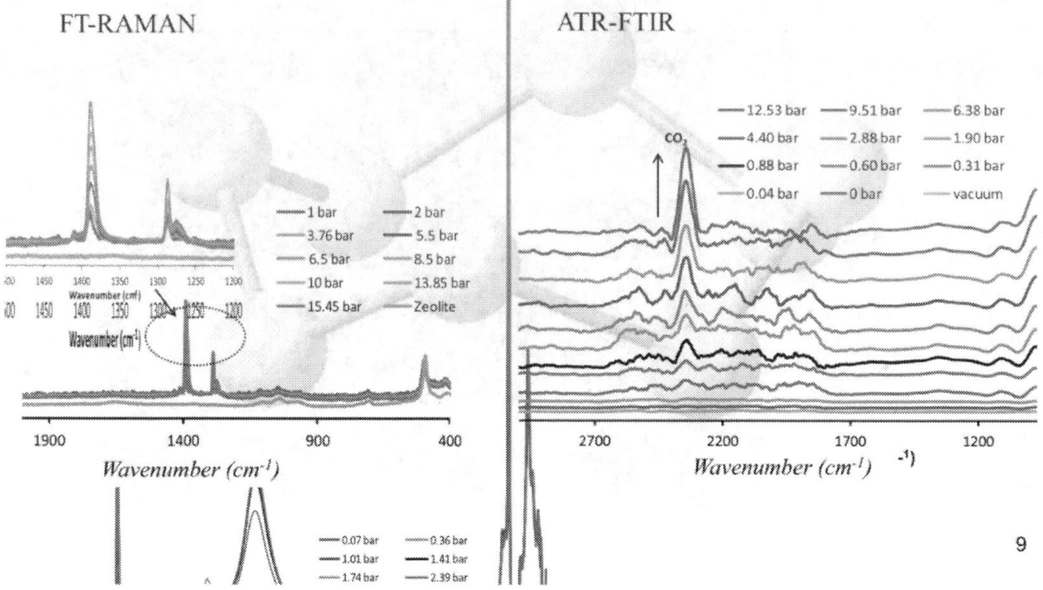


9

# *In-situ* FT-RAMAN ADSORPTION OF $H_2S$ AND $CH_4$ ONTO ZEOLITE 4A AT $T$ = 25°C

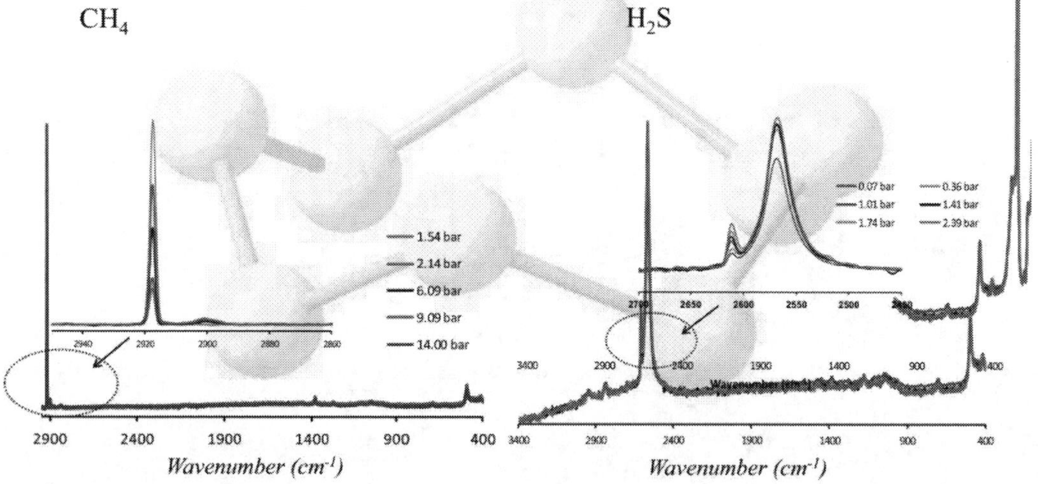


10

# A SMALL CELL ADSORPTION APPARATUS FOR MULTICOMPONENT SOUR GAS ADSORPTOIN

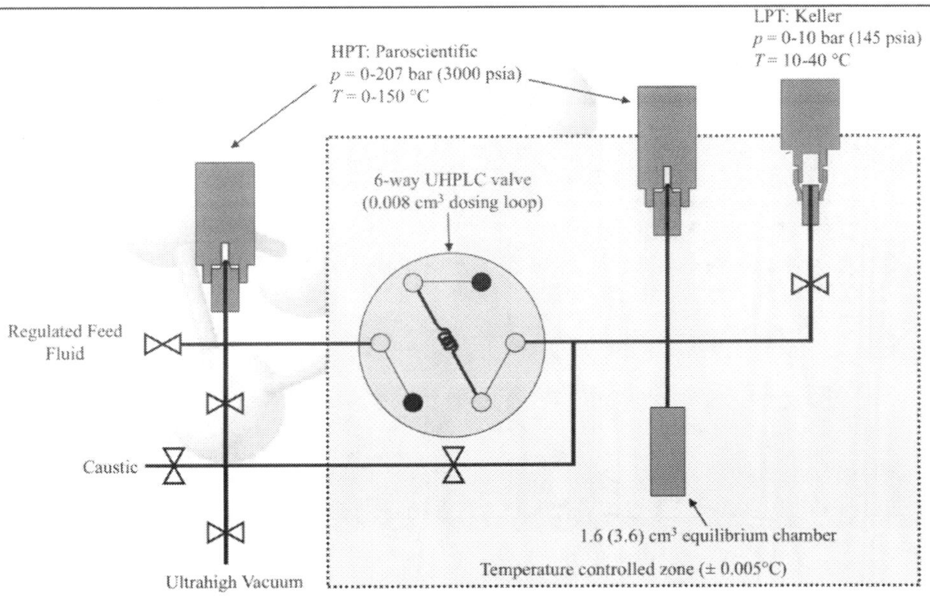


11

# A NEW SMALL VOLUME ADSORPTION APPARATUS

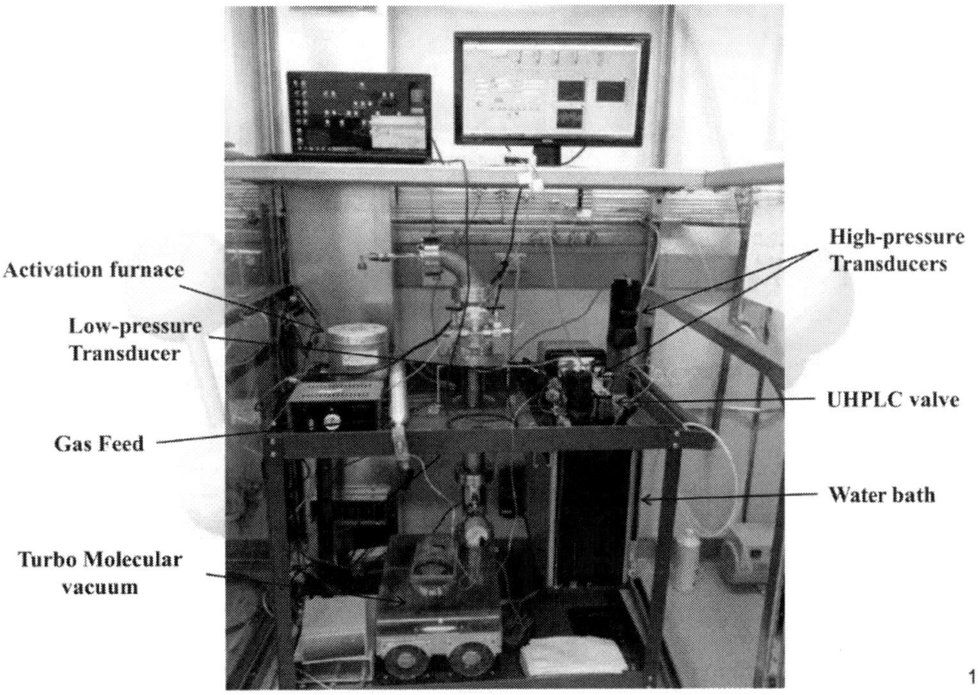


12

# ADSORPTION CONDITIONS FOR PURE HETEROGENEOUS AND HOMOGENEOUS POROUS SOLIDS

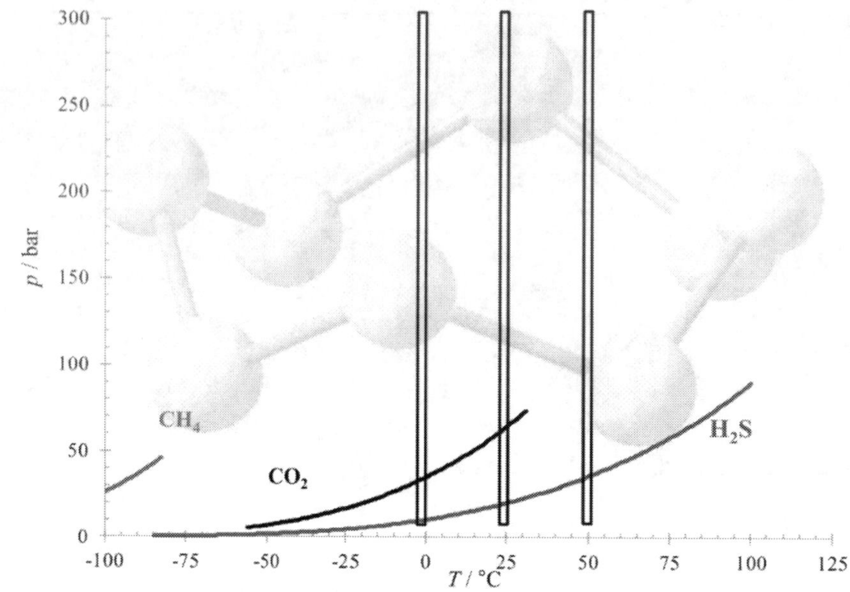


13

# THE EXCESS ADSORPTION OF $CO_2$ ONTO ZEOLITE 4A

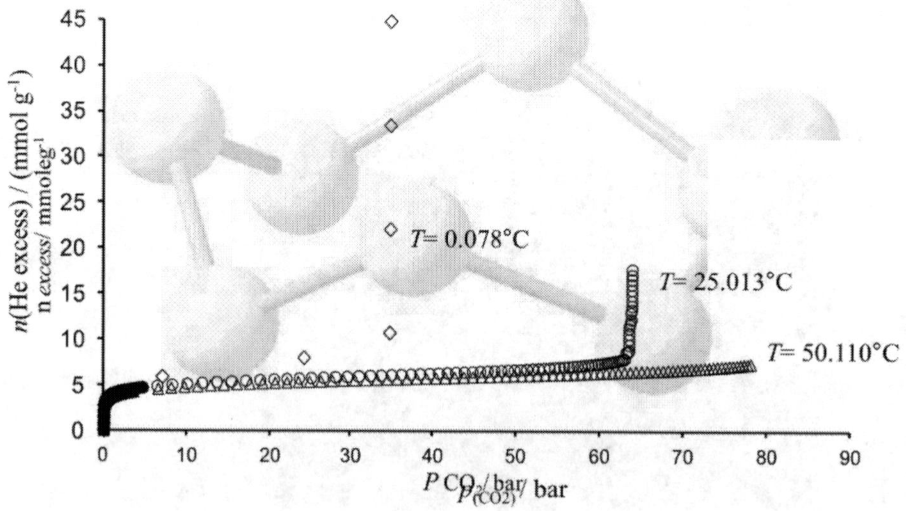


14

# PRELIMINARY $CO_2$ ADSORPTION DATA

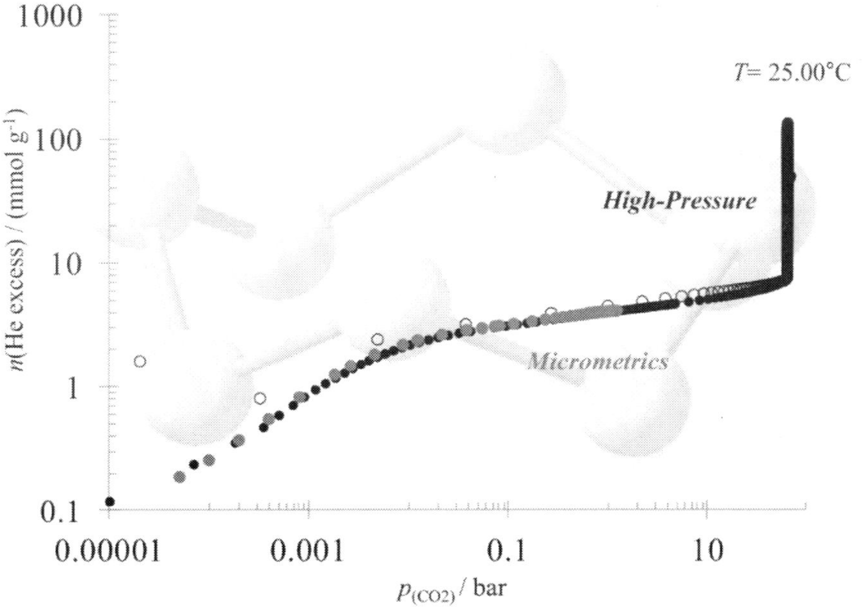


15

# THE EXCESS ADSORPTION OF $H_2S$ ONTO ZEOLITE 4A

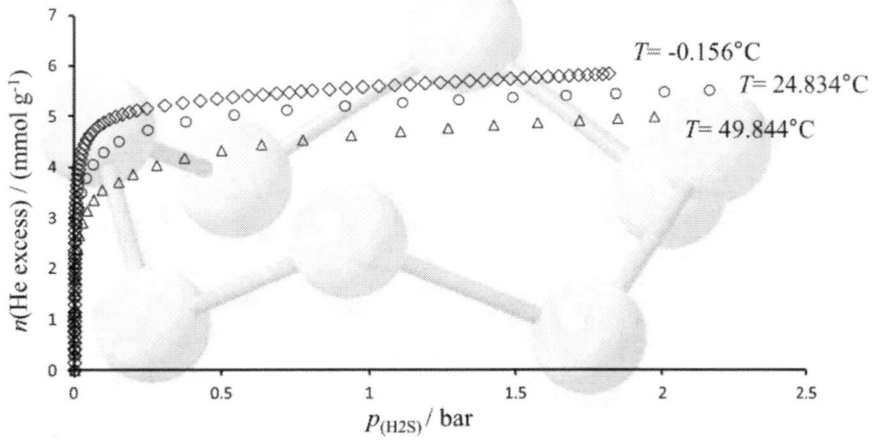


16

# THE EXCESS ADSORPTIONS OF $CH_4$ ONTO ZEOLITE 4A

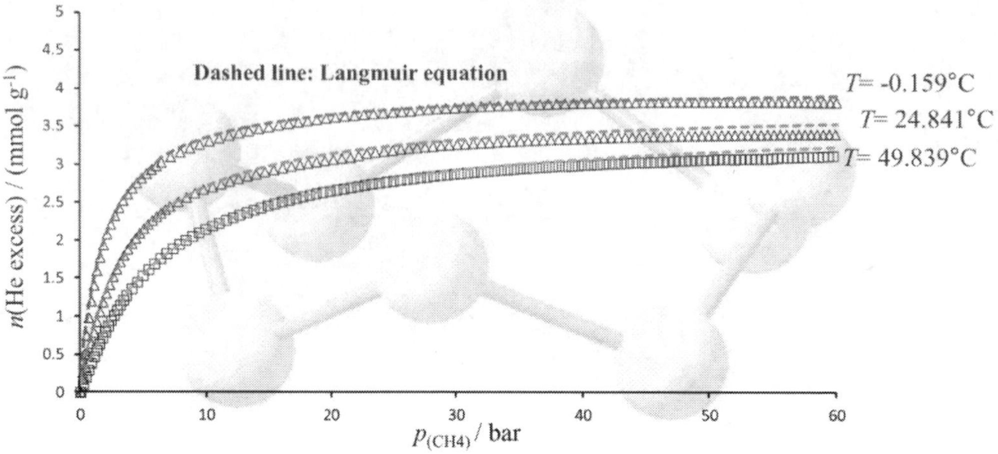


17

# THE DIFFERENCE BETWEEN ABSOLUTE AND EXCESS ADSORPTION

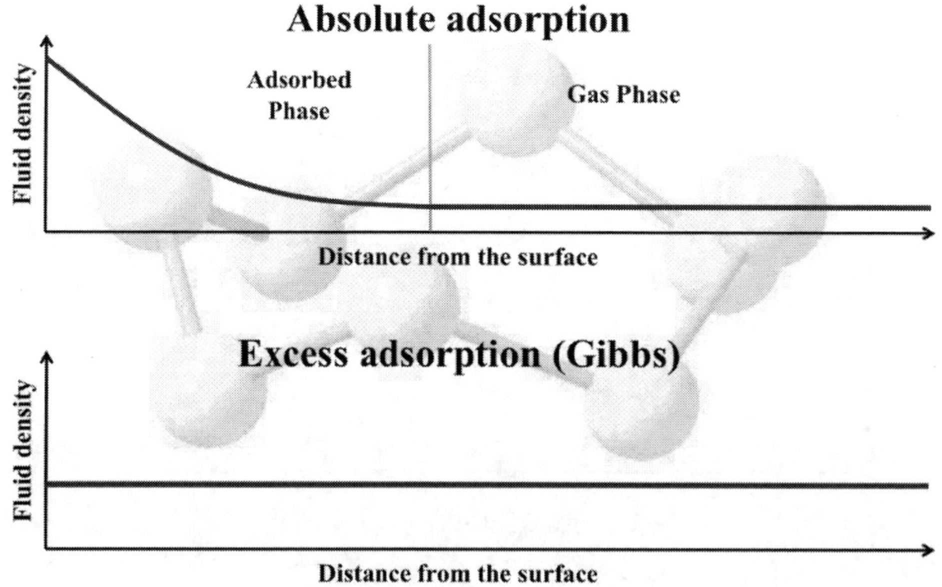


18

## THE EXCESS & ABSOLUTE ADSORPTIONS OF $CH_4$ ONTO ZEOLITE 4A AT $T$ = -0.159°C

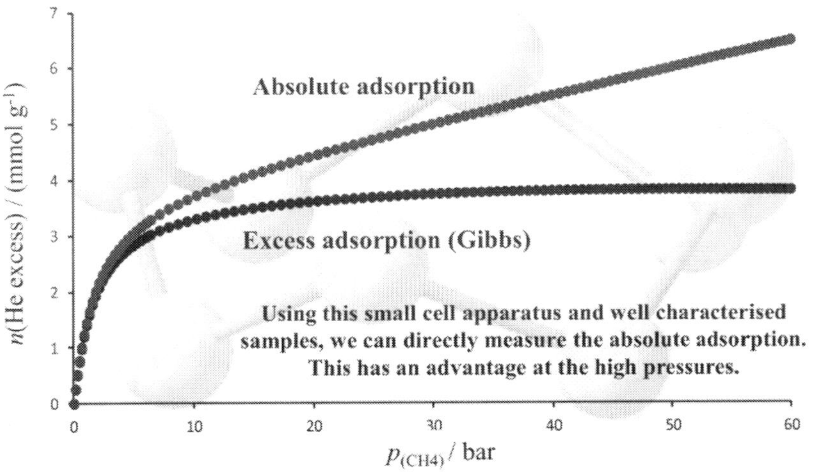


19

## THE HEAT OF ADSORPTION OF METHANE ONTO ZEOLITE 4A AT $T$ = 25°C

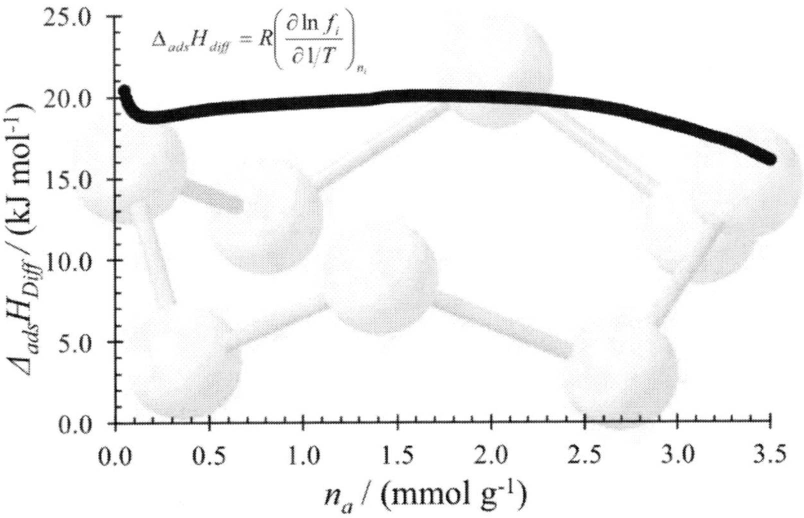


$\Delta_{ads}H_{diff} = 21 KJmol^{-1}$ "Adsorption of Methane on Titanosilicate ETS-10 Zeolite", *J. Chem. Eng. Data* 2005, *50*, 843-848

$\Delta_{ads}H_{diff} = 20.9 KJmol^{-1}$ "Adsorption of Methane on MFI", *Order Porous Solids*, 2011, Science

20

# SUMMARY AND FUTURE WORK

- High-quality Zeolites 4Awas synthesized *via* hydrothermal techniques and characterized by XRD, SEM/EDX, and DLS.
- High-pressure ATR-FTIR and FT-Raman were used to check the purity of the zeolite 4A by the adsorption of $CO_2$, $CH_4$ and $H_2S$ at different pressure (room temperature).
- High-pressure adsorption isotherms of $CO_2$, $CH_4$ and $H_2S$ on the zeolite 4A at 0, 25, and 50 °C using a home-built adsorption apparatus were studied.
- The results can be used to benchmark and compare materials and instruments used for measuring multicomponent sour gas adsorption.
- Future adsorptions models to be parameterized with internally consistent data.
- Future work will include high pressure adsorption isotherms of $CO_2$, $CH_4$, $H_2S$, COS, $SO_2$ and elemental sulfur on the synthesized zeolite 5A, 3A and 13X at different temperatures.

21

***Thank you***

Natural Sciences and Engineering Research Council of Canada
(NSERC)
Alberta Sulphur Research Ltd. (ASRL)
The University of Calgary

# 352233 Design of Units for Mercury Removal from Hydrocarbon Gas Streams

**Abstract:**

**Design of Units for Mercury Removal from Hydrocarbon Gas Streams**

Robert W. Soffel, Selective Adsorption Associates, Inc.

John Markovs, Adsorption Solutions, LLC

Abstract:

The hydrocarbon industry became aware of the detrimental effects of mercury some 40 years ago when the aluminum heat exchangers were damaged by mercury at the Skikda LNG plant in Algeria. Since mercury is widely present in natural gas, subsequent plants have chosen to install Mercury Removal Units (MRU's) to protect heat exchangers and other plant operations. The predominant means for removing the mercury has been the use of fixed bed adsorption.

Review is made of the process design criteria for achieving dependable long-term operation of mercury removal units. Essentially total mercury removal is required, since even part per trillion concentrations of mercury in the gas can damage aluminum heat exchangers. The effects of gas residence time in the MRU as it affects the attainable purity are examined, as well as the effects of the adsorbent particle size. Examples of MRU designs are presented for a variety of feed gas conditions, including wet and dry feed gases. The physical and chemical properties of the four types of commercially-available mercury removal adsorbents are compared. The design of the MRU vessel and its relationship to the process gas pressure drop is examined. The importance of designing single-phase flow protection into the mercury removal process to attaining the desired design performance is discussed. Experience-based recommendations for achieving reliable mercury removal performance and predictable mercury adsorbent life are described.

# Estimation of Calorific Value and Grindability of Colombian Caribbean Coals by Multiple Regression and Artificial Neural Networks.

*Yamid Alí Gómez Rueda, Junior Researcher, Universidad del Atlántico, Barranquilla, Colombia.*

*M.Sc. Ricardo Angulo Mercado, Associate Professor, Universidad del Atlántico, Barranquilla, Colombia.*

**Abstract.**

Coal is one of the most important raw materials in the world; it is used in diverse applications such as water treatment, steel production and molecular sieves; however coal is mainly used as fuel in power generation plants. In power industry the coal must have specific properties to make an affordable, efficient and environmentally responsible generation process, therefore a lot of instrumental analysis are required to determine the value of those properties, spending time, equipment and qualified personal. Between many properties measured, there are two that are fundamental for combustion and grinding processes: Gross calorific value and Hardgrove Grindability index.

Mathematical models of coal properties – including Gross calorific value and Hardgrove grindability index– have been developed, mainly, to reduce the number of instrumental analysis, determining the mathematical relations of certain properties with parameters that are taken from more simple analysis. Although mathematical models generalizing performance of properties are the goal, in empirical models of Coals this is very difficult, due to coal heterogeneity; hence empirical models are often developed for coals of specific locations, where the estimation of properties becomes the main goal.

Due to the above, added to the lack of studies of Caribbean Colombian Coals, in this study we aim to find through multiple regression and artificial neural networks an empirical mathematical model to determine Gross calorific value and Hardgrove grindability index for Colombian Caribbean Coals using the data from proximate and ultimate analysis as regressors.

In the development of this research, we collected data from three different coal mines, we applied parallelly the regression method and the artificial neural network and finally we compared the two mathematical models to determine which of two methods showed a better performance. After that we made a comparison of our final model with another mathematical models published in the international literature applied to studied coals. Finally we established some criteria to determine which procedure – of the two utilized in this study - use when somebody wants to predict certain property.

## Introduction

The behavior of coals are often dependent on the abundance of C,H,N,O,S, also fixed carbon, volatile matter and, in low range coals, humidity and ash content. It is possible to predict a large number of coal parameters using these components. Even if there is no a causation link, useful correlations are allowed by the coalification processes, thus many studies have been published in the last 50 years. Many of them are plots which coal scientists still use, like Seyler's plot. In recent years, the trends are accompanied by equations and artificial neural network models that made significant predictions. In this paper we aim to use traditional multiple regression and Artificial Neural Network (ANN) regression to predict Gross Calorifc Value (GCV) and Hardgrove Grindability Index (HGI) of three Caribbean Colombian Coals for GCV and two for HGI, and the models obtained are compared with others of the scientific literature.

## Methodology

Step 1. Database building. Proximate and ultimate analysis data was collected from a variety of coals, covering a diverse range of volatile matter, mineral matter, and humidity. The data was divided in two groups randomly: the first one, containing 80% of the data, was used to build the model; the second one, containing 20% of the data, was used to validate the model.

Step 2. Variable Selection. Here we aim to minimize the number of variables of the model, using the principle of Occam's razor. We use here correlation analysis and some theory of causation to select the variables.

Step 3. Selection of the equation's relationship and ANN structure. Here we select the type of equation (linear, quadratic, logarithmic) and the ANN architecture and learning method.

Step 4. Choose of the best model. To choose the best model we used tow criteria: Root Mean Squared Error (RMSE) and mean absolute error (MAE)

Step 5. Comparisons with scientific literature models. Here we aim to see the differences between predicting power of published models and their ones.

## Results.

### *Gross Calorific Value.*

A filtering was made and only 9000 BTU/lb and higher Calorific Power Coals were chosen; 773 samples were used to perform the analysis. For GCV we used the data provided by proximate analysis (Fixed Carbon, FC; Volatile Matter, VM; Ash, A) and their combinations as product. The next graph shows the correlations between proximate analysis variables and GCV.

**Graphic 1.** Correlation plots between proximate analysis and GCV

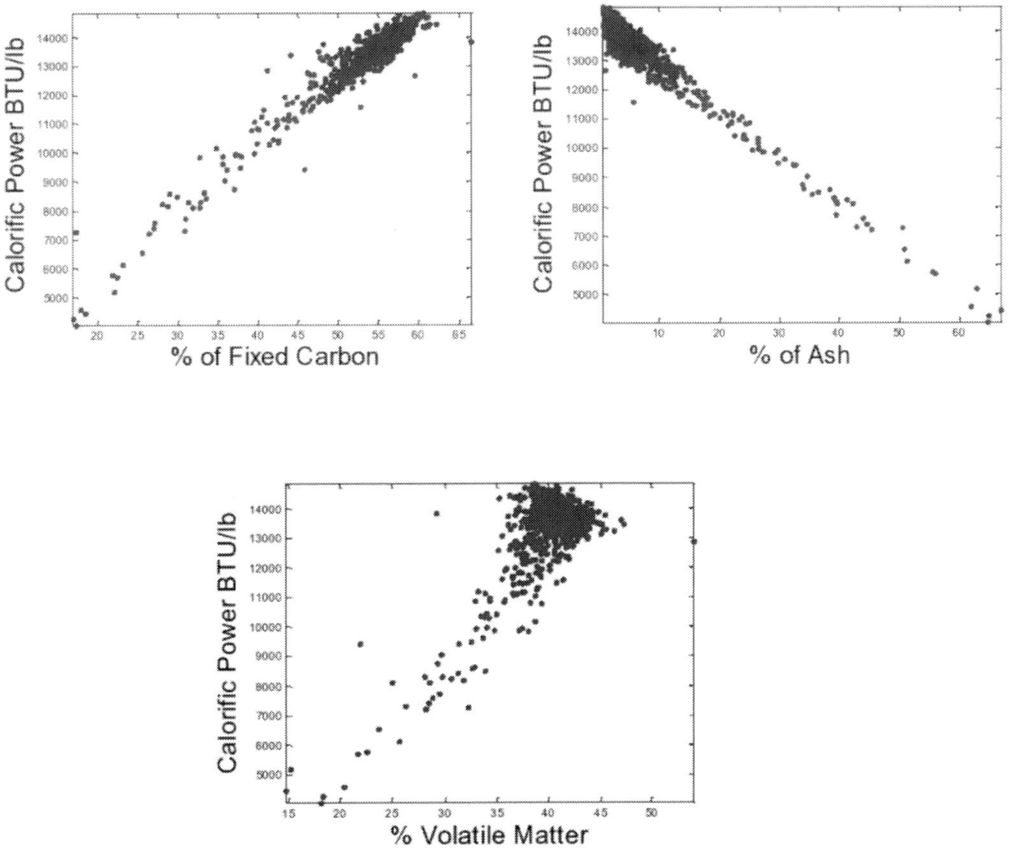


The regression equation that showed the best performance for global data is stated below with $r^2$=0.96

$$GCV(BTU/lb)=-22212{,}9-250{,}13*FC+56{,}908*VM-4{,}641*FC*VM-1{,}229*A*VM$$

On the other hand, two ANN architectures were used. The first ANN is a Multilayer Perceptron with 20 neurons in the hidden layer, using the Levenberg-Marquardt learning algorithm. The second one was a GRNN with a smoothing parameter optimized with a value of 2.

The performance of the regression and ANN models are presented in the Table 1.

**Table 1.** Performance of regression and ANN models

| | r2 | | | RMSE | | |
|---|---|---|---|---|---|---|
| **Data group** | **Regression** | **ANN 1** | **ANN 2** | **Regression** | **ANN 1** | **ANN 2** |
| Guajira 1 | 0.938 | 0.954 | 0,965 | 271,98 | 4,50 | 205,59 |
| Guajira 2 | 0.984 | 0.988 | 0,985 | 270,24 | 16,96 | 150,98 |
| Cesar | 0.988 | 0.991 | 0,990 | 122,69 | 3,82 | 106,23 |
| Global | 0.959 | 0.971 | 0,965 | 294,53 | 14,74 | 168,76 |

According to the table, there is no significant difference between the performance of ANN models and regression models. The best models were obtained on individual mines instead of the Global data of the three mines. In all cases, the best model was ANN 1 which corresponds to a model of Multilinear Perceptron with Levenberg-Marquardt learning algorithm.

***Hardgrove Grindability Index.***

In HGI the regression models shows an almost null relation between HGI and proximate and ultimate analysis data ($r^2$ = 0,164). Then only ANN models were used. We use 4 ANN models which are described in the table 2.

**Table 2.** ANN models for HGI

| | Neurons in hidden layers | | |
|---|---|---|---|
| **MODELS** | **I** | **j** | **k** |
| Model 1 | 20 | 0 | 0 |
| Model 2 | 32 | 0 | 0 |
| Model 3 | 10 | 10 | 0 |
| Model 4 | 10 | 10 | 12 |

The performance of ANN models are presented in the next table.

**Table 3.** Performance of the ANN models for HGI

| | **Global** | **Guajira_1** | **Guajira_2** |
|---|---|---|---|
| ANN Model 1 | 0,61663 | 0,69 | 0,92893 |
| ANN Model 2 | 0,64086 | 0,63 | 0,97 |
| ANN Model 3 | 0,78322 | 0,81635 | 0,95 |
| ANN Model 4 | 0,70392 | 0,79216 | 0,94541 |

The performance is similar to the GCV, showing a better performance for individual mines than Global data. However in the three dataset the best model was the model 3, which corresponds to a model with two hidden layers

### ***Comparison with other models.***

Comparing the results of our study and other published empirical models we can conclude that the models have the same capacity of prediction, even when they were calculated to coals from different sources.

**Table 4.** Comparison of GCV models

| BTU/lb | Banerje et al (2008) | Chelgani (2009) | Patel et al(2006) | This paper |
|---|---|---|---|---|
| $r^2$ | 0,9783 | 0,97 | 0,996 | 0,971 |

For the HGI we chose the Maximum Absolute Error (MAE) as comparison criterion. Due to our prediction parameters it can be said that the maximum error of 10 is admissible, although other empirical models showed maximum errors less than 10, they used prediction parameters more elaborated than proximate and ultimate analysis.

**Table 5.** Comparison of HGI models

| HGI | This paper | Chelgani(2007) | Li et al(2005) | Ozbayoglu et al(2006) |
|---|---|---|---|---|
| MAE | 10 | 4.5 | 5 | 17,46 |

## Conclusions.

A lot of predictive models that relate proximate and ultimate analysis with properties of coal have been developed in the last 70 years. These models have been developed for coals of specific locations and in this study we create two models using multiple regression and artificial neural networks to obtain them. In this study, the ANN models show a better performance than regression models. In the case of GCV, the Multilayer perceptron with 20 neurons, one hidden layer and Levenberg-Marquardt learning algorithm, although the regression models showed a good performance, if is not required much precision, the regression models can be used; otherwise, ANN model can give the best performance.

The HGI prediction was harder, and no regression model was obtained, due to the show practically any correlation between proximate and ultimate analysis and the HGI. The ANN models worked better, and although they did not show an excellent performance, it is enough to make calculations in some processes that required the data quickly. The ANN model that performs the best was a Multilayer Perceptron with two hidden layers and ten neurons in each layer.

## References.


1. AKYILDIZ, Mustafa; URAL; SUPHI. Studies of the relationship between mineral matter and grinding properties for low-rank coals. 2004.
2. ASTM Standards of Coal and Coke. ASTM D121, ASTM D167, ASTM D388 ASTMD409 ASTM D2013, ASTM D2234, ASTMD3172, ASTM D3174, ASTM3176, ASTMD3180, ASTM D3302, ASTM D4749, ASTM D5016, ASTM D5865 ASTM D6518.
3. BANERJEE et al. Development of a new proximate analysis based correlation to predict calorific value of coal. Elsevier. 2008
4. BAGHERIEH et al. Prediction of coal grindability based on petrography, proximate and ultimate analysis using multiple regression and artificial neural networks.
5. CANNAVOS G. Probabilidad y Estadística Aplicación y métodos. Ed. en español Mc Graw-Hill. México.1995.
6. CHANNIWALA; GHOSAL; PARIKH, Jigisha. A correlation for calculating HHV from proximate analysis of solid fuels.Fuel. 2004
7. CHELGANI, Chehreh; HOWER, James; JORJANI; MESROGHLI,Sh. Simultaneous prediction of coal rank parameters based on ultimate analysis using regression and artificial neural network. International Journal of Coal Geology. 2010.
8. CHELGANI, Chereh; JORJANI; MESROHI . Estimation of the gross calorific value based on coal analysis using regression and artificial neural networks.Elsevier. International Journal of Coal Geology. 2009
9. DUNXI Yu et al. Prediction of grindability with multivariable regression and neural network in Chinese coal.
10. KUCUKBAYRAK, Sadriye. Estimation of calorific values of Turkish lignites. Fuel. 1991.
11. HAXHI, Hajri. MARTIN, Duncan; LOLJA, Saimir. Correlations in the properties of Albanian coals. Correlations in the properties of Albanian coals. Fuel. 2002.
12. LI, Peisheng et al. Prediction of grindability with multivariable regression and neural network in Chinese Coal. Elsevier. Fuel Journal. 2005.
13. MATHEWS, Jonathan et al. A review of coal properties with elemental composition.Fuel process technology. 2014, pp. 104-133
14. MONTAÑO MORENO. Redes Neuronales Artificiales aplicadas al análisis de datos. Universidad de les Illes Balears: Tesis Doctoral. 2002.
15. OZBAYOGLU, Gülham. Estimation of Hardgrove Grindability index of Turkish coals by neural networks. International Journal of mineral processing. 2008
16. PATEL, Shagufta. Estimation of gross calorific value of coals using artificial neural networks. Elsevier. 2006.
17. SPECHT, Donald. A General Regression Neural Network. IEEE, transactions in neural networks. 1991.
18. SPEIGHT, James. Handbook of Coal Analysis. Wiley &Sons. 2005
19. The Mathworks Inc. Matlab Help, Neural Network Toolbox, Version 7.9.0.
20. WALPOLE; MYERS. Probabilidad y Estadística para ingeniería y ciencias. Prentice Hall 2000.

# AUTHOR INDEX

# AUTHOR INDEX